CHEMICAL ENGINEERING ECONOMICS

DONALD E. GARRETT

Adjunct Professor, University of California, Santa Barbara
and
Former Executive Vice-President for R & D
Occidental Petroleum Corp.

WITHDRAWN

Printed in the United States of America

Van Nostrand Reinhold
115 Fifth Avenue
New York, New York 10003

Van Nostrand Reinhold International Company Limited
11 New Fetter Lane
London EC4P 4EE, England

Van Nostrand Reinhold
480 La Trobe Street
Melbourne, Victoria 3000, Australia

Macmillan of Canada
Division of Canada Publishing Corporation
164 Commander Boulevard
Agincourt, Ontario M1S 3C7, Canada

16 15 14 13 12 11 10 9 8 7 6 5 4 3 2 1

Library of Congress Cataloging-in-Publication Data

Garrett, Donald E.
    Chemical engineering economics / Donald E. Garrett.
        p. cm
    Includes bibliographies and index.
    ISBN 0-442-31801-4 :
        1. Chemical engineering—Costs. 2. Chemical engineering-
-Estimates. I. Title.
TP155.G35   1989
660.2'8'00681—dc19   88-19187

PREFACE AND ACKNOWLEDGMENTS

There are many excellent books about economics for the process industries, each containing unique features. However, none of them emphasize the basic economic information that most working engineers will need. When the author started teaching chemical engineering economics some years ago at the University of California, Santa Barbara (UCSB), there were no texts available to provide specific instruction on cost estimating, project evaluation techniques and industry background on economics that related to, or could be profitably used by the engineer. Because of these circumstances a series of notes were assembled that gradually led to this book.

The book is intended primarily for the working engineer who is not a cost estimator, upper level manager, or economics specialist, but who needs economics in his or her daily activities for better job performance and advancement. When practicing engineers are asked in surveys about the subjects they wish they had taken in more detail in school, almost all place economics high on their list. This book is designed to fulfill that wish and to be useful in the chemical engineering curriculum.

Because of the importance of economics, and since it is a required course for accreditation, most engineers will have had some economics instruction in college. This book, whether used in conjunction with a design course or not, can provide the fundamental economic background that can be used in day-to-day engineering problems and assignments. Finally, since most chemical engineering professors are highly skilled technical specialists, this book can offer an easier and more comprehensive basic economic guide for the noneconomics instructor teaching the course.

The author wishes to acknowledge the support and assistance given by the UCSB Chemmical Engineering Department, which has made writing this book and the companion teaching assignment highly interesting, enjoyable, and rewarding activities. The friendship and assistance of Professor Jack Meyers, former UCSB Dean of Engineering, have been of great value. And last, but not least, the author wishes to thank his constantly helpful wife Maggie and his secretary Pat Weimer; the former for her patience, encouragement, and for acting as a sounding-board, and the latter who toiled endlessly, cheerfully, and most competently on the book's preparation.

CONTENTS

1

INTRODUCTION

Economics is a subject that almost everyone deals with on a daily basis, since one of its facets is the study or practice of utilizing, controlling, or budgeting money or wealth. Most people earn money, make spending decisions and budgets, and plans for future income, investments or spending. This truly is the practice of economics, but of course, it is a fairly limited part of the subject. Far more economic detail is required of most chemical engineers in the execution of their jobs, and naturally even more for managers, financial staffs, and so on.

Chemical engineering almost by definition involves the practical and economic blending of science, engineering, and mathematics to solve problems and make engineering decisions in the conduct of the employer's business. Every engineer is expected to have some economic knowledge, and most will use this training (at least as background) frequently. The need for chemical engineering economic familiarity might include, among other things: decision making on equipment, projects, or plants; managing or controlling new or old operations, projects, or plants; cost estimating; budgeting, accounting; understanding reports, balance sheets, operating statements; marketing, market research; foreign sales and currency; personal finances, the stock market, funds and other investments; and economic indicators, inflation, national budgets; and so on.

Economics is far from an exact science, and much of what you may need to know about it will be generalities and procedures rather than exact equations or models. Even though "unscientific" it still will have considerable value, and the skillful and common sense use of economics can constantly improve most chemical engineers' capabilities. Unfortunately, this lack of scientific basis, its general simplicity, and the frequent inexactness of the economic techniques makes some engineers or students feel that it has little value and is too easy and general to be interesting. This is largely unavoidable, so it is urged that if this reaction is encountered, try to rise above it, and attempt to master the subject as well as possible. There is no question that it can be useful in your career and in your personal life.

FREQUENTLY USED ECONOMIC STUDIES

Economics is as much a career tool for chemical engineers as any of the unit operations or other engineering subjects. As with any single facet of an education, most engineers will only make detailed economic studies periodically, and for extensive use considerable additional learning of company methods and new subjects will be required. However, early and throughout the career of most engineers a basic knowledge of the major economic procedures used in financial decisions can provide a much better ability to solve problems, as well as provide a ''feel'' for the employer's business, and an ability and confidence to better conduct, and then understand the outcome of engineering assignments. Knowledgeably employing economic principles throughout one's career will generally also increase the probability of engineering success and advancement.

Economics is a required subject in all accredited chemical engineering departments, indicative of the general feeling that economics is an integral part of an engineering education and that all engineers should have a ''practical'' viewpoint, and consider costs and general economics in their decisions and actions. It is also interesting that in most surveys of recent chemical engineering graduates or others concerning what additional courses should have been offered, or which they wished they had taken in their college careers, more economics is always stressed (Basta 1986, 1987; Hughson and Lipowitz 1983; Septenary Committee 1985). The chemical engineering curriculums are over crowded with high priority, new potential subjects, so this is difficult to respond to (*Compressed Air Magazine* 1986; Farrell 1985; Kreiger 1987; Mathis 1986), but it does indicate that a fair percentage of young engineers do find that economics is an important subject in their jobs. This has very much been the author's own experience, not only in his own industrial career, but with most economics courses taught to date, one or more graduating seniors have come back the next year and related how useful the economics course was in their first assignments, and the cudos they had received for their skill with the subject.

Examples of how one might use economics throughout a chemical engineering career are given in Table 1-1. It is seen that for most job categories, preliminary cost estimates, simple profitability analyses, budgets, a general understanding of their industry's activities, corporate operating statements, balance sheets and financial indicators, project management, and personal finances are the economic principles that most chemical engineers will need to know. It is, therefore, this basic information that will be covered in this text. Initial and primary emphasis will be on cost estimation, followed by techniques for making economic decisions and a review of the economics of the chemical industry.

It is usually quite expensive and time consuming to have either an engineering contractor or an in-house engineering department prepare preliminary cost estimates and economic analyses for a small project or a new idea, and consequently it is often not possible to obtain management approval and budgets too early or frequently for such work. Thus, until one is at the authorizing management level an engineer must do most of this estimating himself, if it is to be done at all. In many cases such estimates are very preliminary or for personal information and guidance only, so they do not need to be extremely accurate, but a knowledge of the details in the cost breakdown and factors in reasonably reliable economic analyses is much more useful than an off-the-cuff guess.

It is fairly obvious that medium to upper level management will frequently make economic decisions, and perhaps occasionally use the wide variety of economic analyses covered in advanced economic texts, but this need will be many years in the future for most engineers. The economic fundamentals such as the time value of money taught now will still be valid, but by then all of the financial calculations and procedures will be organized in computers for each company (most large companies are at this point now), and the specific analyses used will be somewhat different from the ones now in vogue. As with most aspects of one's career, there will be a need to constantly learn new and changing technologies, and economics will not be an exception. Consequently, it is felt that to be most useful this book should concentrate on the basic principals of the more widely used economic concepts and knowledge frequently encountered in a working engineer's career, and provide information and techniques that can be used during his advancement period. An attempt will be made to provide such information in the following chapters.

BASIC ECONOMIC SUBJECTS

Priorities

Since economics is such a broad general subject with many specialties and quite divergent detailed areas of interest, even the experts usually do not attempt to master the entire subject. The same general limitations usually apply to engineers, for even within their area of interest there are a rather large array of specialties and fringe economic fields. This text will not attempt to provide the detail that a specialist or highly experienced engineer may need to know, but rather to cover only the more basic components of a wide range of economic subjects that are frequently encountered in engineering assignments. Similarly it is not intended to present a compendium of all of the economic ideas or information that have been presented on each subject, but rather to focus primarily on the most commonly used methods, procedures, and information that are somewhat of a current concensus of opinion.

In this manner it is hoped that the working engineer will be provided with a broad basic economic background which will allow him to better perform and enjoy his work, and assist in his career advancement. Economic knowledge and understanding are highly prized by industry, particularly if they cover a wide range of subjects, such as project evaluation, a knowledge of current industry

Table 1-1
Examples of the Use of Chemical Engineering Economics.*

1. Production or Plant Technical Service
   a. Plant equipment continuously needs repair, replacement, or modernization. The responsible engineer should know roughly what the comparative performance, costs, and payout periods are, even if there is a plant engineering group to do that type of analysis, or later firm price quotations will be obtained.
   b. Plant changes, such as initiated by the ever increasing costs of energy and environmental requirements, necessitates that many energy saving, pollution, and hazardous waste control possibilities must be considered. For the responsible engineer to make intelligent recommendations, he should personally conduct design and cost estimates, pay-back, and economic calculations on the alternatives before making even preliminary recommendations.
   c. Competitors' processing methods, as well as R & D, sales, or management suggested changes must be continuously examined. Supervisors or other groups may be responsible, but the staff engineer can help a great deal by making preliminary cost estimates and economic analyses of the changes to guide his own thinking (and hopefully the group's position).
   d. All engineers should have a feel for their company's business, products, and economics. This requires occasional economic reading, and a basic understanding of company annual reports and general industry economic news.
2. Research and Development
   a. In the process of being creative one thinks of many novel ideas. During the analytical phase of creative thinking many of the ideas will require a quick cost estimation and economic analysis to provide a better idea of their merit. Even though supervisors or others may be assigned to do this work, the chances of conceiving good ideas and having them accepted increases immensely if some economic screening can be done by the originator.
   b. While conducting R & D studies there are always many stumbling points, or alternative directions that may be taken in attempting to solve the problems. Often brief cost estimates and economic analyses will help in deciding which are the most promising directions to pursue.
   c. After an early or intermediate stage of an R & D program has been successful, new funding requests usually are required to continue the study. These requests can always benefit from having potential preliminary economic analyses. Later, in the final stages of a successful project the engineer may be part of a team assigned to provide a more definitive preliminary economic projection and analysis.
   d. In dealing with production, sales, or management personnel one can usually gain more respect, and be considered more practical and less ''theoretical'' by having a reasonable knowledge of the costs and economics of the projects under study, and general industry economics.
3. Sales
   a. A general knowledge of company costs, profits, and competition are very helpful for more effective salesmanship.
   b. Salesmen often recommend new products, improvements, or pricing ideas to their management. A cost and economic estimate for these ideas should be helpful in the proposal report.
   c. Salesmen sometimes perform market surveys. Again, a general economic knowledge of the industries and companies surveyed may be essential, and is always useful.
   d. Salesmen, as the other chemical engineering occupational categories, may move into management, where economic knowledge is a major part of the job.
4. Engineering
   a. Because of the extreme specialization of most engineering companies and many company engineering departments, cost estimating and economics may not be directly required in

Table 1-1 *(Continued)*

many engineering company or department jobs. However, other jobs will deal exclusively with cost estimating and economic analysis, and all will benefit from a good, fluent knowledge of the basic economic procedures. Engineering departments or companies usually have very well-developed in-house methods and data that must be used, but the basics are still applicable.

5. General
   a. All chemical engineers are assumed to know the rudiments of cost estimating, economic evaluation, and the economics of their industry. A high percentage will find this knowledge useful or necessary throughout their careers.
   b. All work situations are more or less competitive, and one means of maintaining the highest advancement potential with most jobs is to convince superiors of your knowledge and interest in management, business, and economics. Associated with this is the demonstration of ability, an interest in accepting responsibility, and ability to communicate. Many companies promote people whom they think can "manage," such as those with MBA (Master of Business Administration) degrees, or with perceived equivalent capabilities ahead of people with a better performance record. Basically, a confidence that you can learn managerial skills as you need them, and a knowledge of economics should make most chemical engineers equal or preferred candidates for advancement.

*(See Table 12-3 for general job classifications).

conditions, and so on. Hopefully the material in this book can provide this base, which will be appreciated by management, directly useful in engineering assignments, and later used as a foundation for further specialization if desired.

The text starts with the basics of making your own cost estimates, first with single pieces of equipment, then with plants, and finally in estimating manufacturing costs. It next uses this information, along with consideration of the time value of money to evaluate the profitability of the equipment, processes, or plants that you have just estimated. These are the basic components of engineering economics. However, the following chapters present somewhat periphery economic information that is just as useful and needed for a complete knowledge and feel for the subject. This starts with a fairly detailed review of the present status of the chemical industry, and the basics of accounting, budgets, and corporate reports. It is followed by the economics of project management, and finally, a brief discussion is given on the more personal aspects of economics—investing and job hunting.

These chapters have been presented in a general sequence of their importance and dependence upon preceding information. Thus, with limited time available, the first six chapters may be considered as the most basic and important. The following chapters examine equally useful economic information, but progress further and further into varied subdivisions of the main or basic engineering economic requirements. Most engineers will profitably use all of this material in their careers, but the first six chapters are absolutely essential, and the others perhaps more optional.

Problems

For each chapter a brief series of questions and problems on the information covered are included in Appendix 4, followed by the answers and often a brief review of how the problem was approached. Hopefully these questions will cover some of the more important information, and provide practice in solving assignments that might be encountered in common engineering situations. The problems were not inserted in the text in order to reduce potential distraction to the reader, but in case there is some uncertainty while reading the text or the calculation methods it may be advisable to turn to Appendix 4 and work through the relevant problems. This practice may also be useful to provide some fluency in the various procedures.

It should be cautioned in considering the problems that in most practical economic situations additional calculations, heat and material balances, and almost always some (or considerable) conversion of units are required to utilize your basic design data in a form that can be applied to the tables and charts or calculation methods. Conversion is one of the constant requirements of engineering problems, and in order to assist in this regard some of the more common conversion factors encountered in these calculations are listed in Appendix 5.

Appendixes

The first three appendixes provide cost estimating charts for chemical engineering equipment, complete processing plants, and manufacturing cost estimates, respectively. Appendix 4 gives questions and answers, while Appendix 5 lists some of the most commonly used chemical engineering conversion factors. The equipment estimating charts of Appendix 1 are intended to allow simple equipment or plant cost estimates of a rough but ballpark accuracy which can be quickly performed when vendor quotations are not available. The complete plant estimating charts provide an estimate of what at least one or a small number of actual plants have cost, thus again allowing a rough estimate of these more complex facilities to produce specific industrial chemicals. The complete plant manufacturing cost data are more limited, but still cover a number of the more common chemicals. In total these charts provide a fairly broad compendium and average of the cost estimating literature in one easy-to-use location.

REFERENCES

Basta, Nicholas, and Mark A. Lipowitz. 1986. Challenging times for chemical engineers. *Chemical Engineering* (Oct. 27):110–135.
— 1987. Chemical engineering education: Are changes really called for? *Chemical Engineering* (June 22):23–27.

Compressed Air Magazine. 1986. Engineering education crisis. (Sept.):8–15.

Farrell, Pia. 1985. Chemical engineering schools, what they're doing to adapt. *Chemical Engineering* (April 15):22–25.

Hughson, Roy V., and Mark A. Lipowitz. 1983. A CE survey, putting colleges back on course. *Chemical Engineering* (Sept.):48–60.

Kreiger, James H. 1987. Report on chemical engineering reflects a profession in flux. *Chemical and Engineering News* (Nov. 30):7–12.

Mathis, J. F. 1986. Chemical engineers meet a changing world. *Chemical Engineering Progress* (July):17–21.

Septenary Committee, Chemical Engineering Dept., University of Texas, Austin. 1985. Chemical engineering education for the future. *Chemical Engineering Progress* (Oct.):9–14.

2

EQUIPMENT COST ESTIMATING

As indicated in Chapter 1, one of the basic economic skills that chemical engineers often need is the ability to make cost estimates. The foundation of most estimates is the cost of individual pieces of equipment, and developing that estimating skill is the purpose of this chapter. All chemical engineers should know how to do this, as it is fundamental to considering costs and economics for engineering decisions and recommendations, as well as being "practical" and "cost conscience" in one's work. On a direct use basis, equipment cost estimates are frequently required in many types of work, and are the first step in more detailed plant cost estimates.

MANUFACTURERS' QUOTATIONS

When an engineer has the time and authorization to obtain actual price quotations from several vendors on process equipment his cost estimates will be fairly accurate for the direct purchased price over a reasonable time period. He will still have to estimate shipping and installation charges, but again, with adequate time and funds, price quotations and existing company records on previous jobs can lead to accurate estimates for the installed equipment. This is the preferable method of making all cost estimates, but if there is not time or funds for such detail, or one's supervisor does not authorize the extensive outside contact that would be required, vendor quotations should be obtained on at least the major or most expensive equipment, if at all possible. Then, even if installation costs and the minor equipment must be estimated, the final results will still be fairly accurate and well documented.

Several methods may be used to obtain vendor quotations. For fairly widely used equipment the local phone book yellow pages, or those from the nearest large city, will generally supply the names of agents for different equipment manufacturers, and they can be called directly. For less common equipment, or more extensive lists of vendors for all types of each piece of equipment, there are excellent directories readily available, such as the *Chemical Week* (1987), *Chemical Engineering* (1987) and *Chemical Processing* (1987) *Buyers*

Guides, the *Hydrocarbon Processing Catalog* (1987) and many more (*American Laboratory Buyer's Guide* 1987, etc.). These volumes are given free to all subscribers of their corresponding magazines because of their advertising content, and are very useful. Many of the technical or news magazines (some of them also free) also periodically publish lists of all manufacturers of selected equipment, along with notations of the equipment's major characteristics. Other technical journals periodically publish review articles discussing different equipment in detail, providing excellent information to use in deciding which types of equipment you should consider, and what the major design variables are.

When asking the manufacturer or sales agent for prices you generally will be speaking to a technical representative, who sometimes is a chemical engineer himself, and frequently is very knowledgeable and helpful in discussing design and equipment type alternatives. In these cases you end up with a great deal of technical information as well as the equipment prices you want, and the conversation will be both a pleasant and profitable one. Unfortunately, however, most inquiries do not go that well. The salesman will often want to know design details that you have not yet considered, and some in a trade jargon where you have no idea of what they want. Usually anything more than basic design parameters are unimportant for preliminary cost estimates, but some sales people will not give quotations without their special information. In this case a number of calls are required before you learn the correct language and have calculated a few more parameters in order to obtain an answer, and even then you may not be successful because the salesman may not feel that you are a bonafide potential customer, or the proper representative from your company. Thus, obtaining price estimates on equipment can be a lengthy and difficult process, or it can be quick and educational. It is difficult to know which it will be in advance, so patience and an open mind are required.

Professional cost estimators with engineering companies or company engineering departments make their estimates in this manner on essentially all jobs except those on the tightest budgets and with well-announced and acceptable lower accuracy ranges. For the normal preliminary cost estimate (that will hopefully be within 20–35% accuracy) professional estimators might only obtain one cost quotation on each major piece of equipment, and estimate the installed costs from their extensive records. For greater accuracy, however, they will obtain several price quotations on the equipment, actual costs or quotations on any other component possible, and then detail labor and materials, along with overhead, for all of the installation costs.

The cost to make such estimates is very high, and it takes considerable company and contractor labor and time. It is rare that an engineering company will make any cost estimate for less than $50,000 unless they are doing it as a fairly sure inducement of being granted a large construction contract. Generally the final cost estimates are taken from the detailed engineering design for the plant, and would cost from 1 to 5% of the total plant cost. Because of the high possibility for overrun, if a contractor is required to give a firm cost quotation on a new plant he will define the scope of work extremely tightly and still add a 10 to 20% contingency factor. With luck this could make the contractor a very good profit on a fixed cost job, but the risks are so great that many of the largest contractors will not even bid on a fixed cost construction project. This further illustrates the difficulty of making accurate cost estimates.

With the advent of the personal computer as a common engineering tool numerous software packages have been developed to assist with cost estimating. These vary from fairly simple programs such as "Price" from McGraw-Hill to elaborate ones such as "Questinate" from the Icarus Corp. Also, groups such as the Icarus Corp. will perform capital cost estimates on a contract basis. These programs basically work with cost versus size data, as will next be discussed, and probably have about the same accuracy as our estimating charts. They may be somewhat more current and extensive, although this is not certain. It is well to know of these software packages and services, but for most cost estimating work to be discussed in this text they are not necessary or too useful. For more extensive cost estimating requirements they may become much more valuable.

ESTIMATING CHARTS

The factors of high cost and lengthy time requirements to prepare cost estimates require that most estimates needed by working engineers must be made by themselves, and often even vendor quotations are not possible. To cover this situation a large number of estimating charts have been published in the literature over the years and are available for much less accurate, but more quickly prepared preliminary costs estimates. There have been a number of excellent compilations of such charts in books and articles in the past, and new estimating charts, tables, and equations are occasionally published. Unfortunately, however, most of them are based upon old data, and consequently may have further decreased in their accuracy. Inflation factors are always required to attempt to correct the charts to a current date, but because of the variability of changing technology and unequal price inflation for different equipment with time, this may not be a consistent correction. Charts are not available for every type of equipment one may need to estimate, but often there are charts for somewhat similar equipment, and at least a rough approximation may be obtained. Even with these limitations, equipment cost estimating charts serve a very useful purpose for most chemical engineers.

A number of charts of this type have been assembled in Appendix 1, with the individual equipment listed in alphabetical order. Most charts are a compilation of data from a number of sources, combined with some current price quotations, so they may be considered to represent a somewhat conservative (although generally higher) median price. The range of actual current price quotations from all vendors with widely differing quality, designs, and specifications can often vary by +100%, −50% from these curves. However, the values still should be in the generally correct range, and represent an approximate average of good-to-high quality equipment. By the time a number of pieces of equipment have been estimated and totaled for a complete design package, the overall estimated accuracy should be somewhat improved.

Size Factoring Exponents

A typical equipment cost curve from the Appendix is presented in Figure 2-1 for a centrifugal compressor. Note that most cost versus size curves are plotted on log–log paper, and that the data are depicted as a straight line. This is the most general situation for much of the cost data, and the slope of this line becomes an independent estimating exponent that allows the cost of one size of equipment (which you may know) to predict the cost of another size by the relationship

$$\text{cost of second size equipment} = \text{known cost of first size}\left(\frac{\text{size of second}}{\text{size of first}}\right)^{\text{exponent}}$$

These exponents (the slope of the cost versus size lines) are given under each chart, and in this example is 0.80. Thus, if a 200-HP reciprocating compressor with a motor drive is known to cost $106,000, a 1,000-HP unit should cost

$$\$106,000\left(\frac{1000}{200}\right)^{0.80} = \$380,000$$

With some log–log graphs the cost data plot as a curve, and if this is the case the slope is constantly changing, and the best that can be realized with sizing exponents is a series of slopes (exponents) that approximate the curve for a short segment. Centrifugal pumps present such a problem, so often three or more exponents are listed for short size ranges. If the cost curve is best represented by a semilog graph or regular coordinates (such as centrifuges), then a sizing exponent has little meaning.

With all of the cost curves there are obviously more than one variable that affect the cost. Because of this some authors have chosen one parameter, such as cubic feet of gas per minute for compressors, while others have chosen another, such as horse power, to act as the primary variable. In general in this

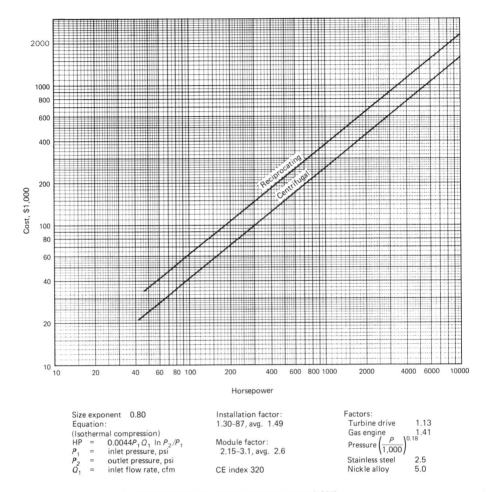

Size exponent   0.80
Equation:
(Isothermal compression)
HP   = $0.0044P_1Q_1 \ln P_2/P_1$
P_1 =   inlet pressure, psi
P_2 =   outlet pressure, psi
Q_1 =   inlet flow rate, cfm

Installation factor:
1.30–87, avg. 1.49

Module factor:
2.15–3.1, avg. 2.6

CE index 320

Factors:
Turbine drive   1.13
Gas engine   1.41
Pressure $\left(\dfrac{P}{1,000}\right)^{0.18}$
Stainless steel   2.5
Nickle alloy   5.0

a. Average of Guthrie (1974); Hall et al. (1982); Happel and Jordan (1975);
   Peters and Timmerhaus (1980); Pikulik and Diaz (1977)

Figure 2-1. Compressors, high capacity and/or pressure[a].
(1,000 psi; electric motor drive; gear reducer; steel)

book, an attempt has been made to choose the more dominant variable, or a combination such as gallons per minute times the head (as pounds per square inch) for pumps. Whenever simple variables exist that can be plotted as a companion graphs (in Figure 2-1, with reciprocating or centrifugal types of compressors) this was done, or if multiplication factors could be used with the charts such as for different materials of construction or higher pressure, this was added as a brief table below the graph.

Inflation Cost Indexes

Another item of information that is needed for each graph is the *Chemical Engineering* Plant Cost Index (CE Index—see Figure 2-2) that existed at the time the graphs were constructed; 320 for early 1987. This is an inflation indicator made specifically for the chemical industry to correct the cost of each piece of equipment to the date of your estimate, by the relationship

$$\text{equipment cost at your date} = \text{chart cost} \times \frac{\text{CE Index, your date}}{\text{CE Index, chart } (=320)}$$

For example, the chart indicates a pump to cost $1,000 at the time when the CE Index was 320, but it is 345 when you make your estimate. You would show the pump as costing

$$\$1,000 \times \frac{345}{320} = \$1,078, \text{ or } \$1,100 \text{ based on the probable accuracy}$$

There are a fairly wide variety of inflation cost indicators that could be used to provide a measure of how the costs of labor, material, supplies, and equipment increase each year. Any one of the factors could be used to update the equipment cost charts, but the oldest and perhaps the best known of these indicators for engineers is the Marshall and Swift Equipment (M & S) Cost Index (Figure 2-2 and Table 2-2). Many chemical engineers prefer to use it, but the one specifically designed for chemical plants is the *Chemical Engineering* Plant Cost Index, called the CE Index. Both are listed each month, along with a 10-year notation of past yearly indexes, in the magazine *Chemical Engineering,* as shown in Figure 2-2. Added to this page are CE Index values back to 1956 (Table 2-1). The CE Index is composed of four components, weighted as follows: equipment, machinery, and supports, 61%; erection and installation labor, 22%; buildings, material, and labor, 7%; and engineering and supervision, 10%. A survey is taken each month of selected manufacturers and contractors in the industry, and the price increases averaged and tabulated to form the index. The yearly figure is established as the average value for that year.

It would be expected that technological improvement, competition, the price of raw materials, labor contracts, and so on would require each manufacturer, product, and cost component to have quite distinct rates of cost escalation, but for the entire industry and larger assemblages of equipment the index should provide a reasonable means of expressing the average escalation of costs.

The curves of Figure 2-3 compare the Engineering News-Record (ENR) Construction Cost Inflation Index with the M & S and CE Plant Indexes, and

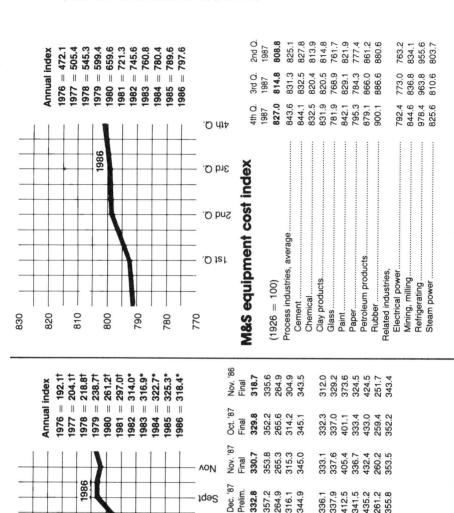

CE plant cost index

Annual index
1976 = 192.1†
1977 = 204.1†
1978 = 218.8†
1979 = 238.7†
1980 = 261.2†
1981 = 297.0†
1982 = 314.0*
1983 = 316.9*
1984 = 322.7*
1985 = 325.3*
1986 = 318.4*

(1957-59 = 100)	Dec. '87 Prelim.	Nov. '87 Final	Oct. '87 Final	Nov. '86 Final
CE plant cost index	**332.8**	**330.7**	**329.8**	**318.7**
Equipment, machinery, supports	357.2	353.8	352.2	335.6
Construction labor	264.9	265.3	265.6	264.9
Buildings	316.1	315.3	314.2	304.9
Engineering & supervision	344.9	345.0	345.1	343.5
Fabricated equipment	336.1	333.1	332.3	312.0
Process machinery	337.9	337.6	337.0	329.2
Pipe, valves & fittings	412.5	405.4	401.1	373.6
Process instruments	341.5	336.7	333.4	324.5
Pumps & compressors	435.2	432.4	433.0	424.5
Electrical equipment	261.2	260.2	259.4	251.7
Structural supports & misc.	355.8	353.5	352.2	343.4

*Revised; productivity factor = 1.75.
†Unrevised; productivity factor = 2.50.

M&S equipment cost index

Annual index
1976 = **472.1**
1977 = **505.4**
1978 = **545.3**
1979 = **599.4**
1980 = **659.6**
1981 = **721.3**
1982 = **745.6**
1983 = **760.8**
1984 = **780.4**
1985 = **789.6**
1986 = **797.6**

(1926 = 100)	4th Q. 1987	3rd Q. 1987	2nd Q. 1987
Process industries, average	**827.0**	**814.8**	**808.8**
Cement	843.6	831.3	825.1
Chemical	844.1	832.5	827.8
Clay products	832.5	820.4	813.9
Glass	831.9	820.5	814.8
Paint	781.9	768.9	761.7
Paper	842.1	829.1	821.9
Petroleum products	795.3	784.3	777.4
Rubber	879.1	866.0	861.2
Related industries,	900.1	886.6	880.6
Electrical power	792.4	773.0	763.2
Mining, milling	844.6	836.8	834.1
Refrigerating	978.4	963.8	955.6
Steam power	825.6	810.6	803.7

Source: *Chemical Engineering* 1987. Excerpted by special permission from *Chemical Engineering*, Feb. 15, 1987. Copyright © 1987, by McGraw-Hill, Inc., New York.

Table 2-1.
Trend of Plant Costs Since 1956 (Base period: 1957–1959 = 100).

	1956	1957	1958	1959	1960	1961	1962	1963	1964
Chemical Engineering Plant Cost Index	93.9	98.5	99.7	101.8	102.0	101.5	102.0	102.4	103.3
Equipment, machinery and supports	92.7	98.5	99.6	101.9	101.7	100.2	100.6	100.5	101.2
Construction labor	95.8	98.6	100.0	101.4	103.7	105.1	105.6	107.2	108.5
Buildings	98.0	99.1	99.5	101.4	101.5	100.8	101.4	102.1	103.3
Engineering and manpower	94.2	98.2	99.3	102.5	101.3	101.7	102.6	103.4	104.2

Year 1965	1966	1967	1968	1969	1970	1971	1972	1973	1974	1975	1976	1977
104.2	107.2	109.7	113.6	119.0	125.7	132.3	137.2	144.1	165.4	182.4	192.1	204.1
102.1	105.3	107.7	111.5	116.6	123.8	130.4	135.4	141.9	171.2	194.7	205.8	220.9
109.5	112.5	115.8	120.9	128.3	137.4	146.2	152.2	157.9	163.3	168.6	174.2	178.2
104.5	107.9	110.3	115.7	122.5	127.2	135.5	142.0	150.6	165.8	177.0	187.3	199.1
105.6	106.9	107.9	108.6	110.9	110.6	111.4	111.9	122.8	134.4	141.8	150.8	162.1

Source: Kohn 1978.

Trend of Equipment Costs Since 1956 (History of the Equipment Component and Major Subcomponents of CE Plant Cost Index. Base period: 1957–1959 = 100).

	1956	1957	1958	1959	1960	1961	1962	1963	1964
Equipment, machinery and supports	92.7	98.5	99.6	101.9	101.7	100.2	100.6	100.5	101.2
Fabricated equipment	92.5	99.5	99.6	100.9	101.2	100.1	101.0	101.7	102.7
Process machinery	92.2	98.1	100.1	101.8	101.8	101.1	101.9	102.0	102.5
Pipe, valves and fittings	94.8	97.9	98.8	103.3	104.1	101.1	100.6	100.7	101.6
Process instruments and controls	91.2	96.7	100.4	102.9	105.4	105.9	105.9	105.7	105.8
Pumps and compressors	90.0	97.5	100.0	102.5	101.7	100.8	101.1	100.1	101.0
Electrical equipment and materials	93.5	98.4	100.6	101.0	95.7	92.3	89.4	87.6	85.5
Structural supports and miscellaneous	92.5	98.0	100.4	101.6	101.9	99.8	99.2	97.3	98.3

Year 1965	1966	1967	1968	1969	1970	1971	1972	1973	1974	1975	1976	1977
102.1	105.3	107.7	111.5	116.6	123.8	130.4	135.4	141.9	171.2	194.7	205.8	220.9
103.4	104.8	106.2	109.9	115.1	122.7	130.3	136.2	142.5	170.1	192.2	200.8	216.6
103.6	106.1	108.7	112.1	116.8	122.9	127.1	132.1	137.8	160.0	184.7	197.5	211.6
103.0	109.6	113.0	117.4	123.1	132.0	137.3	142.9	151.5	192.3	217.0	232.5	247.7
106.5	110.0	115.2	120.9	126.1	132.1	139.9	143.8	147.1	164.7	181.4	193.1	203.3
103.4	107.7	111.3	115.2	119.6	125.6	133.2	135.9	139.8	175.7	208.3	220.9	240.2
84.1	86.4	90.1	91.4	92.8	99.8	98.7	99.1	104.2	126.4	142.1	148.9	159.0
98.8	101.0	102.1	105.7	112.6	117.9	126.6	133.6	140.9	171.6	198.6	209.7	226.0

Source: Kohn 1978. Excerpted by special permission from *Chemical Engineering*, May 8, 1978. Copyright © 1978, by McGraw-Hill, Inc., New York.

Table 2-2.
Marshall & Swift Annual Indexes of Comparative Equipment Costs Since 1956 (Base Period: 1926 = 100).

	1956	1957	1958	1959	1960	1961	1962	1963	1964
Equipment Cost Index	208.8	225.1	229.2	234.5	237.7	237.2	238.5	239.2	231.8
Process industries									
Cement	199.4	216.4	222.8	228.7	232.1	231.1	231.8	232.5	235.9
Chemical	209.1	226.5	232.3	236.5	239.2	237.7	238.0	238.7	241.1
Clay products	193.8	210.2	216.8	222.2	225.7	224.6	225.5	225.8	229.2
Glass	197.5	213.8	219.3	223.2	225.3	224.4	224.7	225.4	227.6
Paint	201.2	217.6	223.2	226.9	229.5	230.0	231.5	232.1	235.0
Paper	210.5	218.2	223.8	227.8	229.9	229.0	229.3	229.9	232.3
Petroleum Products	205.4	222.2	228.0	231.8	234.3	235.0	238.2	238.8	241.8
Rubber	207.9	224.9	230.8	234.6	237.3	237.9	239.2	240.0	243.0
Related Industries									
Electrical power	211.0	229.2	235.2	239.0	241.0	236.3	235.6	234.7	236.8
Mining, milling	210.4	227.9	233.8	237.1	240.6	239.2	239.5	240.1	242.6
Refrigerating	234.3	254.2	260.8	265.1	268.2	268.8	270.4	271.2	274.6
Steam power	197.0	213.0	218.6	222.9	224.7	225.3	226.6	227.2	230.1

Year												
1965	1966	1967	1968	1969	1970	1971	1972	1973	1974	1975	1976	1977
244.9	252.5	262.9	273.1	285.0	303.3	321.3	332.0	344.1	398.4	444.3	472.1	505.4
239.3	249.6	258.1	268.3	280.1	298.3	317.3	328.5	340.1	398.1	454.3	482.2	518.3
243.8	246.1	261.8	271.6	282.8	300.8	319.1	329.7	341.0	400.5	447.6	473.2	507.1
232.6	239.0	250.7	260.7	272.2	289.9	308.4	319.3	330.3	384.5	437.2	466.9	504.2
230.1	240.6	271.1	256.3	266.9	283.9	301.2	311.3	322.0	375.4	425.3	451.3	483.5
238.1	243.9	255.7	267.4	280.3	300.1	320.7	333.3	345.9	406.0	455.0	481.2	515.3
234.8	247.5	252.1	261.6	272.4	289.7	307.5	317.7	328.4	381.7	430.1	457.6	491.2
244.9	253.9	263.4	275.2	288.4	308.8	330.6	343.0	356.0	419.3	470.9	497.7	535.9
236.2	251.2	264.7	276.5	289.9	310.3	331.6	344.7	357.8	415.2	468.5	497.3	532.8
239.4	246.5	257.3	264.9	274.8	290.3	307.4	316.4	327.2	391.7	438.5	465.1	494.1
245.3	253.0	263.5	273.2	284.5	302.6	321.1	331.8	342.9	394.3	451.2	482.9	520.8
278.2	287.1	299.1	312.5	327.5	350.6	374.7	389.5	404.6	472.4	528.1	558.4	595.9
233.0	240.4	250.6	261.8	274.3	293.8	313.9	326.3	338.8	400.8	448.2	475.7	507.7

Source: Kohn 1978. Excerpted by special permission from *Chemical Engineering*, May 8, 1978. Copyright © 1978, by McGraw-Hill, Inc., New York.

show that they do not exactly follow each other, nor maintain the same ratios. However, they are roughly comparable, and since the CE Index applies specifically to chemical plants, it shall be assumed to be more accurate and used as the reference for comparison in this text. The CE Index existing at the time the data in each chart were assembled was 320, corresponding approximately to April, 1987.

Installation Factor

An additional factor shown on many of the charts is an installation factor. In some cost estimates one will be considering only one piece of equipment, or a

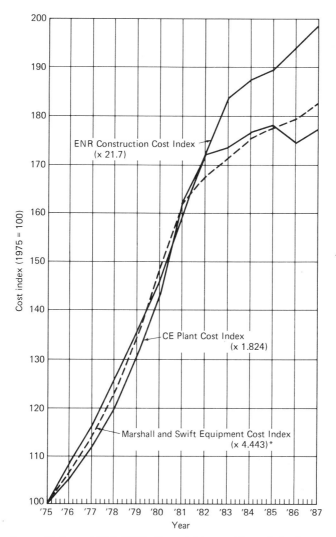

Source: Matley and Hick (1988). Excerpted by special permission from
Chemical Engineering, Apr. 11, 1988. Copyright © 1988, by McGraw-
Hill, Inc., New York.
*To convert to actual index

Figure 2-3. Comparison of inflation indexes.

very small number of pieces, and it is desired to know not only the cost of the
equipment itself, but also the total cost to purchase the equipment, have it
shipped, received, installed, and ready to run. Obviously this adds a large
number of additional factors, and no single installation number will be very
accurate or authoritative for each particular case. However, a rather general,

average figure can perhaps be quoted for a new installation in an existing plant that will give a rough, ballpark estimate, and these are listed below the charts whenever available. As an example, a rotary dryer installation factor is listed at 1.25–96, with the average 1.64. Thus, if the dryer cost is $50,000, the installed cost would usually vary from $63,000 to $98,000, with the average $82,000. This would include freight, crane service, foundations and supports, all labor, electrical, switch gear and controls, gas or oil piping connections, mounting (including the necessary extra material—rollers, bolts, motor couplings, chain drives, etc.), inlet and outlet chutes and hoods, painting, safety guards, and so on. It would not include the burner air compressor (if needed), the conveyor feed and discharge systems, dust collector, fan, ducting, and so on. In other words, this factor would only cover the actual mounting of the equipment itself, ready to run in an existing plant with all of the utilities and services nearby, but without any of its support or auxiliary equipment. The factor is mainly composed of labor, although there are always periphery equipment such as foundations, electrical switch gear, crane service, and so on required. These factors are not meant to be used for anything but single equipment estimates, and not complete plants. Their accuracy is a further step reduced from that of the equipment itself, but again should be in the correct general cost range for very preliminary estimates.

Module Factor

A similar factor to the installation cost is the "module factor" which is defined to include not only the installation cost of the equipment in question, but also the purchase and installation cost of all of the supporting equipment and connections. For the rotary dryers example, it now *would* include the feed and discharge conveyors, the dust collection equipment, the fan, ducting, and everything required to make the entire rotary dryer section of the plant operational. Its primary usefulness is in estimating costs when one major piece of equipment (alone) is to be installed as a new addition to an existing plant, and all of its periphery and support equipment must also be purchased and installed. The smaller "installation factor" is employed more frequently, since replacing or modernizing an existing piece of equipment only requires the installation of the new equipment alone, since all of the support equipment is already in place. Module factors are listed on the estimating charts when available, again with a range as given by various authors, and an average value.

Other additions may be made to the equipment cost estimating charts, such as when an equation has been proposed that describes the cost–size relationship, or pertains to the estimate, it is also given under the chart. The equation may not exactly follow the plotted line. Alternate material, pressure, size, and other variable factors are also included when available.

Estimating Accuracy

The rather poor accuracy of estimating charts has previously been discussed, but it must be remembered on the positive side that just as in purchasing any article of food or clothing, there will be price variations between different manufacturers of equipment that cover the range noted above based upon different features, quality, unusual competition, regional increases, and each manufacturer's costs and profit. Because of this, each chart should actually show a broad band, rather than a line, to indicate the true anticipated price differences. However, a single value is more useful to the estimator, and assuming that the charts try to show an average or typical cost, the user must be even more cautious in stating the accuracy obtainable in his final analysis. As noted earlier, actual vendor price quotations should be obtained whenever possible, but for preliminary estimates, and when actual prices can not be gathered, these charts can still provide excellent information when properly used. In practice this is a very useful function, for in much of an engineer's work approximate estimates are all that is needed.

It is recommended that any engineer interested in economic analysis or ultimate advancement into management keep a notebook of new cost estimating articles encountered in professional reading, and actual price quotations when obtained. In this manner cost estimates can become progressively more accurate, and one will constantly maintain a good general knowledge and feel for costs.

ESTIMATING EXAMPLE

With most of the cost estimating that an engineer may wish to do the problem that actually needs to be answered is somewhat different, and considerably more complex than just looking at the estimating charts. One will often have to go through at least a rough design procedure to establish the size parameter, sometimes convert units from one system to another, and in many cases make an estimate based upon judgment (or a guess) on some of the factors involved. All of this tends to be discouraging as the engineer thinks of the further inaccuracy or uncertainty involved, especially if there is not time or information to adequately study the problem and feel confident of the judgments. However, the point to remember in these very preliminary estimates is that they generally are for guidance and direction only, and that high accuracy is neither possible nor necessary. The engineer needs to be in the generally correct range with the answer, and if reasonable care is taken this will usually be the case. The results will be both useful and quite educational throughout the various steps of the study. One will have learned more about the equipment, its design, and the costs involved. And if the results look promising and management concurs, there will be adequate time later to do the detailed engineering, cost estimating, and prepare the final proposal much more accurately.

An example of a problem that an engineer might encounter in a production department where energy costs are high is as follows:

PROBLEM. You are a junior production engineer in a Green River, Wyoming natural soda ash plant and have been instructed to search for ways to reduce energy costs. The plant produces 1,000,000 t/yr Na_2CO_3, and now consumes 4.0 MM Btu (as steam)/ton Na_2CO_3 in the monohydrate, triple effect evaporator step. It has a 2.4/1 lb water evaporated/lb steam efficiency. You have heard that one of your competitors now uses only 0.8 MM Btu/ton Na_2CO_3 for this step after converting to vapor recompression. Would this be practical for your company? The CE Index is now at 400.

ANSWER. First find an approximate cost for the necessary new vapor recompressors. This requires a rough estimate of the cfm of steam to be "recompressed," which would be (assuming all evaporated water compressed twofold)

$$1,000,000 \text{ t/yr } Na_2CO_3 \times \frac{70\% \text{ H}_2\text{O in sat. soln.}}{30\% \text{ solids}}$$

$$\times \frac{2,000 \text{ lb/ton}}{365 \text{ d/yr} \times 0.95 \text{ on stream efficiency} \times 24 \text{ hr/d} \times 60 \text{ min/hr}}$$

$$\times \frac{359 \text{ cf/lb mole}}{18 \text{ lb/lb mole}} \times \frac{14.7 \text{ psa std. pres.}}{29.4 \text{ psa final pres.}} \times \frac{672°\text{R actual temp.}}{460°\text{R std. temp.}}$$

$$= 1.36 \cdot 10^5 \text{ cfm}$$

Using the horsepower equation, HP = $1.36 \cdot 10^5 \cdot 4.4 \cdot 10^{-3} \times 14.7 \cdot \ln 2$ = 4710 HP (see Figure 2-1). However, it would be easiest to use a separate compressor on each of the three evaporator effects, making the vapor flow from each effect $3.5 \cdot 10^4$ cfm. Using the horsepower equation, HP = $3.5 \cdot 10^4 \cdot 4.4 \cdot 10^{-3} \times 14.7 \cdot \ln 2 = 1570$ HP

From Figure 2-1 this would indicate a cost of about $0.36 million per centrifugal unit and $0.41 with a turbine drive. When installed (1.49 factor) this would be $0.41 × 1.49 × 400/320 = 0.76 MM for each unit, or $2.3 MM for the total cost of the three compressors. This estimate assumes installing the compressors to a well-prepared existing plant with all of the support equipment in place. However, this is a case where a totally new set of equipment (the compressors and their support equipment) is to be installed, so the module factor would be more appropriate for the complete installation. Also, since rather extensive new steam lines and controls will be required, perhaps the higher end of the module factor range should be used. On this basis the three new compressors might cost $0.41 million × 3 × 3.1 × 400/320 = 4.77 million. If constructed of stainless steel the cost would be $11.93 MM.

The potential profitability, assuming the competitor's steam consumption and a cost of $4/MM Btu steam in the existing coal fired boilers would then be 1 MM t/yr × (4.0 − 0.8) MM Btu/ton savings × $4/MM Btu = $12.8 MM savings for a $11.93 MM investment, or 11.93 × 12/12.8 = 11 month payout for the capital investment.

Obviously the situation will be much more complex than these simple calculations, and in reality when all factors are considered, such as the plant's steam-power balance, the downtime for the change over, and other costs that may be required or desirable, the savings may be much less. Later cost estimates based upon complete heat and material balances, equipment designs, and the cost of lost production for the change could be quite different. However, it would appear conclusively from this very quick preliminary economic analysis that the potential for considerable savings exists with such a change, and that the project should be highly recommended for further study.

REFERENCES

American Laboratory Buyers' Guide. 1987. International Scientific Communications, Inc., 808 Kings Highway, Fairfield, CT 06430–5416.

Chemical Engineering. 1988. (Feb. 15):9.

Chemical Engineering Equipment Buyers' Guide. 1987. McGraw-Hill, Inc., 1221 Ave. of the Americas, New York, NY 10020.

Chemical Processing Guide and Directory. 1987. Putnam Publishing Co., 301 E. Erie St., Chicago, IL 606111.

Chemical Week Buyers' Guide. 1987. McGraw-Hill, Inc., 1221 Ave. of the Americas, New York, NY 10020.

Guthrie, Kenneth M. 1974. *Process Plant Estimating, Evaluation and Control.* Craftsman Book Co., Solana Beach, CA:125–180; 334–353; 369–371.

Hall, Richard S., Matley, Jay, and McNaughton, Kenneth J. 1982. Current costs of process equipment. *Chemical Engineering* (April 5):80–116.

Happel, J., and D. G. Jordan. 1975. *Chemical Process Economics.* Marcel Dekker, New York:219–231.

Hydrocarbon Processing Catalog. 1987. Gulf Publishing Co., Box 2608, Houston, TX 77001.

Kohn, Philip M. 1978. Chemical engineering cost indexes maintain 13-year ascent. *Chemical Engineering* (May 8).

Matley, Jay and Hick, Ann. 1988. CE cost indexes. *Chemical Engineering* (April 11):711–73.

Peters, M. S., and K. D. Timmerhaus. 1980. *Plant Design and Economics for Chemical Engineers.* McGraw-Hill, New York:973.

Pikulik, Arkadie, and Hector E. Diaz. 1977. Cost estimating for major process equipment. *Chemical Engineering* (Oct. 10):107–122.

3

PLANT COST ESTIMATES

Once individual equipment costs are known from either manufacturers' price quotations or estimating charts as discussed in the previous chapter, they can be utilized to form preliminary total plant cost estimates. A suggested sequence for obtaining such estimates is as follows:

1. Prepare a flow sheet for the process or operation to be estimated, showing all of the major equipment and whatever of the basic auxiliary or general plant facilities that directly affect the process (i.e., boilers for steam; cooling towers; special electric requirements, unusual storage facilities, transportation equipment, etc.).
2. Prepare heat and material balances around each piece of equipment to the degree of accuracy and detail required to size the equipment.
3. Size all of the equipment with the precision required to obtain the parameters needed for manufacturers' cost quotations or to estimate costs from charts.
4. Analyze the process carefully to determine what "plant cost factors" should be used, and then prepare a detailed breakdown of the total plant cost package. This will be discussed in the following sections.

ACCURACY AND COSTS OF ESTIMATES

The American Association of Cost Engineers (AACE) has published a listing of different types of cost estimates, and the accuracy that such estimates theoretically have when estimated by professionals. It is shown in Table 3-1, along with perhaps a more realistic prediction of the accuracy range of a working engineer or contractor's estimates, and what management might actually count on for all estimates. A graph showing the possible spread in the accuracy of these estimates is shown in Figure 3-1. If it is assumed that what has just been done on equipment cost estimating, and will soon be done for plant costs is the first item, order of magnitude or very preliminary estimates, then the AACE (professional) accuracy figure would appear to be totally beyond the confidence

Table 3-1
Characteristics and Possible Accuracy of Chemical Plant Capital Estimates

Type of Estimate (AACE Name)	Common Names	Usual Basis	Usually Prepared by	Basis for Estimate	Result; Possible Approval for	Possible Accuracy[b]	
						AACE 1974	More Realistic Estimate
Order of magnitude	Very preliminary estimate	Estimating charts; previous cost information	Individual engineer	Basic idea	Inexpensive study	40	40–100
Study	Detailed preliminary, or factored estimate	Some vendor quotations; estimating charts, etc.	Project group	Following initial study	Expensive study	25	30–50
Preliminary	Initial budget, scope estimate	Vendor quotes on all major equipment	Contractor[a] (professional estimating)	Following final study	Detailed design; market research	12	20–35
Definitive	Project control estimate	Detailed quotes, labor, material estimates; not complete drawings	Contractor[a]	Complete design drawings	Construct plant	6	10–15
Detailed	Firm estimate, contractor's estimate	Competitive vendor quotes, complete drawings and specifications	Contractor[a]	Most equipment purchased	Continue construction	3	5–10

[a] Or company engineering department.
[b] ± % of total cost.

23

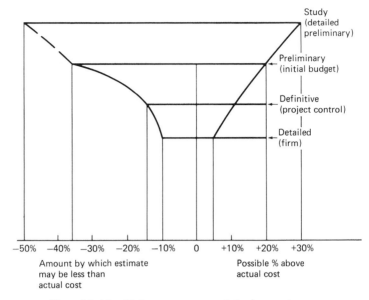

Figure 3-1. Most likely accuracy range of plant cost estimates.

factor of the average engineering estimate. As one looks down the table, this would also appear to be the case for each of their predicted accuracies.

The explanation of this claim for such high accuracy probably comes from several factors: first when these numbers were prepared (apparently in the 1960s) world economics were much more stable. Inflation and interest rates had been historically much lower and less subject to change, labor costs and work rules were simpler, more reliable, and stable, energy costs were much lower, safety rules and environmental permits and pressures were much less, and so on. All of these factors, along with equipment shortages, changed dramatically in the 1970s. Secondly, professional cost estimators may not take responsibility for changes such as the above, or other unforeseen "acts of governments, environmentalists, unions, or God," and claim that their accuracy is only based upon more normal changes and things they can control. Finally, it is extremely rare that a professional estimator would perform the first two types of studies. Their work is too expensive and slow for most companies to be able to afford their services for such early estimating purposes. Usually these estimates are left to the individual engineer or group, and they do not have the experience, exhaustive records, or time to make the estimates with such accuracy.

It is more likely that the initial very preliminary estimate has a range of up to 100% error, hopefully much better, but one cannot be sure that the error is not somewhere between 40 and 100%. This estimate will be made primarily with cost estimating charts or other plant estimating factors at early stages of a project evaluation by the engineer or group involved.

Table 3-2
Possible Cost of Chemical Plant
Cost Estimates

Type of Estimate	Possible Cost, $
Very preliminary	$2,000–5,000
Detailed preliminary	10,000–50,0000
Initial budget	50,000–200,000
Definitive	150,000–700,000
Detailed	1–5%[a]

[a]Percent of total plant cost. The range expands or contracts for small (<$2 million) or large (>$100 million) plants.

The detailed preliminary estimate is usually performed when a project is partly completed, or finished with its development stage. As such the estimate will often be made by a group, or an individual with much more time. Manufacturer's (vendor's) or subcontractor cost quotations will be used whenever possible, but many factors, charts, and estimating methods will still be employed. The estimate's accuracy should be in the 30–50% error range.

The initial budget estimate is usually the first one performed by a contractor or the company's engineering department—professional estimators. For this estimate they will use single vendor quotations on major equipment, and some charts and factors, but theirs will be based upon much more recent experience and far more detailed estimating information, and should be more correct. A 20–35% accuracy might be expected.

The next two estimates come after all competitive equipment bids have been received, and should be quite close; 10–15% before all of the drawings, engineering, and purchasing are complete, and 5–10% afterward. Even with the professional estimator's know-how and experience, however, environmental problems, permit hang-ups, labor trouble, supply problems, unusual weather, and so on can make each of these professional estimates have many fold the anticipated error. Such problems have become more the rule than the exception.

The cost to prepare economic estimates is even harder to predict, but might be somewhat in the range shown in Table 3-2. These figures are based upon typical time and material spent in preparing the estimate, with the actual cost dependent upon the detail desired and the complexity of the plant. Even an engineer's own early estimates are expensive if rigorously accounted for, but many will be done on one's own time for his or her own knowledge, and only an occasional estimate may be authorized or requested by the company.

Cost Overruns

The previous section has dealt with the generalities of cost estimating accuracy. It is interesting to compare this with actual studies on the cost of very large projects. A mid-1987 study (Parkinson 1987) concluded that most large projects were underestimated, and that the principal cause of cost escalation for megaprojects ($500 million to $10 billion) resulted from conflicts between the plant owner and governmental agencies. It is very likely that a similar situation exists for most medium-to-large projects. For the 52 mega projects studied the average cost overrun was 82% from that estimated at the beginning of detailed engineering; only four projects were under their budgets. (An earlier study for projects under $500 million averaged 31% escalation). The average time slippage was 18%. The study found that the owners and contractors went into the projects expecting to experience problems with logistics, labor, equipment, and materials, but in reality the problems were with environmental regulations, opposition from institutions and the public, worker health safety rules, labor practices, and procurement controls.

The study recommended that if costs were to be controlled, project managers and estimators must:

1. Explore the regulations of all concerned governmental agencies with respect to each aspects of a project as an absolutely essential part of project definition. Environmental impact statements (or their equivalent) may not be demanded, but usually are, and should generally be completed in comprehensive detail.
2. Frequent meetings are necessary with the regulators, politicians, and community groups.
3. Laws and regulations must be seen as legitimate by project managers, even if they are privately considered distasteful or wasteful.
4. The government makes the rules; they can change them at anytime—and often do (such as late in the permit stage, or during construction suddenly demanding environmental impact or other studies).
5. A project may provoke changes in rules by generating problems or opposition that politicians will seek to resolve or benefit from.
6. Politicians consider getting elected and staying in office more important than the success of any project.
7. Do not expect bureaucrats anywhere to be more reasonable, understanding, flexible, or quick about their (your) business than they are locally.

PLANT COST ESTIMATING FACTORS

There are many components in a total chemical plant besides the processing equipment itself. The plant occupies space, so there must be land or a platform. Each piece of equipment must be supported by foundations or structural members, and interconnected by piping, conveyors, electrical lines and switch gear, instruments, and so on. There must be buildings for labs, offices, warehouses, maintenance, and perhaps to enclose some or all of the plant. The so-called off-site facilities include boilers, generators, roads, utility service and transportation equipment. Finally, to build a plant requires engineering, construction labor, supervision and contractor's profit. Later there will need to be start-up expenses and working capital. A partial list of some of these plant components and costs are given in Table 3-3.

A large number of these items and many more are included in the cost of a totally new, or "grass roots" plant, while only some of them are required with simpler plant expansions or modernizations. The cost of each component can of course be estimated individually and in detail for the greatest accuracy, but it has been found that simplified estimating factors may also be used to provide a quick and reasonably reliable plant cost estimate.

Table 3-3
Partial List of Components in a Plant Cost Estimate

On-site Facilities
    Process equipment, such as:

Towers, columns	Dryers
Heat exchangers	Pumps
Filters, centrifuges	Evaporators
Reactors	Tanks

    Installation costs:

Labor to install equipment	Insulation
Piping	Utilities
Electrical	Yard improvements
Instruments	(grade, pave, fence)
Foundations	Painting
Platforms	Railing, catwalks
Buildings	Safety equipment
Fireproofing	Inventory, supplies, catalyst
Product storage, handling	Environmental facilities

    Construction expense:

Engineering, design	Tools, temporary facilities
Model; computers, software	
Supervision, overhead	Changes, additions
Construction equipment	Licenses, fees
Accounting; scheduling; planning	Environmental, safety, etc. studies, reports

    Company Costs:

Design, drafting	Research and development
Engineering	Licensing fees
Owner's inspection	Feasibility studies
	Market research

Off-site Facilities
  (See Table 3-6)
Start-up; Working Capital

In its simplest form, the total purchased price of all of the equipment included in the plant can be added and then multiplied by a single factor to estimate the total plant cost. Single multipliers of the equipment cost, used directly and alone have been called "Lang" (1947) factors, and currently vary from about 3.4 to 5.2, depending upon whether the plant processes primarily solids or liquids, and its complexity. Other estimating methods may ultimately come down to such single numbers, but the potential accuracy is much better when using smaller individual factors for the major plant components, such as piping, electrical, and so on. This also allows a more detailed understanding of where the costs are in a total plant cost estimate so that they may hopefully be reduced, and much better controlled.

In considering individual plant component (piping, etc.) factors, different authors have used many different breakdowns and estimating methods. Chilton (1960) used equipment multiplying factors which changed according to the variables most important to that type of plant, such as hazardous, complex and new. Guthrie (1974) used "modules" which encompassed all of the plant cost components related to one major piece of equipment or portion of the plant. Most others, however, have used the simplified "solids" (a small multiplying factor) or "fluids" (a large factor) handling basis, with more or less of the total plant components included. Some have based these multipliers upon the *installed* cost of the basic major equipment, but since this requires an extra conversion calculation to change the purchased cost to installed costs, it is far simpler to use purchased cost directly, so this will be done in this book.

Table 3-4 lists some of the more important components of a plant, and the multiplying factors suggested to be used with the purchase price of all of the equipment. As might be expected, over the years there have been literally hundreds of different, but similar, charts of this type suggested, with an even greater number of multiplying factors (Holland et al. 1974; Klumpar and Slavsky 1985a; Ward 1986). This table represents a reasonable consensus among them, and an attempted simplification. Greater complexity to perhaps provide better accuracy would not appear to be warranted because of the chart's intrinsically limited accuracy (Cran 1981; Viola 1981). The major cost components will be discussed separately below:

Equipment Installation

Equipment installation costs have been mentioned in the previous chapter and are given in various of the equipment cost charts. It is seen that each piece of equipment requires quite a different multiple of its purchased price to pay for its installation. Generally the costs are fairly specific for, and based upon special characteristics of the equipment, with some influence of the local installation conditions for the equipment mounting, and the process, location, and size also helping to determine this cost. The installation cost consists of the freight from

Table 3-4
Plant Cost Estimating Factors

Component	Plant Cost Factor, Fraction of Total Purchased Equipment Cost	Optional Estimating Factors, Fraction of (Subtotal) Plant Costs
Purchased equipment	1.00	
Piping	0.15–0.70	
Electrical	0.10–0.15	
Instrumentation	0.10–0.35	
Utilities	0.30–0.75	
Foundations	0.07–0.12	
Insulation	0.02–0.08	
Painting, fireproofing, safety	0.02–0.10	
Yard improvements	0.05–0.15	
Environmental	0.10–0.30	
Buildings	0.05–1.00	
Land	0–0.10	
Subtotal	1.96–4.80	
Construction, engineering	0.30–0.75	0.10–0.40
Contractors fee	0.10–0.45	0.03–0.10
Contingency	0.15–0.80	0.05–0.20
Total	2.51–6.80	2.31–8.16

Usual limits for total factor		Lang Factor
Minimum (solids processing)	3.00	3.4
Average (mixed processing)	4.00	
Maximum (fluid processing)	5.50	5.2

Other capital requirements	Additional Factor, % of Total Plant Cost
Off-site facilities	0–30
Plant start-up	5–10
Working capital	10–20, or 10–35% of mfg. cost

the factory, the unloading and handling costs, foundations or supports, physically putting the equipment in place and securing it, and connecting it so that it will run (electric switch gear, etc.) and function (connect piping, etc.). A knowledge of this installation cost is necessary if one is considering single pieces of equipment only, or small process additions. Otherwise, however, for complete plant estimates, installation is *not* considered separately and is included indirectly in other items such as the electrical hookup and switch gear, instrumentation, piping or conveyor attachment, and so on, all of which are necessary for the equipment in a new plant to function. The installation cost ranges and averages listed on the equipment charts were taken from many authors, often with a somewhat different basis. The average installation cost for all of the equipment in the charts of Appendix 1 is 63%.

Instrumentation

This factor has traditionally been about 2–18% of the total purchased equipment cost for chemical plants, but rising labor charges and the rapidly increasing use of computer and more complex instrument control has significantly increased its value. It may be more typically 10–35% now, and will continue to increase in the future. Figure 3-2 shows generalized relationships of instrument cost versus equipment cost and the total plant cost. The values on this chart have been corrected to a CE Index of 320.

Piping

Process piping has always represented a major cost in chemical plants. Computer-assisted design and scale-size models are helping to optimize piping runs and plant design (Brooks 1988), but the need for energy saving (larger pipe sizes) and attempts to reduce maintenance with better corrosion and erosion resistant materials is tending to offset this gain and increase piping costs. It varies widely, but a 15–70% factor usually covers the range for most plants. Operations where there is considerable solids handling and conveying equipment would cause the piping to be at the lower end of the range, while plants handling only fluids with considerable recycling and heat exchange would have the highest factors. Some piping component costs (i.e., pipe, valves, etc.) are also given in Appendix 1, if specific pipe cost estimates are needed.

Insulation

As with piping sizes, the need to conserve energy by insulating much more thoroughly is a rapidly increasing factor for both old and new plants. The processing temperatures and heat duty determine the amount of insulation required, but equipment cost multiples of 2–8% may be fairly typical at present. Limited general and piping insulation costs are given in Appendix 1.

Electrical

Electrical hookups, conduits, wiring, switch gear, control panels, transformers, and distribution panels and vaults or centers are all part of the electrical cost factor. This has stayed constant at about 10–15% of the equipment cost. Some direct electrical installation costs are listed in Appendix 1.

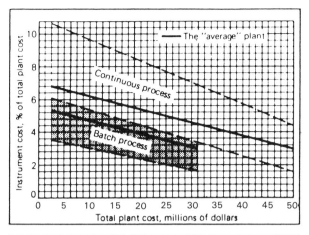

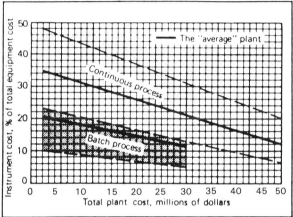

Correction Factors to Reflect
Instrumentation Philosophy

Features	Factor, F*
Localized control	−0.20
Pneumatic instrumentation	0.00
Centralized control	0.00
Sample analysis performed in laboratory	0.00
General-purpose process area	0.00
Explosion-proof process area	+0.10
Graphic panel display	+0.10
Special alloys required for pipeline items	+0.15
Sample analysis by online analyzers	+0.20
Electronic instrumentation	+0.20
Limited-scope optimizer computer included	+0.25
All loops on computer control	+0.45

*Fraction change in total cost.
Source: Liptak 1970. Excerpted by special permission from *Chemical Engineering*, Sept. 1970. Copyright © 1970, by McGraw-Hill, Inc., New York.

Figure 3-2. Instrumentation costs.

Buildings

A certain number of buildings are required for any plant, such as control rooms; offices; shops; laboratory; employee lunch, change rooms, and lavatories; and so on. However, the really important cost of buildings arises when there are large product or raw material storage warehouses or silos required, and when some or all of the plant must be housed in buildings for reasons of weather, safety or environmental control. Although varying widely with the specific circumstances, the building factor may range from 5 to 100% of the initial equipment purchase price. Various direct building estimating costs are listed in Appendix 1.

Environmental Control

Over the past few years environmental control has become a major cost in most new plants. Many governmental jurisdictions are requiring essentially a zero discharge of any contaminant into the air, water, or land, and more will in the future. Other agencies are requiring a close monitoring and control of hazards and hazardous wastes. This usually means extensive pollution control and treating facilities, groundwater monitoring wells, leak detectors and soil sampling surveys, air pollution monitoring, controls and alarms, hazardous waste handling, storage and shipping facilities, and very costly to prepare permits. At present this might total from 10 to 30% of the basic plant equipment, but the figure will be steadily rising. Some direct wastewater treatment costs and air pollution control facilities are listed in Appendix 1.

Painting, Fire Protection, Safety, Miscellaneous

This is a small factor, but an important one to the appearance and safety of the plant. It might average 2–10% of the equipment cost, but with increasing insurance costs and public scrutiny it is increasing.

Yard Improvements

This item includes grading, drainage, roads, rail facilities, parking, fences, landscaping, concrete slabs and walkways, sumps, lighting, as well as all other required nonprocess components of the site or yard in the plant area. It might vary from 5 to 15% of the equipment cost. Some of the yard improvement costs often are considered as part of the off-site facilities.

Utilities

Utilities include the general services required to operate the plant, such as telephone, water, gas, other fuel, steam, compressed air, inert gas, sewage, and bringing electricity into the plant. These items usually represent a major plant cost, as exemplified by the electrical hookup. It may only require bringing in one outside power supply, or hopefully two, or a standby emergency generator, and installing the necessary high voltage transformers and substations throughout the plant. However, it may also require, in addition, installing a power plant or a cogeneration power plant-steam generator facility. Obviously these items would be expensive, but they may save considerably on energy costs and emergency shutdowns, and many plants are now installing them. In a similar manner, large plants often have a complex water system, with drinking water, boiler makeup water, process water, cooling water, sanitary and landscaping water, fire water, and so on. Depending upon the water source (your own wells, river, sea, city, recycled, etc.), its scarcity, and discharge requirements, some or all of these (or more) may be needed.

The other utilities are usually not so complex, but they still can be quite involved and expensive. Costs may range from 30 to 75% of the purchased equipment cost for the total utility package. Some direct utility costs are listed in Appendix 1.

Land

This may not be involved in the plant costs, but if new land is to be purchased it may run from 3 to 10% of the equipment cost.

Construction and Engineering Expense, Contractor's Fee, Contingency

These contractor factors can be quite variable, and each may be in the range of 3 to 40% of the total plant cost (Mendel 1985). Construction and engineering expense is for the detailed engineering required for the plant design, drawings, permits, and managing and supervising construction. These charges are over and above the construction cost itemized in the previous factors. Engineering and supervision is generally charged on a cost plus expenses and overhead basis, so it is quite variable, but it may be about 30–75% of the purchased equipment cost, or 10–40% of the total plant cost. The contractor's profit (fee) is usually negotiable, but usually runs from 10 to 45% of the equipment cost, or 3 to 10% of the total plant cost. The contingency can be set by the owner at any number desired, but is almost always a necessity between 15 and 80% of the equipment cost, or 5 and 20% of the total plant costs to cover the many uncertainties of design, permits, purchasing, and construction. The variability with these items depends mainly on plant size, but also on its complexity and uniqueness. For a simple or standard plant or type that the owner and/or contractor has considerable experience with, the fees and costs can be on the low side of the ranges.

Total Multiplier

By the time all of the factors are added a total is reached which can be used to multiply the total purchased equipment cost, just like the Lang factor. In Table 3-4 the normal variation of these numbers is usually not the sum of each of the maximum and minimums in the columns above, but a more realistic somewhat tempered range as indicated by the "usual limits for total factor" table. For every cost estimate the final multiplier should be compared to these values to double check its reasonableness. Many estimates can have higher or lower values, but they are unusual, and if the estimate is beyond this range there should be a good reason for it.

Each of the total factors is large, showing that there is considerable capital required for a plant besides the basic equipment. In deciding upon these estimating factors in preparing a very preliminary plant cost estimate, an engineer may feel that he does not have the experience or knowledge to estimate the factors meaningfully because of the lack of data available, and that an overall total might be just as accurate. This may very well be the case, but for most estimates, even at an early stage, the itemization and best judgment on each factor will be of value in better understanding the total cost, and as guidance for later development or more detailed estimates. Furthermore, many times the engineer himself later, or a reviewer of the cost estimate may wish to know the value assigned to certain cost components, and suggest changes. With the detailed cost breakdown this can be easily done.

COMPLETE PLANT ESTIMATING CHARTS

Just as there are estimating charts for process equipment as a function of size, there are similar charts for the average cost of complete plants. Typical of this are the extensive charts of Guthrie (1970), *Chemical Engineering* (1974) and the nomographs of Kharbanda (1979). This data have been factored to a CE Index of 320, averaged, and new data added to produce the charts of Appendix 2. They are especially useful for providing a general idea of the cost of these complex, highly evolved basic commodity plants whenever the same product or a very similar operation is needed. Many of these plants are so complex that reasonable estimates can only be obtained from complete plant charts such as these, or by comparison with a similar plant where data are available. However, since most of the charts are fairly old or based upon only one reported plant, and because of rapidly changing technology, new design requirements (environmental controls, more costly fuel or energy, etc.), and the specific designs of each licensor, contractor, or site, many of these estimates have a fairly high potential error. Also, many of the original source charts covered a size range that is somewhat smaller than is now standard installation practice. For example, it would be very rare for any new ammonia plant to be built smaller than with a 1,000 t/d capacity, and 1,500 t/d is a more frequently employed size. The original charts ended at the smaller of these capacities.

Cost per Ton of Product

Plant cost estimates are sometimes made in a very approximate manner by using a ratio of the capital cost of a new plant per annual ton of product. This type of estimate can be used for rough predictions when data are available on exactly designed and sized plants compared to the ratio data, but intrinsically this is an erroneous method. Plant costs usually vary as some exponent of size, so using a linear relationship such as this can be many fold in error. Some numbers of this type are listed in Table 3-5, but it is cautioned that they are at best only accurate for the listed plant size. They were taken from plant cost data of Appendix 2. A typical example of their variability is with polyvinyl chloride, where at 200,000 t/yr capacity the factor is $370 per annual ton, while at 20,000 t/yr the cost is $930 per ton.

Capital Ratio (Turnover Ratio)

A similar type of generalization is based upon the ratio of the yearly value of the product sold to the initial plant capital cost, and is called the capital or turnover ratio. This method may not be quite as inaccurate for approximating the capital cost of a plant as the cost per ton method, and is sometimes quite useful for an entire plant or a totally unknown design. It divides a number into the yearly product value to estimate the original capital requirement. Examples of this ratio are also listed in Table 3-5, again estimated from the plant costs in the charts and the product sales value listed in the *Chemical Marketing Reporter* (CMR). For many products the ratios vary from 0.4 to 3, although it actually can vary between the range of 0.2 to 5.0 and greater. It is interesting to note, however, that for the chemical industry as a whole their plants' asset value is about equal to the yearly value of the product sold (see Table 7-4). Since the average plant is about 50% depreciated, this implies that the original plant cost averaged about two times the yearly product sales value, or a turnover ratio of 0.5. Table 3-4 would not indicate anywhere near this low an average product turnover ratio, which demonstrates that: (a) the plants are not operating at capacity; (b) the CMR product sales prices probably are higher than actually realized; and (c) the plant costs of Appendix 2 do not include start-up, auxiliaries, and working capital, which are a significant part of the total investment. Because of this industry-wide consistent ratio of 0.5, however, it becomes a reasonable number to use if no other information is known about a plant's cost.

A number of other simplified plant cost estimating methods have been proposed over the years (Cran 1981; Ward 1986), such as "functional or operating steps" (Viola 1981) where plant capacity and the number of major

Table 3-5
Plant Cost per Annual Ton of Product, and Turnover Ratio (Annual Sales Value per Original Plant Cost) Numbers
(Estimated from Appendix 2 and Mid-1987 CMR Prices)

Compound	Capacity, M t/yr	$ Sales/$ Plant	$ Plant/annual ton
Acetalehyde	50	1.8	410
Acetic acid	20	1.7	440
Acetone	200	3.4	140
Acrylonitrile	300	1.4	560
Alumina	100	1.9	430
Aluminum sulfate	25	1.5	130
Ammonia	330	0.63	130
Ammonium nitrate	300	4.6	28
Ammonium phosphate	250	2.9	48
Ammonium sulfate	300	3.7	22
Benzene	260	8.1	51
Butadiene	250	1.8	140
Butanol	100	1.4	480
Caprolactam	45	1.6	1,100
Carbon tetrachloride	30	1.1	420
Chlorine	300	1.0	190
Citric acid	3.3	0.50	4,800
Cyclohexane	100	9.0	61
Diphenylamine	10	2.0	1,250
Ethanol	30	0.14	2,500
Ethanolamine	25	6.1	360
Ethylbenzene/paraxylene	20	0.63	700
	8.5	0.35	1,000
Ethyl chloride	15	0.51	940
Ethyl ether	35	5.7	160
Ethylene dichloride	25	0.57	600
Ethylene oxide	200	1.0	700
Glycerine	35	2.2	810
Hydrogen peroxide	200	2.5	180
Isopropyl alcohol	150	2.5	240
Maleic anhydride	50	5.4	200
Methanol	330	0.93	110
Melamine	70	0.94	810
Methyl chloride	10	2.2	230
Methyl isobutyl ketone	2.5	1.8	400
Methyl isobutyl carbonyl	10	2.1	530
Nitric acid	200	4.1	46
Para xylene	20	0.24	1,500
Phenol	200	2.1	280
Phosphorous	100	1.7	1,100
Phosphoric acid	20	2.2	270
Phathalic anhydride	200	2.7	200
Polyethylene	20	0.38	1,800

Table 3-5 (Continued)

Compound	Capacity, M t/yr	$ Sales/$ Plant	$ Plant/annual ton
Polypropylene	20	0.32	2,800
Polyvinyl chloride	200	2.7	370
Propylene	20	1.9	180
Sodium carbonate (Mine)	500	0.40	210
Sodium carbonate (Brine)	400	0.18	460
Sodium metal	20	1.10	1,170
Styrene	500	5.0	110
Sulfur	15	0.24	490
Sulfuric acid	330	0.60	83
Titanium dioxide	50	0.58	2,800
Tolune disocyanate	12.5	0.69	2,900
Urea	200	2.40	84
Vinyl acetate	200	1.9	420
Vinyl chloride	500	3.3	320

processing component steps are considered the basis of plant cost. Most of these methods are not too easy to use and may require additional arbitrary judgments. Some can increase the accuracy of preliminary estimates, but with much additional complexity. This provides no significant advantage, so they shall not be considered further in this book.

Factoring Exponents

Both individual pieces of equipment (as noted before) and entire plants may be roughly cost estimated by knowing the cost of one size and an exponent that relates cost with size. Since most of the complete plant cost versus size charts are straight lines, or can be approximated by a series of straight lines on log-log plots, a sizing exponent for a limited range (the slope of this line) can reasonably factor the costs. These exponents are listed in Appendix 2 under each plant cost chart. However, the likelihood of having accurate data on one plant size, or the sizing exponent, and the high probability of technical or process changes, coupled with each owner having quite different specifications for the plant, make such estimates still of only marginal usefulness. One of the most frequent uses for the sizing exponents, however, is to examine the effect of plant size on profitability, and to examine plant costs as part of the "sensitivity analysis." This will be considered in Chapter 6. If not specifically available, an exponent of 0.6 is considered to be a good general average for most chemical or related plants and equipment, although the average actual exponent for the 40 plants listed by Guthrie in Appendix 2 was 0.64.

PLANT MODIFICATIONS

The preceding discussions primarily concern cost estimates for complete plants. The same principles apply to the perhaps more common objective of plant modification. Because of environmental obstacles in siting a totally new grass roots chemical plant, and the general public's fear of unknown "toxics" from chemicals in general and their plants in particular, for many locations in the United States it is far easier to expand or modify an existing facility than it is to build a totally new plant. Also, there is the normal need to modernize and make older plants more efficient and environmentally acceptable if they are to remain competitive. Modification often becomes the most cost effective and practical procedure, and this, along with the difficulty of obtaining permits for totally new plants make additions or modifications the most common estimating need for the average engineer.

The procedures for estimating capital costs for plant modifications are exactly the same as previously discussed, except that some of the cost factors are either eliminated or greatly reduced. Land costs, for instance, are eliminated, and utility costs greatly reduced. Sometimes demolition, facility removal, and tie-ins require major additional expenses. Lost production from the old facility and arrangements to cause the minimum disruption to the plant often can also result in considerable scheduling problems and extra cost.

OTHER COMPONENTS OF TOTAL CAPITAL INVESTMENT

The total capital investment for any project is usually far greater than the cost of the "battery limits," or direct processing plant alone, and may include any of the following items.

Off-Site Facilities

A typical condensed checklist for off-site or additional facilities to be considered with any new plant estimate is given in Table 3-6. Some of these items relating to auxiliary, "off-site" or nonprocessing or production facilities may have been included in the previous capital cost estimate. If the auxiliary equipment will only be used by the production plant being cost estimated, then their cost should be itemized in the detailed battery limits estimate. However, if the plant is to share these facilities with other plants, or the facilities are quite general and removed from the new plant, then they will probably have to be considered separately. This might include assuming some (or all) of the cost of headquarters buildings; research and development facilities; engineering and plant technical service departments; safety, security; environmental control or hazardous waste facilities; power plants, utilities and sewage facilities; shipping facilities; and so on. The list of possible off-site requirements can be quite extensive, and should always be carefully reviewed to see which, if any, is

Table 3-6
Partial List of Possible Auxiliary Facilities

Typical Utilities
    Boilers; condensate and makeup water systems
    Generators (including cogeneration)
    Standby generators or battery assemblies
    Main power transformer stations
    Fuel storage and distribution facilities (oil, coal, gas, etc.)
    Plant-wide air conditioning facilities
    Plant-wide paging, emergency communication system
    Sewage collection (and treatment)
    Inert gas systems
    Fire fighting equipment and systems
    Flares, stacks, waste gas treatment
    Compressed, instrument air
    Cooling towers, distribution systems
    Refrigeration; hot oil systems
Service Buildings and Related Facilities
    Employee lockers, showers, time-card, lunch, restrooms
    Office building (management, sales, accounting, administration)
    Engineering, technical service facilities
    Laboratory (analytical), R & D, environmental
    Shipping, receiving office; supply warehouse
    Inventory (raw materials, products, supplies) storage
    Maintenance buildings, shops
Product Sales
    Packaging facilities
    Loading, unloading rail spurs, docks, forklifts, loaders, etc.
    Warehouses
    Shipping equipment (trucks, railcars, ships, barges, etc.)
    Distant storage, reshipping facilities
Environment
    Air, water, and ground monitoring equipment and controls
    Water treating and reuse facilities
    Incineration equipment
    Solid or liquid waste processing, handling equipment
    Solid or liquid waste shipping facilities

applicable, or perhaps not included in other sections of the cost estimate before totaling the capital investment. The cost for these facilities may be estimated directly (see Appendix 1), or they may be factored, usually as 0–30% (see Table 3-4) of the total plant cost.

Distribution Facilities

For some products the only way that they can be sold is for the company to install some, or extensive packaging and distribution facilities. This might include warehouses; branch offices; cars, trucks, ships, barges, rail cars;

shipping, receiving, or reshipping (i.e., rail to truck, etc.) facilities; packaging machinery; and so on. Such capital items may not be involved in a particular estimate, or if they are, perhaps they can be shared with many other products, but some new plants cannot be considered complete until these facilities have been included as part of the total capital investment. In some cases this investment may be optional and/or evaluated on its own separate merits with its own value-added economics, but for all estimates it should be investigated and itemized if appropriate. For this equipment the costs must be estimated separately and not merely factored from the plant costs because of the unusual and unique demands of the products or plants under consideration.

Research and Development, Engineering, Licensing

For many plants there is a sizable capital cost that must be added to a new project for the prior expense of research and development that was done for it, licensing if a lump sum payment (with or without royalties) is required, and for preliminary engineering design, review, and analysis before the new plant was approved for detailed engineering and construction. As will be seen in later chapters, the long time period that may be required for the research and development and engineering before the project is approved can have a considerable impact upon the ultimate profitability of the operation. The government tax laws do not allow these expenses to be considered part of the prior operating cost, and if a plant is built they must be "capitalized" and added to the plant capital.

Working Capital

Every plant has a requirement for a certain amount of capital to be available to pay the bills and sustain the operation before product is sold and payment is received. Some of this capital is needed prior to the first months' operation, and all of it is needed throughout the plant's life. It should be totally recovered at the end of the operating period when the plant is finally shut down. Its value generally consists of:

1. Operating capital equal to the operating expense for:
   a. the average length of time the product is being manufactured and in storage in the plant, plus shipping and storage time in other locations (if any) before it is sent to the customer, and
   b. the average length of time that it takes to collect for the merchandise sold (accounts receivables). This later figure generally averages about 60 days, although for very efficient companies it may be as low as 30 days, and companies selling to seasonal markets (i.e., agriculture) may run 120–180 days or longer.

2. Cash for wages, fringe benefits, local taxes, and other current obligations.
3. Inventories of raw materials, maintenance, and operating supplies.

Raw materials often require about a one month's supply, and maintenance and operating supplies about 2–4% of the total plant cost. However, for remote locations this must be much larger, and for urban areas it may be smaller. The current "just-in-time" supply and raw materials policy of some manufacturers (auto industry, etc.) is an attempt to greatly reduce this supply inventory (and warehousing) cost by precisely scheduling these materials to arrive exactly as needed. It would be very difficult to schedule in such a manner for the chemical industry and in fact, chemical companies servicing the just-in-time plants generally have to increase their inventory in proportion to the customer's decrease.

Working capital is often 10–20% of the plant's total cost (see Table 3-4), or perhaps 50% of the installed equipment costs. Such values may be acceptable for very rough estimates, but calculating a more precise figure by means of itemizing the manufacturing cost components is generally simple, and should be done if possible. A fraction of the yearly manufacturing cost is a much more relevant method of establishing working capital than using a factor times the installed plant cost, and might typically be 10–35% of the yearly operating cost.

Start-up Expenses

The final item of possible extra capital expense that should be considered is the cost of starting the plant and bringing it to full production. Usually there is little saleable, good quality, product generated during this period, so the start-up expense represents an additional increment of capital cost. Labor, materials, and overhead expenses during start-up time would be required, plus often extra engineering and considerable minor equipment, piping, controls, and so forth, modification. There is no way to precisely estimate this capital, or the start-up time, but for many plants the costs might run from 5 to 10% of the total plant cost. (See Table 3-4.)

FOREIGN LOCATIONS

The cost estimates in this text are based upon average U. S. prices and locations. It can be readily visualized that the same plant built in a foreign location would cost a somewhat different amount because of the difference in local labor costs, governmental requirements and taxes, varying transportation costs and availability of services. Often some of the construction costs will be cheaper, and some more expensive, but usually an overall factor can be applied to the U. S. costs to provide a first approximation of the costs in a foreign location. Table 3-7 gives an estimate of such factors. As with all the other estimating methods

Table 3-7
Foreign Location Factors

Australia		1.3	Malaysia		0.8
Austria		1.0	Middle East		1.1
Belgium		1.0	Newfoundland		1.2
Canada		1.15	New Zealand		1.3
Central Africa		~2.0	North Africa	(imported element)	1.1
Central America		1.0		(indigenous element)	0.75
China	(imported element)	1.1	Norway		1.1
	(indigenous element)	0.55	Portugal		0.75
Denmark		1.0	South Africa		1.15
Eire		0.8	South America (North)		1.35
Finland		1.2	South America (South)		2.25
France		0.95	Spain	(imported element)	1.2
Germany (West)		1.0		(indigenous element)	0.75
Greece		0.9	Sweden		1.1
Holland		1.0	Switzerland		1.1
India	(imported element)	1.8	Turkey		1.0
	(indigenous element)	0.65	U.K.		0.9
Italy		0.9	U.S.		1.0
Japan		0.9	Yugoslavia		0.9

Notes:
1. Increase a factor by 10% for each 1,000 miles, or part of 1,000 miles, that the new plant location is distant from a major manufacturing or import center, or both.

Source: Bridgwater 1979. Excerpted by special permission from *Chemical Engineering*, Nov. 5, 1979. Copyright © 1979, by McGraw-Hill, Inc., New York.

this is perhaps an oversimplification of the problem and not too accurate, but it should nevertheless provide a somewhat more realistic estimate of plant costs in other countries.

REFERENCES

Bridgwater, A. V. 1979. International construction cost location factors. *Chemical Engineering* (Nov. 5).

Brooks, Kenneth. 1988. A suite of software to aid plant engineers. *Chemical Week* (March 2):32–35.

Chemical Engineering. 1974. *Sources and Production Economics of Chemical Products*. McGraw-Hill, New York:121–180

Chilton, C. H. 1960. *Cost Engineering in the Process Industries*. McGraw-Hill, New York.

Cran, John. 1981. Improved factored method gives better preliminary cost estimates. *Chemical Engineering* (April 6):65–79.

Guthrie, Kenneth M. 1970. Capital and operating costs for 54 chemical processes. *Chemical Engineering* (June 15):140–156.

—— 1974. *Process Plant Estimating, Evaluation and Control*. Craftsman Book Co., Solana Beach, CA:125–180; 334–353; 369–371.

Holland, F.A., F.A. Watson, and J. K. Wilkinson. 1974. *Introduction to Process Economics*. John Wiley & Sons, London:1–346.

Kharbanda, O. P. 1979. *Process Plant and Equipment Cost Estimation*. Craftsman Book Co., Solana Beach, CA:213–225.

Klumpar, Ivan V., and Steven T. Slavsky 1985*a*. Updated cost factors: Process equipment. *Chemical Engineering* (July 22):73.

—— 1985*b*. Commodity materials. *Chemical Engineering* (Aug. 19):76–77.

—— 1985*c*. Installation labor. *Chemical Engineering* (Sept. 16):85–87.

Lang, H. J. 1947. Cost data correlated. *Chemical Engineering* (Sept.):130–133.

Liptak, B. G. 1970. Cost of process instruments. *Chemical Engineering* (Sept).

Mendel, Otto 1985. Estimating engineering cost. *Chemical Engineering* (Dec. 9 & 23):117–118.

Parkinson, Gerald 1987. New study provides clues to project-cost overruns. *Chemical Engineering* (Aug. 17):33–37.

Viola, J. L., Jr. 1981. Estimate capital costs via a new shortcut method. *Chemical Engineering* (April 16):80–86.

Ward, Thomas J. 1984. Predesign estimates of plant capital cost. *Chemical Engineering* (Sept. 17):121–124.

—— 1986. Cost estimating methods. American Institute of Chemical Engineers Modular Instruction Series, *Plant Design and Cost Estimating*, Vol. 1:12–21.

4

MANUFACTURING COST

DETAILED ESTIMATES

Of equal importance to the capital cost estimate in an economic evaluation is the operating or manufacturing cost. It predicts the expense of producing the desired product, and thus, together with the capital cost and sales realization, allows the profitability and potential attractiveness of an operation to be evaluated. The manufacturing cost can be determined by gathering exact data on those factors which can be definitely established (taxes, insurance, utility needs, etc.), and making detailed estimates for the other factors involved based upon manning charts, local wage scales and manufacturer's expected maintenance schedules. Alternately, a factoring method may also be used.

Manufacturing costs are generally broken down into two broad categories: variable or controllable costs, and fixed costs. The plant manager has some ability to control the former items; the later are determined by the plant itself or other groups and are essentially "fixed." Table 4-1 gives a partial checklist of some of the components in each category. Naturally every plant has many items that are unique to it alone, and it is often debatable as to how controllable certain items are. For instance, operating labor is fairly well fixed for a certain number of hours of production per year, and wages may be specified by union negotiation. However, presumably overtime decisions, extra labor for special projects and the give-and-take of labor and automatic controls still allow the plant manager some flexibility in this area. Also, most of the service function jobs are more directly controllable, since they are not absolutely essential to production, and there is some leeway upon when and if they get done.

On-Stream Efficiency

No plant is capable of running all of the time, since mechanical breakdown, maintenance, power disruption, shortages of feed materials or sales, cleaning or catalyst change, and so forth, cause it to periodically shut down. Under ideal conditions many plants schedule major maintenance, or "turnaround" periods

Table 4-1.
Partial Checklist for Manufacturing Costs.

Variable Costs (Controllable)
  Raw materials, additives, catalysts
  Utilities: fuel, electricity, water, steam, air, telephone, sewage
  Labor: operating, supervision, maintenance, technical service, engineering, safety, environ-
    mental, laboratory, clerical, accounting, legal, security, etc.
  Indirect labor charges; fringe benefits such as:
      health insurance, retirement, social security, workman's compensation, disability
      insurance, vacation, holidays, sick leave pay, payroll taxes, overtime, bonuses, etc.
  Maintenance: material, services, contract maintenance
  General: office, plant, safety, lab supplies; books, subscriptions, dues, memberships; outside
      legal, accounting; consultants; travel, meetings; environmental; miscellaneous
  Transportation, freight
  Distribution, packaging, storage and sales expense
  Donations, public relations
  Research and development
Fixed Costs
  Depreciation
  Taxes, business and licensing fees, insurance
  General and administrative expenses, corporate overhead
  Patents and royalties
  Interest

---

of about two weeks per year. If this were their only downtime it would result in an on-stream efficiency (ose) of $50/52 = 96\%$, which is about as good as is obtainable. However, the current percent of design capacity operation for the entire industry is only about 80% because of limited sales and normal overcapacity. Based primarily upon maintenance considerations, standard preliminary design practice is to assume an optimistic 90 to 95% on-stream efficiency (or about 330–345 days per year), and a design rate of exactly the estimated sales quantity. If it turns out that the market actually is larger, it is usually assumed that modest "de-bottle necking" or increasing the capacity of only a few rate limiting pieces of equipment can provide a 10–20% capacity increase for a reasonable cost at a later time.

It is often difficult to know what manufacturing costs continue when the plant is not operating. Raw materials, catalysts, and chemical additives consumption is generally tied directly to production. Utilities and fuel are also mainly tied to production, although some usage continues during down time, but often this is small. The sales related costs actually do continue, but they are estimated based upon sales, so for estimating purposes their costs are also assumed to be proportional to production. All other costs continue, except perhaps occasionally labor if there are other jobs the staff can be shifted to and that will pay their salaries, but this is relatively uncommon.

Several of the manufacturing cost factors must be determined separately whether a detailed or factored estimate is made. They are discussed in the following sections.

Raw Materials

The raw materials required by the process may be calculated from the stoichiometry and a material balance for the process with an allowance for extra materials because of the plant's inevitable inefficiencies and losses, estimated from laboratory or pilot plant data, prior experience, or related processes. Included with the raw materials should be all major additives, treating agents, catalyst, filter aids, and so forth, that are required to complete the process. For many operations these additional raw material components can be a significant cost factor. The initial process requirements, such as catalyst charge to the system must be treated as a capital cost, but catalyst makeup, regeneration, and replacement are operating charges. The inventory of all raw materials, in-process, and finished products in storage are considered part of the working capital cost.

The *Chemical Marketing Reporter* (1988) is a weekly newspaper that prints the current list price of most chemicals, and provides a reasonable estimate of the cost of the raw materials and the sales price that might be obtained for the product. Often there are major discounts available on many of the commodities listed, and regional or local competition can dramatically change the price. Also, on occasion it is very difficult to find a manufacturer who will sell as cheaply as listed in the *Chemical Marketing Reporter* (CMR). A recent example of the uncertainties in chemical costs (for California) are: (1) sulfuric acid in mid-1987 was listed in CMR for the West Coast at $85/ton, virgin, 100% basis, tank cars, FOB ("Free on Board"), meaning pure acid, the price for equivalent 100% H_2SO_4 (even if sold at a lower concentration) in tank car quantities, and the price at the manufacturer's plant: (2) smelter sulfuric acid, 100% basis, tank cars FOB Arizona, and so on, was listed at $20/ton as by-product acid from the large western copper smelters. In this and many cases one must know where the manufacturer's plant is located and the freight charges to the desired location before the actual delivered price can be calculated for the manufacturing cost estimate. This adds additional uncertainty and inaccuracy, but it is an unavoidable part of the cost estimating process.

As a second example, an attempt was made to purchase carload quantities of another chemical by calling every producer listed in the *Chemical Buyer's Directory* (1987), and then calling CMR to check their sources. The closest price obtained was 30% greater than that listed. CMR may have listed imported material which was hard to track down, the listed price may have been for large customers only and not available to the occasional purchaser, it may have recently increased, or it may have been in error.

Various free publications, such as the *Chemical Buyers Directory* and others (*Chem Cyclopedia* 1987) list most of the manufacturers and distributors for a large number of chemicals, which will provide locations and telephone numbers in case one wishes to write or call for additional information on price. The manufacturer or distributor might also be able to estimate freight costs, but if not, your company's purchasing department should be able to provide a good estimate (and perhaps on the purchase price as well). As a last resort one can contact trucking companies or the railroad for freight estimates. Once this information is available you can then compare the purchase price plus freight from different locations to estimate the most favorable listed price. Later, as the project nears completion, the company's purchasing department may be able to obtain a more favorable price.

Utilities

The cost of utilities has now become one of the larger segments of a chemical plant's operating cost, and where there is often the greatest potential to economize. When most older plants were designed and the processes developed, energy was very cheap, so only moderate energy savings were attempted as the most economical balance between savings in the operating cost for energy, and the additional capital and operating cost for greater energy saving equipment. Now, however, this balance has been considerably shifted, and new equipment can be justified to greatly reduce energy consumption.

In manufacturing cost estimates energy requirements must be itemized and estimated for each plant, with the simplest method being to tabulate motor horsepower, steam consumption, fuel requirements, cooling water needs, and so on, directly from the flow sheet and heat and material balance. Adding these together, estimating other utilities that may not be listed (compressed air, room heating or air conditioning, telephone, etc.), and then applying reasonable on-stream factors should give a close approximation to the actual utility needs. The local price of the utilities should be readily available from either the accounting department or the company records, but if not, typical costs in the United States for the most common utilities are shown in Table 4-2. These numbers may vary widely and in a few plants the actual costs, based upon old purchase contracts or other favorable factors, may be much lower than the listed values.

Operating Labor

Another manufacturing cost that always must be itemized is the operating labor required to run the plant. In the factoring methods this does not include maintenance, supervision, analytical, clerical, or other types of totally necessary labor, since these staff costs will later be estimated from the operating labor or the plant capital cost. The simplest means of estimating the operating labor is to

Table 4-2.
Typical Common Utility Costs
(Southern California, 1987; CE Index 320)

Electricity	$0.08/kW hr
Gas	$7/MM Btu
Fuel oil, low sulfur	$7/MM Btu ($20/bbl)
Steam, 250 psig (Jones 1987)	$12/MM Btu
Cooling tower water	$0.05/M gal
Process water: city water	$0.20/M gal
well water	$0.10/M gal
Recycled process water	$0.25/10³ gal
Recycled cooling tower water	$0.20/10³ gal
Softened water	$0.55/10³ gal
Demineralized water	$5/10³ gal
(or condensate) for high	
pressure boiler makeup	
Instrument air (dry)	$0.30/M scf
Inert gas, low pressure	$1/M scf
Nitrogen, purchased	$2.20/10³ scf
Refrigeration (ammonia to 30°F)	$1.50/ton-day (288,000 Btu removed)

predict the labor requirements for each major piece of equipment or section of the flow sheet. Chemical plants do not require many operators, but instruments, controls, analyses, and operations must be frequently checked, and some parts of the plant actually require a number of physical manipulations. Usually each major piece of equipment and its supporting facilities and controls needs one operator, but if equipment is grouped together where a central control panel services much or all of one plant, the number can be considerably reduced. If the equipment is by itself or some distance from other facilities a second operator must be added for safety purposes.

To illustrate the labor requirements, in a boric acid extraction plant without extensive instrumentation and a central control room, one operator might be assigned to the incoming feed streams and the borate extraction mixer–settlers. A second would handle the solvent stripping mixer–settlers and the acid and solvent makeup. A third would watch the crystallizer and its steam source and cooling tower; a fourth the centrifuge, dryer elevator, and screens; and perhaps two would be required for product and raw material storage, loading, packaging, and shipping. This would make a total of six operators. With the normal modern (extensive) instrumentation and a single control room, the same operation could be handled by four operators (still two for shipping; two for operation). Maintenance expenses would be increased somewhat, however, to handle the routine maintenance (lubrication, etc.) and emergencies that were previously partly taken care of by the extra operators.

Various estimating charts, formulas, and tables have been suggested for the labor requirements of different equipment or plants, as indicated by Figure

4-1. Such a graph is useful to check the plant section-by-section estimate, and when no other data are available, but it is only a broad generalization, so it should be used with caution. A partial checklist of manpower that might be required in a new plant is given in Table 4-3.

Number of Shifts. If a plant operates around the clock, every day of the year, it requires four or five crews to be able to staff each shift and the days off for an approximately 40 hours per week schedule. This usually involves a rotating shift arrangement, with some overtime or plant downtime with the four-shift schedule to balance the total number of operating days each year. Figure 4-2 shows an example of a four-crew schedule, and Figure 4-3 illustrates a five-crew schedule. If each crew worked 40 hours per week, then around-the-clock continuous operation would require $365 \times 24/(40 \times 52) = 4.21$ crews, or four crews working 42.1 hours per week. Assuming 10 working days vacation, nine holidays, and 10 sick days per year, then $365 \times 24/40(52 - 29/7) = 4.58$ crews are required for complete coverage. On this later basis a four-crew operation would require $0.58/4 = 14\%$ overtime (or 45.8 hours per week) or plant downtime. The actual industry average work-week for 1986 was 42.09 hours (*Chemical and Engineering News* 1987). Alternatively, five crews would only

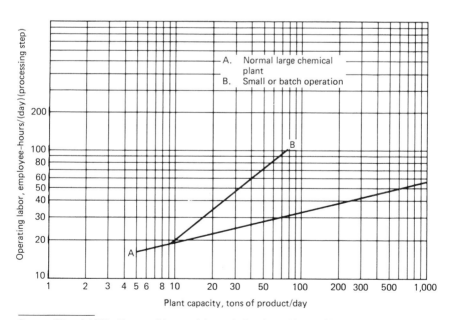

Figure 4-1. Chemical plant labor requirements.

Table 4-3.
Typical Checklist to Staff a New Production Operation

Department	Job Title or Function	Department	Job Title or Function
Administration	Manager	Technical Service	Director
	Administrative director		Engineers
	Safety director		Secretary
	Legal counsel (?)	Environmental	Director
	Purchasing agent		Engineers
	Secretaries, clerks		Secretary
	Nurse (?)	Laboratory	Chief chemist
Accounting	Financial manager		Chemists
	Accountants		Samplers
	Clerks		Secretary
	Secretaries	Operations	Superintendents
Sales Department	Manager		Shift foremen
	Technical sales manager		Operators
	Salesmen		Helpers, laborers
	Clerks		Secretary
	Secretaries	Shipping	Director
Maintenance	Superintendent		Shift foremen
	Craftsmen		Operators, packagers
	Helpers		Loaders, helpers
	Warehouse staff		Dispatcher
	Secretary, clerk		Secretary, clerk
Engineering	Director	Security	Guards
	Engineers		Fire protection
	Draftsmen		Emergency co-ordinator
	Secretary		

	M	T	W	Th	F	S	Su	M	T	W	Th	F	S	Su	M	T	W	Th	F	S	Su	M	T	W	Th	F	S	Su
Crew A	1	1	1	1	1	1	X	X	X	3	3	3	3	3	3	3	X	X	2	2	2	2	2	2	2	X	1	1
Crew B	2	2	2	2	X	1	1	1	1	1	1	1	X	X	X	X	3	3	3	3	3	3	3	X	X	2	2	2
Crew C	3	3	X	X	2	2	2	2	2	2	2	X	1	1	1	1	1	1	1	X	X	X	X	3	3	3	3	3
Crew D	X	X	3	3	3	3	3	3	3	X	X	2	2	2	2	2	2	2	X	1	1	1	1	1	1	1	X	X

[1] First shift of the day (normally dayshift)
[2] Second shift of the day (swing or afternoon shift)
[3] Third shift of the day (night or graveyard)
[X] Off-shift

Source: Rickey 1987. Excerpted by special permission from *Chemical Engineering*, May 11, 1987. Copyright © 1987, by McGraw-Hill, Inc., New York.

Figure 4-2. Typical continuous operation shift schedules, four crews.

	Su	M	T	W	Th	F	S
Crew A	3	3	X	X	1	1	1
Crew B	1	1	1	1	X	X	X
Crew C	X	X	X	X	X	2	2
Crew D	2	2	2	2	2	X	X
Crew E	X	X	3	3	3	3	3

1 First shift of the day   3 Third shift of the day
2 Second shift of the day   X Off-shift

Source: Rickey 1987. Excerpted by special permission from *Chemical Engineering*, May 11, 1987. Copyright © 1987, by McGraw-Hill, Inc., New York.

Figure 4-3. Shift schedules, five crews.

work an average of $4.58 \times 40/5 = 36.6$ hours per week. The assumption of five shifts has now become the most commonly accepted procedure for large companies when they estimate their manpower requirements because of the flexibility in scheduling, and the more assured coverage of illness and requested absence. Often the pay is maintained at an assumed 40-hour work-week. This convention should be assumed for estimating purposes unless the merits of four-shift or less operation is clear.

Salary. The average salary for the production operators varies widely with the job skill, responsibility, and hazard, as well as the presence or absence of a union, the section of the country, and other factors. The local wage rates should be readily available for estimating purposes, but if not, the national average is published periodically in *Chemical Week, Chemical and Engineering News,* and elsewhere. In 1986 it averaged $11.97 per hour (*Chemical and Engineering News* 1987) for the chemical industry. Shift differential (increased) pay is often about $0.30 per hour for the swing shift (early evening such as 3 to 11 P.M.) and $0.50 per hour for the graveyard shift (late evening, early morning, such as 11 P.M. to 7 A.M.).

FACTORING METHOD

As with capital costs estimating, detailed breakdowns and item-by-item accurate manufacturing costing are lengthy and expensive procedures, so for most preliminary estimates a more abbreviated method is required. Often this involves the use of estimating factors such as given in Table 4-4. In this table the items

Table 4-4.
Manufacturing Cost Estimating Factors

1. Raw materials	Itemize
2. Utilities	Itemize
3. Operating labor	Itemize
4. Interest (on loans, if any)	Itemize
5. Labor related costs	
A. Payroll overhead	22 –45% of labor[a]
B. Supervisory, miscellaneous labor	10 –30% of labor
C. Laboratory charges	10 –20% of labor
D. Total	42 –95% of labor
(Typical total)	60% of labor
6. Capital related cost	
A. Maintenance	2 –10% of plant cost[b]
B. Operating supplies	0.5– 3% of plant cost
C. Environmental	0.5– 5% of plant cost
D. Depreciation	5 –10% of plant cost
E. Local taxes, insurance	3 – 5% of plant cost
F. Plant overhead costs	1 – 5% of plant cost
G. Total	12 –38% of plant cost
(Typical total)	26% of plant cost
7. Sales related costs	
A. Patents and royalties	0 – 5% of sales
B. Packaging, storage	0 – 7% of sales
C. Administrative costs	2 –10% of sales
D. Distribution and sales	2 –10% of sales
E. R & D	0.5– 4% of sales
F. Total	4.5–37% of sales
(Typical total)	20% of sales

[a]Operating labor only
[b]Total plant cost, including start up and off-site facilities, but not working capital

have been listed for estimating convenience and not for the more logical or useful sequence desired by future managers of the potential operation. The cost components are shown in four general groupings. The first represents items that are totally specific to the process under study, and must be estimated directly from heat and material balances, heat or cooling loads, horsepower takeoffs from the flow sheet, operating labor estimates, and so on. Each requires individual, detailed estimates as noted in the previous sections; the remaining items may be either factored or estimated in detail, as desired.

The second category is labor related costs, in which operating labor (only) is used to estimate other labor and manufacturing costs that depend directly or indirectly (sometimes only vaguely) upon it. As with each of the other cost categories, many additional items (other labor requirements, etc.) could be added, but these are the more important ones. Note that a typical total is also shown for each group to provide guidance that each individual estimate is similar to the norm, or if not, that there is a good reason for the difference.

The next grouping of costs are related to the total plant capital, generally based upon all of the plant costs, including start-up, auxiliary, or off-site facilities, but not working capital. An exception occasionally is made to not include some of the auxiliary equipment if their operation and maintenance are not part of the basic plant, and their costs are included in overhead charges. Many of the items in this category are exactly tied to the plant cost, such as depreciation, taxes, and insurance, while others are only related to capital in a more indirect manner.

The final category is sales related costs, where again, some items are directly tied to sales (royalties, packaging, etc.) and others such as the overhead items are only indirectly related to sales. These items are included, however, since many companies allocate their overhead costs to the various plants based upon the sales value of the products. A discussion of the manufacturing cost components that are usually factored follows.

Payroll Overhead

For each dollar spent on the direct payroll a certain fraction is spent on overhead or ''fringe benefits.'' Some of this money is paid directly to the government for the employer's share of Social Security and disability insurance; some is paid for workman's compensation and the company's health insurance plan, pension funds, and other welfare programs; and finally some is paid for holidays, vacation, and sick leave pay. The total can be no less than about 22%; 38.3% was the 1986 average (Holzinger 1988); and with rising health care and benefit costs, many companies pay 40–45% of the total labor cost as fringe benefits.

Supervisory and Other Labor

Every plant requires supervisory, engineering, clerical, and other support labor to assist with the operation. Often this can be determined by listing the specific staff required and estimating their salaries. However, experience also has shown that these salaries as a group often run from 10 to 30% of the cost of the operating labor. For smaller operations the percentage is usually greater than for larger plants where it might be shared and spread over more of the operations.

Maintenance

The maintenance required for any plant varies considerably with the operating environment, age of the installation, management's commitment to do ''preventative'' maintenance (in advance of actual breakdowns) and most importantly, the original decisions and design engineer's skill in balancing initial cost with the quality and corrosion or wear resistance of the equipment. Normally the later factor would be based upon the most economic equipment over the life of the plant, but often limited initial capital, unavailability of the best equipment, or in the opposite direction, a strong desire for minimum operating costs will influence the equipment selection. Thus, maintenance expenses can vary widely, but usually are in the range of 2 to 10% of the original total plant cost per year, with about one-third of this expense often being in materials, and two-thirds in labor. As noted earlier, greater instrumentation, with the use of less operating labor, can raise maintenance costs to the top of, or beyond, this range. The type of plant is also a major factor, with more solids handling or less corrosive or severe processes requiring much less maintenance.

Examples of the maintenance requirements for various companies is presented in Table 4-5. It is seen that of the 38 companies listed maintenance varied from 1.2 to 13.9% per year of the original new equipment cost, and averaged 5.8%. Based upon the current depreciated equipment cost, the maintenance averaged 10.5%. This data would indicate that specialty chemical companies had 17% less maintenance expense, and diversified companies 24% more. However, this

Table 4-5.
Typical Maintenance Costs.

	Maintenance Cost	Maintenance Cost as a Percentage of Plant, Property and Equipment	
	1986, million $	*At Cost %*	*Depreciated %*
Industrial Chemical Companies			
Dow Chemical	$756.0	5.9	14.1
Union Carbide	384.0	4.5	8.8
Monsanto	363.0	5.7	12.5
American Cyanamid	142.9	5.3	11.3
Hercules	119.9	5.4	10.4
Morton Thiokol	80.8	9.1	13.6
BF Goodrich	118.7	8.1	13.9
Olin	113.3	5.9	15.7
Ethyl	79.5	6.6	13.4
Pennwalt	44.5	5.2	10.2
Georgia Gulf	38.0	21.8	31.9
Witco	34.8	4.5	9.5
Liquid Air	24.9	3.1	5.2
Reichhold Chemicals	21.3	6.9	10.5
A. Schulman	3.9	4.9	8.9
Diamond Crystal Salt	6.3	8.5	14.0
Average		6.0	11.8

Table 4-5. (Continued)

	Maintenance Cost	Maintenance Cost as a Percentage of Plant, Property and Equipment	
	1986, million $	At Cost %	Depreciated %
Diversified Companies (Chemical Sales 25–50% of Total Sales)			
Du Pont	1,440.0	3.6	7.3
Allied-Signal	394.0	7.7	11.2
W.R. Grace	143.8	4.4	9.0
Borden	116.9	4.6	8.1
Koppers	115.9	11.6	31.0
Vulcan Materials	96.7	9.1	19.7
Nat'l Dist. and Chem.	81.8	4.2	5.9
Cabot Corp.	56.6	4.7	8.8
Engelhard	51.4	7.2	13.8
GAF Corp.	45.8	13.9	22.0
Tyler	28.8	10.8	20.8
Gulf Resources & Chem.	14.1	5.1	11.2
Average		7.2	11.7
Specialty Chemical Companies			
Lubrizol	34.5	5.4	11.9
Ferro	28.7	9.1	18.7
Int'l Flavors & Frag.	17.4	5.2	9.3
Dexter Corp.	13.5	4.5	7.7
Nalco Chemical	10.5	1.9	3.7
Petrolite	5.5	2.7	6.2
Sigma-Aldrich	5.5	4.6	6.4
Average		4.8	9.1
Biotechnology Companies			
Cetus	1.8	3.2	5.0
Biogen N.V.	0.7	1.9	3.7
Alza	0.6	1.2	1.6
Average		2.1	3.4
Total average		5.8	10.5

Source: Brooks 1987. Excerpted by special permission from *Chemical Week*, July 15, 1981. Copyright © 1987, by McGraw-Hill, Inc., New York.

wide difference is not always the case, and an overall industry average of about 6% is more nearly the normal maintenance value.

Additional survey information on industry maintenance experience is listed in Table 4-6. These general figures will allow estimates to be made of the size and makeup of maintenance staffs, and the amount of replacement and service

Table 4-6.
Typical Maintenance Experience (1982 Through 1986).

1. Maintenance spending as a percent of total new capital investment—116%. Six to seven new maintenance staff per $10 million in new capital equipment.

2. Salaried maintenance employees: maintenance managers (38.6% of total); engineers (12.6%); planners (8.1%); others including some project engineers (40.7%). One employee per $260,000 spent on maintenance.

3. Hourly maintenance employees: one employee per $62,000 spent on maintenance. ($78,000 per "skilled" hourly employee). There was 14.1% overtime.

4. Maintenance material inventory value: $310 for each $1,000 spent per year on maintenance; 1.3% of the equipment replacement value; $354,000 per maintenance "stores" employee.

5. Amount of contractor maintenance used—6.7% of the total.

Source: Nolden 1987. Reprinted from *Plant Engineering*, July 1987. © 1987 by Cahners Publishing Company.

equipment needed in inventory. It also shows that during the period from 1982 to 1987 the chemical industry spent 16% more for maintenance than it did for new equipment purchases. This clearly demonstrates the high cost of maintenance and the need for excellent maintenance management, planning, and scheduling.

The longer-term record suggests that maintenance productivity is improving. In most chemical plants maintenance has been subsidiary to operations and engineering; it primarily reacted to breakdowns and what engineering and operations told it to do. Now maintenance departments often find themselves in the reliability business, not the fix-it business. Some have been turned into departments that focus on predicting and avoiding equipment failures. This requires that work be performed not according to a generalized schedule but according to the actual condition of the machine or process unit. Implicit in this concept is a need for practical and convenient ways to monitor the equipment operating performance and to store and analyze the data. Many companies now keep the performance records for each piece of machinery, and every time a machine fails or gets worked on it is noted. These performance and cost records are used to gauge reliability, work out a preventative maintenance plan, and select between competitive machines with new purchases. It includes noting the nature of an equipment failure, the reason for the failure, the amount of down time, the cost of the downtime, the time between breakdowns, and the cost of repairing or replacing the failed unit.

More and better training also is improving the performance of maintenance departments. Many companies try to develop craftsmen with multiple skills in order to have workers always available who can tackle all the required aspects of a maintenance job in the most rapid and effective manner. The cost saving can be significant, but good training programs and union cooperation are required.

Operating Supplies, Laboratory Charges, Royalties

These items are all fairly self-descriptive, and their average values are listed in Table 4-4. Operating supplies are somewhat related to the plant cost, and thus are factored from it. They include all of the supplies not separately listed and that are needed to run the plant. Laboratory charges include lab supplies, as well as the labor required to perform all of the plant's quality and process control by means of analysis or testing. The cost is vaguely related to the total operating labor in the plant and thus is factored from it. Royalties, if involved (and not paid as a lump sum) are often charged on a yearly or tonnage production basis. They might vary from 0.25 to 2% of the total sales value, usually being on the lower end of this scale, and are factored from the total plant sales.

Packaging, Storage

All products require some expense for storage, and most a packaging expense for a least part of the production (i.e., the rest may be shipped in bulk). Multi-layered paper bags in mid-1987 cost about $10–15 per ton of product; 55 gallon steel drums about $25 per drum, and plastic or lined drums about twice that (stainless steel was about $450 per drum) (see Appendix 1). Add to these packaging material costs the expense of storage, rehandling for shipment, bagging or filling drums, and so on, and the cost is often appreciable. For other products where entirely bulk shipments are involved, this cost can be very small and is included in the cost of sales or other items. Packaging and storage costs must be detailed if they are important in the total manufacturing cost.

Environmental

Almost every plant must have environmental, safety, and hazardous waste engineers or groups to handle these problems and interface with the regulatory enforcement staffs of governmental agencies and the public. There are endless meetings, training programs, inspections, citations, reports, and paperwork required by the government, plus the normal company desire for good public relations and the plant to be a safe, clean, nonpolluting activity. This has become an extremely demanding requirement for every chemical plant, and a very expensive one. It is estimated that it might currently involve a yearly cost of 0.5–5% of the plant capital, but very likely this figure will soon be many fold that for plants with any appreciable degree of hazard or visibility to the government and environmental groups. This area, incidentally, is one of the major fields that new chemical engineers are being employed for, or where they find some of their first industry assignments. The technical/regulatory problems facing the chemical industry are very demanding, with highly hostile environmental groups and government agencies, and great difficulty in solving the technical problems in a reasonably economical and government approved manner.

Depreciation

In computing income taxes on the profit of an operation the government allows a deduction for a fraction of the initial cost of the plant as a hypothetical "expense" to be subtracted from the actual gross profit for income tax calculations. This deduction, called depreciation, may be considered as a fund to allow eventual replacement of the plant, or the equivalent of acknowledging that the plant has worn or is becoming somewhat obsolete or less competitive with the passing of time. Land is not allowed to be depreciated. The Internal Revenue Service accepts various means of computing depreciation, but there is *only* one method to use for preliminary manufacturing cost estimates: straight line depreciation over the anticipated life of the plant. The government in mid-1987 allowed this to be 5 to 15 years on much of the process equipment and 25 to 45 years on most buildings. These depreciation period numbers have changed several times in the 1980s, and probably will change again. However, if one can prove that the life expectancy of a piece of equipment or a process is less, a shorter period may be used. The government has published an extensive list of allowable depreciation rates for many pieces of equipment and other items, and your company's accounting or financial departments will know what rates are generally acceptable for the type of equipment or processes used in the plant. The accounting department will almost always keep separate depreciation records on each piece of equipment for income tax and property tax assessment purposes. Depreciation credit stops when the total government-allowed useful life period has been used up, even if the equipment is still in service, but the asset value of the equipment may only be removed from the company books when the equipment has been removed from the premises. In other words, if equipment is worn out or obsolete and placed in a "boneyard" or scrap pile it theoretically must still be kept on the books as an asset.

Straight line depreciation provides for the same deduction each year over the (IRS-allowed) life of the item:

$$\frac{(\text{original plant cost} - \text{salvage value}), \$}{\text{plant or component life, years}} = \text{depreciation}$$

Often the salvage value is unknown or zero, since the cost to dismantle and sell the partially or totally worn out equipment is generally about equal to the value that would be received from a dealer for the sale of the old equipment, making the yearly straight line depreciation 20% (5-year life) to 6.7% (15-year life) of the initial plant cost per year. Ten percent is a good working average if more exact depreciation rates are not known.

For very precise manufacturing cost estimates, the company's accounting department (almost never the engineer) may use one of the IRS allowed so-called "rapid" depreciation procedures. They will usually use the method providing the greatest present value of the depreciation funds over the life of the plant. This might be the *"double declining balance"* method, where the first year's depreciation is two times the straight line depreciation, the second year is two times the straight line depreciation of the remainder, and so on, or

$$\frac{2 \times (\text{original plant cost } - \text{ previous depreciation } - \text{ salvage value})}{\text{life of plant or equipment}}$$

Other multiples, such as 1.25, 1.5, 1.75, and so on, are allowed for certain equipment or buildings.

Another accelerated depreciation method is the *sum-of-years digits* depreciation. Here the numbers for each year of the plant life are totaled (i.e., for a 10-year life, $1 + 2 + 3 + \ldots + 10 = 55$), and each year the depreciation is: first year 10/55, second year 9/55, third year 8/55, and so on, to the tenth year which is the remainder. This procedure has the greatest present worth for plants with greater than five-year life, and thus will show the highest rate of return on investments, and should usually be preferred by the financial departments. However, most companies frown on any but straight line depreciation being used by the estimating engineer, and usually list income tax saving by a depreciation method other than straight line as "deferred taxes" on their balance sheet. A comparison of depreciation allowances by these methods is listed in Table 4-7, along with the present value of these depreciation allowances.

Local Taxes, Fees, Insurance, Interest

Every locality charges taxes (or business license fees) of various types, and there are many governmental permits, yearly charges, assessments, miscellaneous licenses, and so on. Also, each plant generally is required by its customers to carry product liability insurance, and other insurance (particularly personal and property liability) is almost a necessity, while fire, theft and property damage is desirable unless the company wishes to be self-insured. The total cost of such taxes and insurance is typically about 3% of the total plant cost, even though the personal and product liability insurances are rigorously based upon the total labor cost and product sales value, respectively. With a hazardous product, plant, or waste material the insurance cost can be many times this value, if available at all. Currently most chemical plants (and particularly small ones) are having great difficulty in obtaining liability insurance, and hazardous materials or waste liability insurance is essentially unavailable. A number of

Table 4-7.
Comparison of Depreciation Allowances by Various Methods
(Equipment Costing $100.00; 10-Year Life; No Salvage Value).

Year	Straight Line, $	Sum of Digits, $	Double Declining Balance, $
1	10.00	18.18	20.00
2	10.00	16.36	16.00
3	10.00	14.55	12.80
4	10.00	12.72	10.24
5	10.00	10.90	8.19
6	10.00	9.09	6.55
7	10.00	7.27	5.24
8	10.00	5.45	4.91
9	10.00	3.64	3.65
10	10.00	1.82	3.42
Total	100.00	100.00	100.00
Present value, 8% interest	67.87	75.54[a]	68.99
Present value, 8% interest; 5-year depreciation	79.85	83.94	84.22[a]

[a]Most profitable.

industry group insurance companies are being formed to provide basic insurance for their member companies. Liability insurance has become a major industry (and national) problem because of the very high court awards for "damages."

The interest charges on borrowed capital must be treated individually with each project. Most companies borrow some or much of the funds for all new plants, so the management must tell the engineer what fraction of the plant capital will be from loans, and what the interest rate will be. If such information is not available, then assume *no* borrowed capital (this is generally advisable in most cases), and carefully state this in the economic presentation. If part of the capital is borrowed, the engineer is faced with the decision of considering only the company's actual capital input and ignoring the loans as part of the capital requirement when making the final profitability analysis. Again, however, unless instructed otherwise, there is no question but that he should always use the total capital investment, since all of this is at risk with the project (and the company is usually liable for it). However, in some cases the company policy will be to only consider their portion of the capital (i.e., employ "leverage"), and then the estimator must comply.

Administration, Sales, and Research and Development

On every project a portion of the company's corporate and general overhead costs must be assigned to the new plant, and this is usually prorated to each project on the basis of its capital, sales value, or some other formula. Also, each plant requires a certain amount of operational expense to distribute and

Table 4-8.
Examples of General and Administrative, Sales, and R & D
Overhead Costs.

Company	Type of Business	Total Sales, Million $ 1969	S & A (Including R & D), % Sales	S & A Costs (Excluding R & D), % Sales
Related Process Industries				
Alcoa	Metals	1,580	9.2	
Texaco	Petroleum	6,270	9.1	
U.S. Gypsum	Building materials	470	15.9	
General Foods	Consumer products—food	2,060	28.0	
Procter & Gamble	Consumer products—soap	2,730	24.5	
Celanese	Textiles and chemicals	1,250	15.1	
General Electric	General equipment	8,550	18.9	
	Average		17.2	
Chemical Industry				
Stauffer	Chemicals (mainly Industrial)	500	11.3	8.8
Monsanto	Chemicals (some specialties)	1,900	16.6	11.4
Hercules	Chemicals (some specialties)	800	13.9	11.2
Union Carbide	Chemicals (some specialties)	2,900	13.1	10.6
American Cyanamid	Chemicals (considerable specialties)	1,200	20.2	16.5
DuPont	Chemicals (some specialties)	3,600	12.0	9.9
Nalco	Specialty chemicals	165	22.5	18.3
Eastman Kodak	Specialties and chemicals	2,750	16.9	11.5
	Average		15.8	12.3
	Total Average		16.5	

Ohsol 1971. Excerpted by special permission from *Chemical Engineering*, May 1971. Copyright © 1971 by McGraw-Hill, Inc., New York.

sell the product. Both of these values often range from 2 to 10% of sales, but can be considerably higher. Finally, general research and development usually runs from 0.5 to 5% of sales, and this amount must also be allocated to each new project. Table 4-8 lists some older data on administration, sales, and research and development expenses for a few chemical (or related) companies, showing a wide range of costs, but averaging about 16% of sales for these three overhead items. The average was well over 20% in 1987. Figures 4-4 and 4-5 show 1972–1986 research and development spending (alone) for industrial chemical producers, indicating their average to be about 4.5% of sales. However, the average for the entire chemical process industry (CPI) is only 1.5–1.6%, (Table 4-9), primarily because of the low research and development spending by the petroleum industry (for petrochemicals, etc.) and other sections of the CPI. Much of this "research" budget is actually spent for technical service work.

Total Manufacturing Cost

The sum of the individual charges described above are added to provide an estimate of the total manufacturing cost. It is seen that besides the itemized costs (raw materials, operating labor, utilities), a typical plant might accrue yearly operating expenses equal to about 165% of the cost of operating labor, plus 26% of the original total plant cost, and 20% of sales value of the plant's product. These items add a great deal to the manufacturing expense, but they

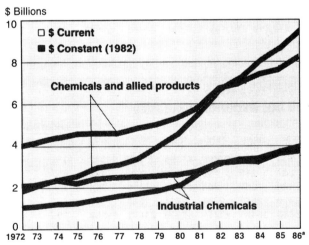

Source: *Chemical and Engineering News* (1987). Reprinted with permission from *Chemical and Engineering News*, June 8, 1987. © 1987, American Chemical Society.

Figure 4-4. R & D spending, $.

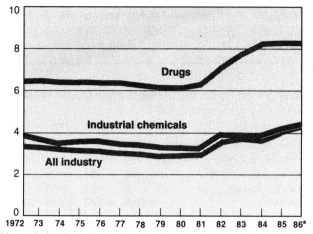

Reprinted with permission from *Chemical and Engineering News*, June 8, 1987. © 1987, American Chemical Society.

Figure 4-5. R & D spending as % of net sales.

Table 4-9.
R & D Spending by Industry.

	Percent of Sales	
	1986	1987
Aerospace	18.7	19.8
Instruments	9.6	9.5
Electrical machinery	8.7	8.6
Nonelectrical machinery	5.4	5.5
Auto, trucks and parts	4.1	NA
Iron and steel	1.5	1.5
Fabricated metals	0.5	0.5
Chemical process industries		
Chemicals	4.5	4.4
Rubber and plastic	1.8	1.8
Petroleum	1.7	1.9
Nonferrous metals	1.3	1.2
Stone, clay and glass	1.2	1.1
Paper and pulp	0.8	0.8
Food and beverage	0.3	0.3
Textiles	0.3	0.3
CPI total	1.6	1.5
All manufacturing	3.5	3.5

Source: Spalding 1987. Excerpted by special permission from *Chemical Week*, July 8, 1987. Copyright ©1987, by McGraw-Hill, Inc., New York.

are realistic, and must be taken into account. The relative importance of each of these cost categories varies widely with the product, as illustrated by Table 4-10, so individual attention to each item is necessary. As with capital cost estimates the individual cost components should almost always be itemized rather than using only these three broad components: total labor, capital, and sales related costs. It is informative to see the individual costs, and many supervisors and reviewers will wish to see various items to compare with their knowledge of what these costs should be. Also, whenever sufficient detail is available each of the factors should be replaced by more complete and accurate estimates. The overall accuracy of such factored operating cost estimates should be similar, but usually somewhat greater, than the corresponding capital cost estimates previously made from Table 3-4.

ESTIMATING CHARTS AND TABLES

For a number of basic commodities, and some individual operations, complete tabulations and charts are available to estimate the manufacturing costs, as listed in Appendix 3. Some of this data on raw materials and utilities for large and complex plants are probably far more accurate than one could personally estimate. Many of these total manufacturing cost charts and tables, however, must still be considered only first approximations because of their age, changing technology, and variably increasing costs of labor and utilities. Nevertheless they have considerable value as a guide for the total product price of complex plants.

Many other manufacturing cost estimating procedures have been suggested, such as the generalized production rate versus fixed operating costs (all costs

Table 4-10.
Examples of Manufacturing Costs Breakdown

| | % of Total Production Costs | | | | |
| | Commodities | | | Specialties | |
	TiO_2	VAC^a	$GPPS^b$	$DLTDP^c$	$CPVC^d$
Raw materials	40	64	76	37	49
Utilities	17	15	1	1	3
Labor related costs	18	4	7	30	14
Capital related costs	18	11	6	14	15
R & D, sales, G & A	7	6	10	18	19

Source: Kyle 1986. Reproduced by permission of the American Institute of Chemical Engineers.
[a]Vinyl acetate.   [c]Dilauryl thiodipropionate.
[b]General-purpose polystyrene.   [d]Chlorinated polyvinylchloride.

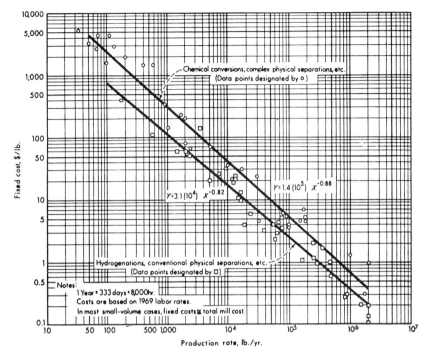

Source: Sommerville 1970. Excerpted by special permission from *Chemical Engineering*, May 1970. Copyright © 1970, by McGraw-Hill, Inc., New York.

Figure 4-6. Generalized relationship of fixed operating costs with production rates.

except raw materials and process energy, in this case) of Figure 4-6. Such correlations can be quite useful as rough approximations, but as with most such estimating methods, their accuracy and usefulness are quite limited.

REFERENCES

Brooks, Kenneth. 1987. Maintenance costs. *Chemical Week* (July 15):45.

— 1986. *Chemical Week* (July 16):36.

Chem Cyclopedia. 1987. American Chemical Society, 1155 16th St., N.W., Washington, D.C., 20036; *Chemical Sources-U.S.A.*, Directories Publishing Co., Inc. P. O. Box 1824, Clemson, S.C. 29633.

Chemical and Engineering News. 1987. Facts and figures for the chemical industry. (June 8):24–76.

Chemical Marketing Reporter. 1988. Schnell Publishing Co., Inc., 100 Church St., New York, NY 10007-2694.

Holzinger, Albert G. 1988. The Real Cost of Benefits. *Nation's Business* (February):30–32.

Jones, Kenneth C. 1987. Steam pricing. *Plant Engineering* (July 9):34–40.

Kyle, H. E. 1986. Effective use of technoeconomic resources. *Chemical Engineering Progress* (Aug.):17–20.

Nolden, Carol. 1987. Plant maintenance costs. *Plant Engineering* (July):38–42.

OPD Chemical Buyer's Director. 1987. Schnell Publishing Co., Inc., 100 Church St., New York, NY 10007–2694.

Ohsol, Ernest O. 1971. Estimating marketing cost. *Chemical Engineering* (May).

Peters, M. S., and K. D. Timmerhaus. 1980. *Plant Design and Economics for Chemical Engineers.* McGraw-Hill, New York.

Rawls, R. L. 1987. Facts and figures for chemical R & D. *Chemical and Engineering News* (July 27):32–62.

Rickey, P. 1987. Shift schedules. *Chemical Engineering* (May 11):69–73.

Sommerville, Robert F. 1970. New method gives accurate estimate of distillation cost. *Chemical Engineering* (May).

— 1970. Estimating mill costs at low production rates. *Chemical Engineering* (April).

Spalding, B. J. 1987. Chemical makers hike R & D spending. *Chemical Week* (July 8):18–19.

Wessel, Henry E. 1952. New graph correlates operating labor data for chemical processes. *Chemical Engineering* (July):209.

5

INTEREST CALCULATIONS; PRELIMINARY PROJECT EVALUATION

In previous chapters methods have been developed for estimating the cost of equipment, additions, modifications, and complete plants, and for the expense of manufacturing a product. This chapter will attempt to provide the initial procedures for evaluating the hoped-for profitability of chemical plant investments, and of estimating the relative attractiveness of different projects and capital expenditures. The first step involves three additional pieces of information that are needed: the product sales value, the company income tax requirements, and a knowledge of the effect that compound interest has upon investments. This later subject is not only important for the money that may be borrowed on a project, but even more so to examine how the project investment might compare with an alternate investment (including a purely financial one, such as bonds, CDs, etc.). This topic shall be reviewed first in the following section.

INTEREST: THE COST OF MONEY

From time immemorial the loaning of money has had a charge, or cost, called interest. Presumably the lender could have invested the money elsewhere and made a profit had he not loaned it, so the interest is his compensation for the otherwise lost profit. The amount of interest depends upon the scarcity of money at the time of the loan and what alternative investments might have yielded. It also depends upon the risk the lender feels that he is taking that the money might not be repaid, or what collateral or security may be pledged to the lender that has an equal or somewhat greater value. Some limited sums may be lent without such a tangible guarantee when the credit worthiness of the borrower is well established, but usually the lender wants to be assured that the money will be paid back, along with interest, using such collateral as the pawn shop's deposit of merchandise, the bank's taking title to property, stocks, or bonds upon default, or obtaining a more creditworthy cosigner.

In the business world the need for borrowing is even more demanding and generally required than with one's personal financing. Few companies can get started or maintain a reasonable vigor without debt. Also, tax laws, inflation, and profit potential generally make borrowing a very prudent option, up to a point. Collateral is still required, either directly for bank loans, or indirectly for the company's ability to sell bonds, debentures, or preferred stock. Interest is paid on these bonds or loans, with the rate depending upon the normal business factors existing at that time, and the creditworthiness of the company. Several financial services rate the potential repayment strength (for loans) of all the larger companies, using a nomenclature from triple A downward. These ratings have a major effect upon the company's ability to sell bonds and the interest rate that is paid.

Most companies maintain a mixture of types of debt, including bank loans that are usually short term (such as for one year or less), and which often are at least partially "rolled over" at the end of their term to give them a longer life. The interest rates on these loans are usually higher than the then-current long-term bonds, and the rate is generally variable and tied to the bank's current "prime rate." This is an interest rate that is some amount (i.e., 2%) higher than the bank can borrow funds from the Federal Reserve at their "discount rate." The banks might loan funds to a very few large, highly solvent firms at this prime rate, but for most business customers the interest charged would be 0.5 to 2.5% above prime, and would vary each time the prime changed.

Companies also sell bonds that are usually in denominations of $1,000, pledging the bond (loan) to the company's assets as collateral, and promising to pay a fixed percentage interest rate (usually semiannually) until a date 10 to 20 years in the future when the bond would be repaid. When a company has issued more bonds than their assets can conservatively guarantee, then subordinated debentures can be sold. They are identical to bonds, but have a secondary position to the bonds (or perhaps no security at all) in being guaranteed by company assets in case of a bankruptcy. They generally pay a slightly (i.e., 0.5 to 1%) higher interest rate than bonds.

The last step in loans with reduced security is a class of bonds called "junk bonds." These bonds are just like regular bonds, but are issued in very uncertain or high debt-to-equity conditions. They are often used to raise money in takeover, merger, or leveraged buy-outs of a company, but they can also be employed in exceptional cases for financing on-going businesses. They often command 2 to 4% additional interest, and are considered to be quite risky. To date, however, their failure rate has not been as high as their enhanced income would cover, and they have been very successful in helping companies become established which would not have been able to obtain normal credit.

INTEREST CALCULATIONS

There are many ways that money can be loaned and repayed, and an equal number of ways that money can accumulate, earn interest, and be dispensed. The mathematics of these procedures is usually fairly simple, even though often complex. In the following chapter some of the standard nomenclature and a few of the more basic calculations will be reviewed. Tables and computer programs are available for some of them, and many references cover their derivation and use. Hopefully with the background given here most engineers can derive any others they may later need.

Simple Interest

When money (capital or principal) is loaned the amount of money earned by a unit of principal in a unit of time (i.e., usually one year) is called the rate of interest. For example, if 10% were the interest agreed to for a loan of $1,000 for a period of one year, the principal would be $1,000 and the interest payment would be 0.10 × 1,000 = $100.00 per year. "Simple interest" such as this requires yearly payment at a constant interest rate times the original principal. If $1,000 were loaned for five years at an interest rate of 10% and no payment was made on the principal, the simple interest earned would be $1,000 × 0.1 × 5 = $500.

In this book P will represent the principal, or original capital, n the number of yearly interest periods, i the interest rate for that interest period, and the amount of simple interest I earned during n interest periods. Thus, for simple interest,

$$I = Pin \qquad (5\text{-}1)$$

The principal must be repaid eventually, so the entire amount S of principle plus simple interest due after n interest periods is

$$S = P + I = P(1 + in) \qquad (5\text{-}2)$$

When borrowing money care must be taken to understand the interest charging basis. Banks have learned that there is a great deal of money to be made by controlling how interest is to be paid. For instance, when it is in their favor and an interest period of less than one year is involved, they may claim that the "ordinary" way to determine simple interest is to assume the year consists of twelve 30-day months, or 360 days. When they owe you, however, the "exact" method is often employed based upon the fact that there are 365 days in a normal year. In a related ploy, they will often delay crediting the deposit of checks into your account for a day or so on local checks, and perhaps as much as a month on out of state or foreign checks. With the advent of electronic fund transfers they usually have received a credit for the money almost immediately, and earn interest on the funds for themselves until they finally credit your account. With a little friendly discussion your banker can usually make these practices considerably more equitable.

Compound Interest

Simple interest is generally paid at the end of each time unit or at agreed upon intermediate intervals. However, interest has an important time value, and when ever the interest is paid the receiver can immediately put the interest to work if he desires and earn additional interest. This is called "compound interest" which assumes that the interest received is not withdrawn, but added to the principle, and interest is received upon this enlarged principal during the following period. Thus, an initial loan of $1,000 at an annual interest rate of 10% would require payment of $100 interest at the end of the first year. When compounded, the interest for the second year would be ($1,000 + $100) (0.10) = $110, and the total amount due after two years would be $1,000 + $100 + $110 = $1,210. The compound amount (S_n) due after any discreet number of interest periods can be determined as follows:

$$\text{first year:} \quad P + Pi = P(1 + i)$$
$$\text{second year:} \quad P + Pi(1 + i) = P(1 + i)^2$$
$$n\text{th year:} \quad P(1 + i)^n$$

Therefore, the total amount of principal plus compounded interest due after n interest periods, is

$$S_n = P(1 + i)^n \tag{5-3}$$

Normally the interest rate i is quoted on a yearly basis, and the time period n is the number of years involved. However, interest payments are actually often made on a more frequent basis, such as semiannually, quarterly, monthly, or even daily. To determine the compounded amount earned on this basis let m equal the number of interest periods per year, and Equation (5-3) then becomes

$$S = P(1 + i/m)^{mn} \tag{5-4}$$

Obviously, the more frequent the compounding period, the greater the return from the interest payments. The ultimate would be if the interest were paid and compounded continuously. It is difficult to imagine how this would occur in practice, but in the case of a multinational company (as most chemical companies now are), with interest being obtained from the funds received from sales around the world and deposited in local banks as received, the total interest that accrued might appear to be fairly continuous. Conversely, in considering the return on a production plant investment, with the bookkeeping on an "accrual" basis where every transaction, including credit for production, was entered on the books when it occurred regardless of when the cash was paid (for purchases) or received (for sale), then with around-the-clock production the profit and equivalent rate of return, or "interest" would be truly continuous.

Mathematically, if earnings are accrued on a continuous basis, the differential increase in the compounded amount dS, over a differential time interval dn is given by

$$dS = Sr \, dn \qquad (5\text{-}5)$$

where r is the continuous interest rate, and n is the time.

Holding r constant with respect to time, and integrating this equation from time zero (the reference point) to time n, one obtains

$$\int_P^S \frac{dS}{S} = \int_O^n r \, dn$$

$$\ln S/P = rn$$

$$S = Pe^{rn} \qquad (5\text{-}6)$$

The following table shows a comparison of the total amounts due at different times for the cases where simple interest, discreet compound interest, and continuous compound interest are used. Assume that $100 was borrowed for 1 or 10 years at 10% interest, compounded either annually, semiannually, quarterly, daily, or continuously. The repayment, S, of interest plus principal would be

Period	Annually	Semiannually	Quarterly	Daily	Continuously
1 yr	$110.00	110.25	110.38	110.52	110.52
10 yr	259.37	265.33	268.51	271.79	271.83
% increase from annual		2.30	3.52	4.79	4.80

It is seen that the compounding period can make a small but important difference in the growth of the principal, but that daily and continuous compounding are nearly the same. Thus, in future project economic evaluations such as "discounted cash flow" (DCF) or "present worth" it can be assumed that the cash received from sales and deposited into the bank would be at least on a daily basis, so either daily or continuous compounding would be acceptable and almost identical to each other. Actually, in profitability calculations any mathematic relationship can be used, just so that consistency is employed.

Present Worth (or Present Value)

In project profitability calculations it is often necessary to determine the present value of funds that will be received at some definite periods in the future. If it is assumed that the principal and interest payments were not withdrawn, then interest compounding's effect upon the amount must be taken into consideration. The present worth (or present value) of a future amount on this basis is considered the present principal which if deposited at a given interest rate and compounded would yield the actual amount received at a future date. The relationship between the indicated future amount and the present worth is calculated by an equivalent "discount factor."

In Equation (5-3), S_n represents the amount available after n interest periods if the initial principal is P and the discreet compound interest rate is i. In other terms, S_n is the amount that would be received during year n. Therefore, the present worth can be determined by merely rearranging Equation (5-3):

$$\text{present worth} = P = S_n(1 + i)^{-n} \tag{5-7}$$

The term $(1 + i)^{-n}$ is referred to as the annual interest discount factor. It would become $(1 + i/m)^{-mn}$ for the more frequent period discount factor. Similarly, for the case of continuous interest compounding, Equation (5-6) gives

$$\text{present worth} = P = S_n^{-rn} \tag{5-8}$$

and e^{-rn} is the continuous interest discount factor.

Annuities

An annuity is a series of constant payments occurring at equal time intervals. Sequences such as this can be used to pay off debt (such as a house or car purchase), accumulate a desired amount of capital (such as an IRA savings plan), or pay periodic installments on a life insurance policy. Engineers often encounter annuity-type formulas in both personal investing and in financial analyses, such as the cost of replacement of equipment or calculations for certain cases of project profitability analyses.

One type of annuity involves payments which occur regularly at the end of each interest period, or year. Interest is received for both the capital and accumulated interest, making the interest compounded on each payment period. In this case the annuity term is the time from the beginning of the first period until the last payment is made. The amount that has been accumulated in the annuity is the sum of all the payments plus compounded interest (assuming no withdrawal) for the annuity term.

To calculate the annuity, let C represent the uniform periodic payment made during n years. The interest rate for the payment period is i, and S is the future accumulated total value of the annuity. The first payment of C is made at the end of the first period and will bear interest for $n - 1$ periods. Thus, at the end of the annuity term, this first payment will have accumulated according to Equation (5-3) to an amount of $C(1 + i)^{n-1}$. The second payment of C is made at the end of the second period and will bear interest for $n - 2$ periods giving an accumulated amount of $R(1 + i)^{n-2}$, and so forth. The amount of the annuity is the sum of each year's accumulated value, or

$$S_n = C(1 + i)^{n-1} + C(1 + i)^{n-2}$$

$$+ C(1 + i)^{n-3} + \cdots + C(1 + i) + C \qquad (5\text{-}9)$$

To simplify Equation (5-9), multiply each side by $(1 + i)$ and subtract Equation (5-9) from the result. This gives

$$S_n = C \frac{(1 + i)^n - 1}{i} \qquad (5\text{-}10)$$

As an example of an annuity calculation, if an engineer invested \$2,000 per year in an IRA retirement fund for 25 years, earning 10% interest per year compounded quarterly, how much would he have upon retirement?

$$S = \$2,000 \frac{[(1 + 0.10/4)^{4 \times 25} - 1]}{0.10/4}$$

$$= 2,000 \frac{[(1.025)^{100} - 1]}{0.025}$$

$$= 2,000 \frac{(11.8137 - 1)}{0.025}$$

$$= 2,000 \times 432.55 = \$865,000$$

The expression for the case of continuous cash flow and interest compounding, equivalent to Equation (5-9) may be developed as follows.

Define C such that the differential change in S with n is equal to C, which is the constant contribution during the year. When the amount due to interest is added, $dS/dn = C + rS$. This expression can be integrated as follows to give

$$\int_0^S \frac{dS}{C + rS} = \int_0^n dn \tag{5-11}$$

$$\ln \frac{C + rS}{C} = rn \quad \text{or} \quad S = C\left(\frac{e^{rn} - 1}{r}\right) \tag{5-12}$$

Present Worth of an Annuity. The present worth of an annuity may be calculated by the discount factors that have previously been developed in Equations (5-7) and (5-8). If P represents the present worth of an ordinary annuity, combining Equation (5-3) with Equation (5-10) gives the present value for annual compounding

$$P = C \frac{(1 + i)^n - 1}{i(1 + i)^n} \tag{5-13}$$

As an example of the sum that would have to be deposited at this time in a bank to be equivalent to the value of an annuity, what is the present-day value of the 25-year annuity calculated above.

$$P = \$2,000 \frac{[(1.025)^{100} - 1]}{0.025 \, (1.025)^{100}}$$

$$2,000 \frac{(10.8137)}{0.025(11.8137)}$$

$$2,000 \times 36.614 = \$73,228$$

For the case of continuous cash flow and interest compounding, combining Equations (5-4) and (5-12) results in the following equation for the present value under these conditions

$$P = C \frac{e^{rn} - 1}{re^{rn}} \tag{5-14}$$

Capitalized Cost

A method sometimes used to compare the total cost or value of competitive equipment considers the hypothetical cost of the unit if it were replaced perpetually. (Assuming a constant cost.) This type of calculation can take into account different equipment costs, service life, and even annual maintenance expense. As an example, if the equipment cost $10,000, had a useful life of 10 years and no salvage value, and interest rates were 10%, then a fund P would be required to generate $10,000 every 10 years. If the fund were to perpetuate itself it would have to be equal to $10,000 (the amount removed each 10 years) plus P, or S = $10,000 + P = $P(1 + i)^n$. Thus, if compounded annually the fund would need to be $10,000 = $P(1 + 0.10)^{10} - P$; $P = \$10,000/1.1^{10} - 1 = \$6,274.55$. At this interest rate the fund would amount to $6,274.55(1 + 0.10)^{10}$ = $16,274.55 after 10 years. At the end of 10 years the equipment could be replaced for $10,000 and $6,274.55 would remain in the fund and the cycle could be repeated indefinitely. The total capital determined in this manner is called the capitalized cost.

The amount required for the replacement is earned as compounded interest over the life of the equipment. With P the amount of present principal (i.e., the present worth) which would accumulate to an amount S_n during n interest periods at interest rate i, $S_n = P(1 + i)^n$. If R is the equipment cost, combining Equations (5-3) and (5-10) gives

$$P = \frac{R}{(1 + i)^n - 1} \tag{5-15}$$

$$\text{capitalized cost} = \text{original cost} \left(1 + \frac{1}{(1 + i)^n - 1}\right) \tag{5-16}$$

With this brief introduction to compound interest calculations we can now return our attention to project profitability analysis, and the additional information that is required for such calculations.

PRODUCT SALES VALUE

In order to determine the income for any new project an estimate must be made of the quantity of product that can be sold, its FOB sales price, the exact product specifications, and any freight allowances or other charges that must be considered. Such information is generally obtained by studies of the existing and potential markets by means of "market research."

Market Research

Most large companies have organized market research groups of specialists that gather and project the sales information that is available on each new or changed product under consideration (Mackenzie and Thomas 1986). They then make presentations and recommendations to management and the sales department for their decision on the proposed new plant size, the product quality, specifications, inventory required, packaging, sales price, discounts, freight equalization, marketing methods, and so on. With other companies the sales department, product managers (or teams), and sometimes even members of the technical staff are assigned this market research responsibility. For most marketing studies a survey of all the available literature sources is first made to obtain as much information as possible on the product, including the amount sold, its growth pattern, the industries sold to, the end use, the size and location of competitors' plants, the price history and trends, shipment method, containers and product specifications. Some of this information is available from the Department of Commerce (1987) or other governmental surveys (Bureau of Mines 1987 on metals and minerals, etc.), trade magazine (*Chemical Marketing Reporter* 1987; *Chemical Week* 1987; etc.) articles or surveys, consulting companies [such as Stanford Research Institute (1987) compilations], and many other sources (International Trade Commission 1987). A second step then involves interviewing as many of the potential customers as possible, asking the same questions about their purchasing needs and whether they would buy the new product, and if so, under what conditions.

For well-established products this information can provide a fairly exact picture of the product's sales potential, and the sales department and management will have a good basis to establish the proposed new plant size and product specifications. They can then estimate the average net FOB price on various tonnages sold for the life of the plant. For newer or specialty products or markets there are much less data available, and sales and customer acceptance becomes a much greater uncertainty, but the estimates still must be made. If management or the sales department does not provide such detailed sales projections, then the estimating engineer will have to make the best guess possible based upon somewhat discounted or related prices from the *Chemical Marketing Reporter,* and assuming a gradual penetration into the market before the total plant capacity is sold.

The lower the fraction of the total market for a product that is assumed can be sold from a new plant, the lower the risk. Ten percent of an established market would normally be a maximum estimate unless there were some unusually favorable conditions for the new plant. For all economic estimates a number of *sensitivity analyses* should be made on possible different sales prices and amounts sold. This often indicates that below a certain capacity the project is not economically attractive, which helps explain the growing trend toward the construction of only very large plants.

INCOME TAX

After the product sales value and the manufacturing costs have been estimated, the difference between the two equals the *gross income* for the operation. This amount is the basis for determining the income tax that will be assessed by the state and federal governments. These taxes change frequently, but for the 1987 period an example is shown below:

1. California income tax (franchise tax board) 9.6% of gross income. Other states vary from zero to slightly higher numbers.
2. U. S. Internal Revenue Service income tax. (Rates for corporate years beginning July 1, 1987 and later.)

Taxable Income	Rate
Not over $50,000	15%
Over $50,000 but not over $75,000	25%
Over $75,000	34% plus 5% on excess over $100,000, or $11,750, whichever is less

3. Maximum corporate capital gains rate increased from 28 to 34% for years beginning after December 31, 1986.

The federal government allows all other taxes, including state income tax to be considered as an expense, while some states (e.g., California) may not accept other taxes as deductions. The state (e.g., California) also may not allow losses from previous years being carried over, and usually have many more restrictions. However, assuming the same gross income, the combined California and IRS tax in 1987 would be

Gross Income, $	Combined Total Tax, %
0–54,800 = 50,000 + 0.096 × 50,000	23.16 = 0.15 (income − income × 0.096) + income × 0.096 = income [0.15(1 − 0.096) + 0.096]
54,800–82,200	32.20
Over 82,200	40.34

For economic estimating purposes a combined maximum income tax figure of 40% is probably a good average value to use currently for the chemical industry, but the total amount of income tax varies with each state, and with each company. It is recommended (to be conservative) that any new project consider the maximum tax that may be due, unless management has specifically instructed estimators to use a different value. In actual practice there are many deductions that may be applied to taxes, and very few companies pay anywhere near the total amount that should be due on gross profit. The chemical industry prior to 1987 averaged well under 25% total taxes on gross profits, and the petroleum industry under 5–10% as shown in Table 5-1. Many chemical (and other CPI) companies pay no taxes at all, as shown for a few companies in

Table 5-1.
Tax Rates of Major U.S. Industries in 1982.

Automobiles	48%	Airlines	16%
Trucking	46	Metal manufacturing	10
Pharmaceuticals	36	Utilities	9
Electronics/appliances	29	Aerospace	7
Food processing	27	Petrochemicals	5
Industrial/farm equipment	24	Crude oil	3
Retailing	23	Commercial banks	2
Oil/refining	19	Railroads	8
Diversified financial	17	Paper/wood	14

Source, Reich 1985; Copyright, 1985, *Los Angeles Times*. Reprinted by permission.

Table 5-2. With the tax law of 1987 it is estimated that there will be slight changes in the amount of taxes paid, with heavy industry (including chemicals) somewhat increasing. The oil and gas industry might increase from a 1987 average of 8.2% to about 12%, and so on (Mendes and Serever 1986).

For companies with international sales a major deduction results from receiving a foreign tax credit for having paid income taxes to foreign governments. Often such foreign income "taxes" are somewhat hypothetical, and in the United States would be considered as normal operating expenses, but the

Table 5-2.
Chemical Companies That Paid No Taxes in 1985.

	Tax Credits, $ Million			Tax Rate Overseas, %	U.S. Taxes
	Investment Credit	ACRS[a]	Other		
1. American Cyanamid	10.05	5.97	19.81	43.70	0
2. DuPont	4.43	12.41	4.14	81.07	0
			51.73		
3. Goodyear Tire & Rubber	64	44			(20.7)
4. Hercules	8.18	20.86	22.88	33.78	(4.86)
5. International Mineral and Chemical	3.01	4.37	22.50	54.00	(5.29)
6. Pennwalt	59.30		(25.36)	33.99	0
7. W. R. Grace	2.47	29.92	(5.52)	47.84	0
8. Air Products & Chemical					0
9. Celenese					0
10. Sun Chemical					0
Total	139.91	117.53	38.45		0
Average	20.0	19.6	6.4	49.1%	0

Source: Katzenberg 1986; Excerpted by special permission from *Chemical Week* October 1, 1986 Copyright © 1986, by McGraw-Hill, Inc., New York. *Chemical and Engineering News* 1987. Reprinted with permission from *Chemical and Engineering News*, July 21, 19878; p. 15; © 1987, American Chemical Society.
[a]Accelerated cost recovery system. The IRS name for accelerated depreciation tax credit.

foreign tax designation allows a significant U. S. tax saving. For companies owning and producing mineral resources (such as oil), the reserves are considered to have a limited life which is being "depleted," and thus 5–28% of the mineral value that was sold each year may be deducted from gross profits as a depletion allowance. These and many other deductions greatly reduce the actual income tax payment, but as noted above, they should *not* be considered in preliminary cost estimates by the engineer unless it is the specific policy of the company.

Another exception to the use of a flat 40% income tax might be in the first year of operation, where various energy and other credits may apply. In prior years equipment or plants with a life of greater than three years received an investment credit of 10% of the capital cost. This amount was deducted directly from the federal income tax (i.e., it was credited much more than as an operating expense) for the year the capital investment occurred (but was not credited for the State tax). Various energy credits were also allowed, but they were, and now are more complicated, and some only apply to the state income tax, so should not be used unless one knows the precise details of the deduction. For instance, in 1985 the state of California allowed a 25% investment credit on many energy saving projects, and the federal government, 15%. These credits were very important to the profitability of some "alternate energy" projects', and when used needed to be carefully noted in the investment analysis presentations. Both have now been (almost completely) repealed.

RETURN ON INVESTMENT (ROI)

After the income taxes have been deducted from the gross or pretax income, the remainder, or *net income* is the amount that belongs free and clear to the corporation and may be used for paying dividends, reinvesting, or spent for any other purposes. This amount is also the basis for determining the simplest measure of the profitability of an investment, the rate of return that the investment is generating:

$$\frac{\text{net income per year}}{\text{total investment}} = \text{rate of return (after taxes)} = \text{ROI}$$

(total plant, including start-up, auxiliaries,

working capital, i.e., plant + working capital)

This rate of return may be compared with alternative investments, including putting the same amount of money in the bank or investing it in bonds, and gives a first approximation of how attractive the new project may be. Since the company management or others may wish to adjust this figure, one should always state what income tax rate and depreciation schedule were used, as well as note any interest on borrowed capital, the amount of product sold, and the sales price. With this data anyone can very simply convert the ROI to any basis they want, such as before taxes, a different depreciation method or period, other sales value, and so on. In general, ROI is a very simple concept that is easy to understand and apply, and for many estimates that are in an early stage of development, or quite simple, it may be all that is warranted. Every project should have an ROI estimate made on it, even if the profit varies from year to year. In this case an average may be assumed, or the profit after steady-state, or the hoped-for income a few years after start-up, is achieved.

PAYOUT PERIOD

An equally general, but perhaps more simplistic, piece of information that is also desirable for every project is how rapidly the project will pay for itself, or return the original investment. This "payout period" may be calculated as follows:

$$\frac{\text{total plant investment less working capital}}{\text{cash flow (= net yearly income after taxes plus depreciation)}}$$

$$= \text{payout period, years}$$

This equation assumes that the working capital is returned at the end of the project life, and that there is no salvage value. Both of these assumptions are usually correct. Thus, the shorter the payout period, the more attractive a project. It also should be estimated for every project, no matter now complex.

REFERENCES

Bureau of Mines, 1987. U.S. Department of the Interior, Mineral Commodity Profiles; Mineral Facts and Problems; Mineral Yearbook; Mineral Industry Surveys.

Chemical Economics Handbook. 1987. Stanford Research Institute, (SRI) International, Menlo Park, CA.

Chemical and Engineering News. 1987. Top corporations avoid paying taxes. (July 21):15.

Chemical Marketing Reporter. 1987. Schnell Publishing Co., Inc., 100 Church St., New York, NY 10007-2694.

Department of Commerce. 1987. U.S. Bureau of the Census; U.S. Imports-Exports; Current Industrial Reports.

International Trade Commission. 1987. U.S. Synthetic Organic Chemical; U.S. Production and Sales.

Katzenberg, Daniel. 1986. Tax reform. *Chemical Week* (Oct. 1):18–19.

Labich, Kenneth. 1987. Tough times living with tax reform. *Fortune* (Nov. 9):113–121.

Mackenzie, Alan K., and Arthur L. Thomas. 1986. Chemical market analysis. *AIChE Plant Design and Cost Engineering* 1:1–11.

Mendes, Jushua, and Andrew E. Serever. 1986. New tax law. *Fortune* (Sept. 1).

Reich, Robert B. 1985. Industrial policy is shaped by unfair tax code. *L.A. Times* (May 12):V3.

6

PROFITABILITY ANALYSIS; DISCOUNTED CASH FLOW (DCF)

The evaluation method most frequently used in the chemical industry for the primary measure of a project's economic attractiveness is called either discounted cash flow (DCF) or internal rate of return (IRR) (even sometimes interest rate of return—IRR). These terms are used interchangeably, even though the words say that one is a cash flow and the others are rates of return. Sometimes to avoid semantic confusion the former is called discounted cash flow return on investment, DCF-ROI. They are a measure of the profitability of a project that takes into account the time value of money. This concept implies that receiving a dollar in income today is worth more than receiving a dollar next year, because today's dollar could be invested and next year be worth one dollar plus the interest that it earned. With any project the investment may be made over a period of time (i.e., plant construction usually takes from one to three years), and the income generated will be over a longer period of time (i.e., 10 to 30 years) and often be quite variable. This method relates all aspects of the cash flow (after tax profit plus depreciation; capital spending; etc.) to a common time basis. Competitive projects, with their quite different spending and income flow can then be compared equally.

Specifically, the discounted cash flow rate of return is a hypothetical interest rate such that when it is used to calculate the present value of all of the income cash flow (income after tax plus the depreciation) for each year or period (these are positive numbers) plus all of the capital expenditure or loss cash flows (these are negative numbers), the present value is zero. In other words, it is the interest rate that would be received if the same capital investment funds were to be placed in a bank for a given period (the life of the plant) and earn the same amount as the cash flow produced by the plant.

CASH FLOW

It has previously been noted that in the typical chemical process industries plant, operation profits (and thus cash flow) are received on a nearly continuous basis.

Most corporate financial departments accept and disperse money daily, and depreciation is shown in the books on at least a quarterly basis and often more frequently. Thus, with all capital investments regarded as a negative cash flow, and the after-tax profit plus depreciation considered as positive cash flows, the effective project earnings are similar to an interest rate which is continuously, or at least frequently, compounded. Such an assumption (or one taken on any other interest compounding basis) will allow calculations to be made on the time value of money in evaluating capital investment alternatives.

It has also previously been recommended that the project engineer understand the accelerated depreciation rules and procedures but that usually only the straight line method for determining depreciation be used unless otherwise instructed by management. In other words, it should normally be assumed that the value of property decreases linearly with time over the IRS allowed depreciation period. The various accelerated depreciation techniques tend to increase project profitability through maximizing cash flow in the early years of the project, and any project's profitability will be improved, just as with the (equally not recommended for engineers) assumption of considering only company funds when borrowed capital is used as part of the capital requirements. Often much of the financing for a new project is through debt, but if a project of marginal profitability were improved to an acceptable level by these techniques, management would not be seeing a consistent and basic analysis. However, if management or the financial department later wish to consider such changes, that is their prerogative and they will be fully aware of the potential risks and the affect on the corporate balance sheet.

DISCOUNTED CASH FLOW CALCULATIONS

The procedure for calculating discounted cash flow is to first tabulate all of the cash flows involved with the project and then discount them to a present (or a reference time) value by an assumed interest rate. These discounted present values are added together and the process repeated with different interest rates until the sum of the discounted cash flow values is zero. That interest rate is the discounted cash flow (DCF) or internal rate of return (IRR) for the project. The sum of each interest rate's total discounted cash flows obtained in reaching the final DCF value is the *present worth* or *present value* of the project at that interest rate (or *net* present worth or value, so named because of the addition of numerous positive and some negative cash flows).

This trial and error calculation is based upon establishing appropriate compound interest discount factors for each future cash flow to bring its value to the present time. Some of the formulas for these discount factors may be the same as previously calculated; others will be derived in the following section. It will be remembered that the discount factor, F_A, is defined as $F_A = P_0/S_n$, where P_0 is the initial or present value, and S_n is the value at time n after the equivalent compound interest was added to the initial value. For cash flow considerations, the discount factor times the cash flow for that year, period, or instant equals the present value, or

$$P_0 = S_n F_A \qquad (6\text{-}1)$$

where S_n is now the future cash flow (in later calculations the cash flow may also be called C_n), and F_A is the appropriate discount factor for the compounding method decided upon and the nature of the cash flow.

The mathematics of a few of the discount factors that might commonly be used with capital expenditures or cash flows are as follows.

Instantaneous Cash Flows

For the compounding of simple periodic interest payments we have previously derived that $S_n = P_0(1 + i)^n$. If earnings were accrued on a continuous basis, and r is the continuous interest rate, the equivalent equation is $S_n = P_0 e^{rn}$. The corresponding discount factors are

$$F_A = (1 + i)^{-n} \text{ (Interest compounded annually)} \qquad (6\text{-}2)$$

$$F_A = e^{-rn} \text{ (Interest compounded continuously)} \qquad (6\text{-}3)$$

These factors may be used to discount a lump sum cash flow resulting from an instantaneous event, such as the recovery of salvage value or the working capital of a plant at the end of the useful project life to a present worth basis. Either of these equations may be used to calculate the discount factor; Table 6-1 contains continuous compounding discount factors calculated by using Equation (6-3) in perhaps a more convenient form. For example, if a lump sum is gained, or capital spent in an instant five years after the reference point, and a uniform continuous interest rate of 10% is applicable, then the corresponding discount factor F_A calculated from Equation (6-3) and shown in Table 6-1 is 0.6065.

In a similar manner, if the interest is compounded periodically and discounted by means of Equation (6-2) the discount values corresponding to the previous example for the salvage funds after five years at 10% interest are:

a. Yearly: $(1 + 0.10)^{-5} = 0.6209$
b. Quarterly: $(1 + 0.10/4)^{-5 \times 4} = 0.6103$
c. Monthly: $(1 + 0.10/12)^{-5 \times 12} = 0.6078$
d. Daily: $(1 + 0.10/365)^{-5 \times 365} = 0.6066$

Table 6-1
Instantaneous event discount factors

These continuous-interest discount factors give present worths for cash flows that occur in an instant at a point in time after the reference point.

$100rn^*$	0	1	2	3	4	5	6	7	8	9
0	1.0000	0.9901	0.9802	0.9704	0.9608	0.9512	0.9418	0.9324	0.9231	0.9139
10	0.9048	0.8958	0.8869	0.8781	0.8694	0.8607	0.8521	0.8437	0.8353	0.8270
20	0.8187	0.8106	0.8025	0.7945	0.7866	0.7788	0.7711	0.7634	0.7558	0.7483
30	0.7408	0.7334	0.7261	0.7189	0.7118	0.7047	0.6977	0.6907	0.6839	0.6771
40	0.6703	0.6637	0.6570	0.6505	0.6440	0.6376	0.6313	0.6250	0.6188	0.6126
50	0.6065	0.6005	0.5945	0.5886	0.5827	0.5770	0.5712	0.5655	0.5599	0.5543
60	0.5488	0.5434	0.5379	0.5326	0.5273	0.5220	0.5169	0.5117	0.5066	0.5016
70	0.4966	0.4916	0.4868	0.4819	0.4771	0.4724	0.4677	0.4630	0.4584	0.4538
80	0.4493	0.4449	0.4404	0.4360	0.4317	0.4274	0.4232	0.4190	0.4148	0.4107
90	0.4066	0.4025	0.3985	0.3946	0.3906	0.3867	0.3829	0.3791	0.3753	0.3716
100	0.3679	0.3642	0.3606	0.3570	0.3535	0.3499	0.3465	0.3430	0.3396	0.3362
110	0.3329	0.3296	0.3263	0.3230	0.3198	0.3166	0.3135	0.3104	0.3073	0.3042
120	0.3012	0.2982	0.2952	0.2923	0.2894	0.2865	0.2837	0.2808	0.2780	0.2753
130	0.2725	0.2698	0.2671	0.2645	0.2618	0.2592	0.2567	0.2541	0.2516	0.2491
140	0.2466	0.2441	0.2417	0.2393	0.2369	0.2346	0.2322	0.2299	0.2276	0.2254
150	0.2231	0.2209	0.2187	0.2165	0.2144	0.2122	0.2101	0.2080	0.2060	0.2039
160	0.2019	0.1999	0.1979	0.1959	0.1940	0.1921	0.1901	0.1882	0.1864	0.1845
170	0.1827	0.1809	0.1791	0.1773	0.1775	0.1738	0.1720	0.1703	0.1686	0.1670
180	0.1653	0.1637	0.1620	0.1604	0.1588	0.1572	0.1557	0.1541	0.1526	0.1511
190	0.1496	0.1481	0.1466	0.1451	0.1437	0.1423	0.1409	0.1395	0.1381	0.1367
200	0.1353	0.1340	0.1327	0.1313	0.1300	0.1287	0.1275	0.1262	0.1249	0.1237
210	0.1225	0.1212	0.1200	0.1188	0.1177	0.1165	0.1153	0.1142	0.1130	0.1119
220	0.1108	0.1097	0.1086	0.1075	0.1065	0.1054	0.1044	0.1033	0.1023	0.1013
230	0.1003	0.0993	0.0983	0.0973	0.0963	0.0954	0.0944	0.0935	0.0926	0.0916
240	0.0907	0.0898	0.0889	0.0880	0.0872	0.0863	0.0854	0.0846	0.0837	0.0829
250	0.0821	0.0813	0.0805	0.0797	0.0789	0.0781	0.0773	0.0765	0.0758	0.0750
260	0.0743	0.0735	0.0728	0.0721	0.0714	0.0707	0.0699	0.0693	0.0686	0.0679
270	0.0672	0.0665	0.0659	0.0652	0.0646	0.0639	0.0633	0.0627	0.0620	0.0614
280	0.0608	0.0602	0.0596	0.0590	0.0584	0.0578	0.0573	0.0567	0.056i	0.0556
290	0.0550	0.0545	0.0539	0.0534	0.0529	0.0523	0.0518	0.0513	0.0508	0.0503
300	0.0498	0.0493	0.0488	0.0483	0.0478	0.0474	0.0469	0.0464	0.0460	0.0455
310	0.0450	0.0446	0.0442	0.0437	0.0433	0.0429	0.0424	0.0420	0.0416	0.0412
320	0.0408	0.0404	0.0400	0.0396	0.0392	0.0388	0.0384	0.0380	0.0376	0.0373
330	0.0369	0.0365	0.0362	0.0358	0.0354	0.0351	0.0347	0.0344	0.0340	0.0337
340	0.0334	0.0330	0.0327	0.0324	0.0321	0.0317	0.0314	0.0311	0.0308	0.0305
350	0.0302	0.0299	0.0296	0.0293	0.0290	0.0287	0.0284	0.0282	0.0279	0.0276
360	0.0273	0.0271	0.0268	0.0265	0.0263	0.0260	0.0257	0.0255	0.0252	0.0250

Table 6-1 (Continued)

100rn*	0	1	2	3	4	5	6	7	8	9
370	0.0247	0.0245	0.0242	0.0240	0.0238	0.0235	0.0233	0.0231	0.0228	0.0226
380	0.0224	0.0221	0.0219	0.0217	0.0215	0.0213	0.0211	0.0209	0.0207	0.0204
390	0.0202	0.0200	0.0198	0.0196	0.0194	0.0193	0.0191	0.0189	0.0187	0.0185
400	0.0183	0.0181	0.0180	0.0178	0.0176	0.0174	0.0172	0.0171	0.0169	0.0167
410	0.0166	0.0164	0.0162	0.0161	0.0159	0.0158	0.0156	0.0155	0.0153	0.0151
420	0.0150	0.0148	0.0147	0.0146	0.0144	0.0143	0.0141	0.0140	0.0138	0.0137
430	0.0136	0.0134	0.0133	0.0132	0.0130	0.0129	0.0128	0.0127	0.0125	0.0124
440	0.0123	0.0122	0.0120	0.0119	0.0118	0.0117	0.0116	0.0114	0.0113	0.0112
450	0.0111	0.0110	0.0109	0.0108	0.0107	0.0106	0.0105	0.0104	0.0103	0.0102
460	0.0101	0.0100	0.0099	0.0098	0.0097	0.0096	0.0095	0.0094	0.0093	0.0092
470	0.0091	0.0090	0.0089	0.0088	0.0087	0.0087	0.0086	0.0085	0.0084	0.0083
480	0.0082	0.0081	0.0081	0.0080	0.0079	0.0078	0.0078	0.0077	0.0076	0.0075
490	0.0074	0.0074	0.0073	0.0072	0.0072	0.0071	0.0070	0.0069	0.0069	0.0068

100rn	0	10	20	30	40	50	60	70	80	90
500	0.0067	0.0061	0.0055	0.0050	0.0045	0.0041	0.0037	0.0033	0.0030	0.0027
600	0.0025	0.0022	0.0020	0.0018	0.0017	0.0015	0.0014	0.0012	0.0011	0.0010
700	0.0009	0.0008	0.0007	0.0007	0.0006	0.0006	0.0005	0.0005	0.0004	0.0004
800	0.0003	0.0003	0.0003	0.0002	0.0002	0.0002	0.0002	0.0002	0.0002	0.0001
900	0.0001	0.0001	0.0001	0.0001	0.0001	0.0001	0.0001	0.0001	0.0001	0.0001
1000	0.0000									

*r = nominal interest compounded continuously; percent/100; n = number of years.
Source: Leibson & Trischman 1971. Excerpted by special permission from *Chemical Engineering*, December 1971. Copyright © 1971, by McGraw-Hill, Inc., New York.

As in previous examples it is seen that continuous and frequent interest compounding give very similar results, and the accuracy (or lack of accuracy) between compounding periods is probably less than the other inaccuracies in the profitability calculations. However, there is enough difference that a consistent calculation method should be used, and the compounding period specified along with the results. If a calculator is used for discounted cash flow or present value calculations, and the compounding method is not stated, then a simple present worth calculation such as the example above will quickly determine the formula programmed into the machine. Small factory programmed hand-held calculators appear to usually employ periodic (annual) compounding.

Income Cash Flows

The case in which a project (continuously) produces a different cash flow each year of its life is a common one. The cash flow may be assumed to be uniform during the year, but the cash flow from one year to the next may not necessarily be equal. For this case, within any year (n), the differential rate of appreciation of the compounded amount (S) is

$$\frac{dS}{dn} = C_n + rS \tag{6-4}$$

where C_n is the cash flow occurring in year n.

Holding r constant with respect to time (n), and integrating this equation from $(n - 1)$ to n, the following is obtained

$$\int_0^{S_n} \frac{dS}{C_n + rs} = \int_{n-1}^{n} dn$$

$$\frac{1}{r} \ln \frac{C_n + rS_n}{C_n} = 1$$

$$C_n + rS_n = C_n e^r$$

$$S_n = C_n \left(\frac{e^r - 1}{r}\right) \tag{6-5}$$

Equation (6-5) calculates the compounded amount for a uniform, continuous, yearly cash flow with continuous interest compounding. The corresponding present worth of cash flow C_n converted from the year n to the reference point is obtained by combining Equation (6-3) with Equation (6-5).

$$P_0 = S_n e^{-rn} = C_n \left(\frac{e^r - 1}{r}\right) e^{-rn} \tag{6-6}$$

The corresponding discount factor (F_B) is

$$F_B = \left(\frac{e^r - 1}{r}\right) e^{-rn} \tag{6-7}$$

This may be used to discount cash flows that occur continuously and uniformly over one-year periods after the reference point to give the corresponding present worth. Table 6-2 contains discount factors calculated by use of Equation (6-7).

For example, for the cash flow for the fifth year of a project $(n = 5$, or from year number 4 to year number 5 after the reference point), with 10% interest $(r = 0.10)$, the corresponding discount factor (F_B) calculated from Equation (6-7) and shown in Table 6-2 is 0.6379.

In a similar manner, the corresponding compounding of periodic cash flow for each year is $C_n(1 + i)$ for a single cash flow received at the beginning of the year, C_n for a single cash flow received at the end of the year, and

$$\frac{(C_n/m)[(1 + i/m)^m - 1]}{i/m} = \frac{C_n[(1 + i/m)^m - 1]}{i} \tag{6-8}$$

for each cash flow received for m periods within the year, but at the end of each period. Each of these three periodic cash flow and compounding possibilities would be converted to the present by dividing by $(1 + i/m)^{nm}$. In the previous example for the period from year 4 to 5, 10% interest, with daily cash flow and compounding, the discount factor would be

$$\frac{(1 + 0.1/365)^{365} - 1}{0.1} = \frac{(1.000274)^{365} - 1}{0.1} = \frac{1.10516 - 1}{0.1}$$

$$= 1.0516; \frac{1}{(1.00274)^{5 \times 365}} = 0.60654$$

The discount factor is thus $1.0516 \times 0.60654 = 0.6378$.

It is to be noted that in this case periodic daily compounding and continuous compounding again give results that are not much different from each other, just as in the previous calculation with instantaneous cash flows.

Initial Plant Investment

Most major plant construction projects require from one to three years for completion. The shape of the cash spending curve usually varies to some extent, depending on the arrangements with the contractor (e.g., lump sum, fixed payment schedule, cost plus, or other) and the type of facility involved. The cash spending curve is typically S-shaped for most projects.

If the spending schedule is forecast or preset by contract, then plant investment capital can be discounted to reflect its value at the reference point (i.e., plant start-up, initial construction, etc.) by using the appropriate discount factors. However, such calculations are usually not easily accomplished because the capital spending pattern can seldom be rigorously defined at the time the economic justification is being prepared. Thus, the period of capital spending is often neglected in profitability evaluations of alternative investment opportunities. Time zero is frequently set when the plant starts production, with no discount factor used on the initial capital investment. Such a practice often causes less than 5% error in evaluating project alternatives, which hopefully is

Table 6-2
Uniform one year period discount factors

These continuous-interest discount factors give present worths for cash flows that occur uniformly over one-year periods after the reference point.

Year*	1%	2%	3%	4%	5%	6%	7%	8%	9%	10%
0–1	0.9950	0.9901	0.9851	0.9803	0.9754	0.9706	0.9658	0.9610	0.9563	0.9516
1–2	0.9851	0.9705	0.9560	0.9418	0.9278	0.9141	0.9005	0.8872	0.8740	0.8611
2–3	0.9753	0.9512	0.9278	0.9049	0.8826	0.8608	0.8396	0.8189	0.7988	0.7791
3–4	0.9656	0.9324	0.9004	0.8694	0.8395	0.8107	0.7829	0.7560	0.7300	0.7050
4–5	0.9560	0.9140	0.8737	0.8353	0.7986	0.7635	0.7299	0.6979	0.6672	0.6379
5–6	0.9465	0.8959	0.8479	0.8026	0.7596	0.7190	0.6806	0.6442	0.6098	0.5772
6–7	0.9371	0.8781	0.8229	0.7711	0.7226	0.6772	0.6346	0.5947	0.5573	0.5223
7–8	0.9278	0.8607	0.7985	0.7409	0.6874	0.6377	0.5917	0.5490	0.5093	0.4726
8–9	0.9185	0.8437	0.7749	0.7118	0.6538	0.6006	0.5517	0.5068	0.4655	0.4276
9–10	0.9094	0.8270	0.7520	0.6839	0.6219	0.5656	0.5144	0.4678	0.4254	0.3869
10–11	0.9003	0.8106	0.7298	0.6571	0.5916	0.5327	0.4796	0.4318	0.3888	0.3501
11–12	0.8914	0.7946	0.7082	0.6312	0.5628	0.5016	0.4472	0.3986	0.3553	0.3168
12–13	0.8825	0.7788	0.6873	0.6065	0.5353	0.4724	0.4169	0.3680	0.3248	0.2866
13–14	0.8737	0.7634	0.6670	0.5827	0.5092	0.4449	0.3888	0.3397	0.2968	0.2593
14–15	0.8650	0.7483	0.6473	0.5599	0.4844	0.4190	0.3625	0.3136	0.2713	0.2347
15–16	0.8564	0.7335	0.6282	0.5380	0.4608	0.3946	0.3380	0.2895	0.2479	0.2123
16–17	0.8479	0.7189	0.6096	0.5169	0.4383	0.3716	0.3151	0.2672	0.2266	0.1921
17–18	0.8395	0.7047	0.5916	0.4966	0.4169	0.3500	0.2938	0.2467	0.2071	0.1739
18–19	0.8311	0.6908	0.5741	0.4772	0.3966	0.3296	0.2740	0.2277	0.1893	0.1573
19–20	0.8228	0.6771	0.5571	0.4584	0.3772	0.3104	0.2554	0.2102	0.1730	0.1423
20–21	0.8147	0.6637	0.5407	0.4405	0.3588	0.2923	0.2382	0.1940	0.1581	0.1288
21–22	0.8065	0.6505	0.5247	0.4232	0.3413	0.2753	0.2221	0.1791	0.1445	0.1165
22–23	0.7985	0.6376	0.5092	0.4066	0.3247	0.2593	0.2071	0.1653	0.1320	0.1054
23–24	0.7906	0.6250	0.4941	0.3907	0.3089	0.2442	0.1931	0.1526	0.1207	0.1954
24–25	0.7827	0.6126	0.4795	0.3753	0.2938	0.2300	0.1800	0.1409	0.1103	0.0863
25–30	0.7596	0.5772	0.4386	0.3334	0.2535	0.1928	0.1466	0.1115	0.0849	0.0646
30–35	0.7226	0.5223	0.3775	0.2730	0.1974	0.1428	0.1033	0.0748	0.0541	0.0392
35–40	0.6874	0.4726	0.3250	0.2235	0.1538	0.1058	0.0728	0.0501	0.0345	0.0238
40–45	0.6538	0.4276	0.2797	0.1830	0.1197	0.0784	0.0513	0.0336	0.0220	0.0144
45–50	0.6219	0.3869	0.2407	0.1498	0.0933	0.0581	0.0362	0.0225	0.0140	0.0087

not too significant compared with the potential error in other inputs (e.g., sales price and volume, capital cost, profit, etc.), especially if all projects are consistently evaluated on this basis. For the most rigorously accurate economic evaluations, however, all cash flows prior to the plant start-up should also be discounted.

The effect of land value, research and development, and prior engineering

Table 6-2 (*Continued*)

Year*	11%	12%	13%	14%	15%	16%	17%	18%	19%	20%	21%
0–1	470	0.9423	0.9377	0.9332	0.9286	0.9241	0.9196	0.9152	0.9107	0.9063	0.9020
1–2	8483	0.8358	0.8234	0.8112	0.7993	0.7875	0.7759	0.7644	0.7531	0.7421	0.7311
2–3	600	0.7413	0.7230	0.7053	0.6879	0.6710	0.6546	0.6385	0.6228	0.6075	0.5926
3–4	808	0.6574	0.6349	0.6131	0.5921	0.5718	0.5522	0.5333	0.5150	0.4974	0.4804
4–5	6099	0.5831	0.5575	0.5330	0.5096	0.4873	0.4659	0.4455	0.4259	0.4072	0.3894
5–6		0.5463	0.5172	0.4895	0.4634	0.4386	0.4152	0.3931	0.3721	0.3522	0.3334	0.3156
6–7		0.4894	0.4588	0.4299	0.4029	0.3775	0.3538	0.3316	0.3108	0.2913	0.2730	0.2558
7–8		0.4385	0.4069	0.3775	0.3502	0.3250	0.3015	0.2798	0.2596	0.2409	0.2235	0.2074
8–9		0.3928	0.3609	0.3314	0.3045	0.2797	0.2569	0.2360	0.2168	0.1992	0.1830	0.1681
9–10	0.3519	0.3201	0.2910	0.2647	0.2407	0.2189	0.1991	0.1811	0.1647	0.1498	0.1363
10–11	0.3152	0.2839	0.2556	0.2301	0.2072	0.1866	0.1680	0.1513	0.1362	0.1227	0.1105
11–12	0.2824	0.2518	0.2244	0.2000	0.1783	0.1590	0.1417	0.1264	0.1126	0.1004	0.0895
12–13	0.2530	0.2233	0.1970	0.1739	0.1535	0.1355	0.1196	0.1055	0.0932	0.0822	0.0726
13–14	0.2266	0.1981	0.1730	0.1512	0.1321	0.1154	0.1009	0.0882	0.0770	0.0673	0.0588
14–15	0.2030	0.1757	0.1519	0.1314	0.1137	0.0984	0.0851	0.0736	0.0637	0.0551	0.0477
15–16	0.1819	0.1558	0.1334	0.1143	0.0979	0.0838	0.0718	0.0615	0.0527	0.0451	0.0387
16–17	0.1629	0.1382	0.1172	0.0993	0.0842	0.0714	0.0606	0.0514	0.0436	0.0369	0.0313
17–18	0.1460	0.1225	0.1029	0.0864	0.0725	0.0609	0.0511	0.0429	0.0360	0.0303	0.0254
18–19	0.1308	0.1087	0.0903	0.0751	0.0624	0.0519	0.0431	0.0358	0.0298	0.0248	0.0206
19–20	0.1171	0.0964	0.0793	0.0653	0.0537	0.0442	0.0364	0.0299	0.0246	0.0203	0.0167
20–21	0.1049	0.0855	0.0697	0.0568	0.0462	0.0377	0.0307	0.0250	0.0204	0.0166	
21–22	0.0940	0.0758	0.0612	0.0493	0.0398	0.0321	0.0259	0.0209	0.0169	0.0136	
22–23	0.0842	0.0673	0.0537	0.0429	0.0343	0.0274	0.0218	0.0175	0.0139	0.0111	
23–24	0.0754	0.0596	0.0472	0.0373	0.0295	0.0233	0.0184	0.0146	0.0115	0.0091	
24–25	0.0676	0.0529	0.0414	0.0324	0.0254	0.0199	0.0156	0.0122	0.0095	0.0075	
25–30	0.0492	0.0374	0.0285	0.0217	0.0165	0.0126	0.0096	0.0073	0.0056	0.0043	
30–35	0.0284	0.0205	0.0149	0.0108	0.0078	0.0057	0.0041	0.0030	0.0022	0.0016	
35–40	0.0164	0.0113	0.0078	0.0054	0.0037	0.0025	0.0018	0.0012	0.0008	0.0006	
40–45	0.0094	0.0062	0.0041	0.0027	0.0017	0.0011	0.0008	0.0005	0.0003	0.0002	
45–50	0.0054	0.0034	0.0021	0.0013	0.0008	0.0005	0.0003	0.0002	0.0001	0.0001	

*Year indicates one-year period in which cash flow occurs.
Source: Leibson and Trischman 1971. Excerpted by special permission from *Chemical Engineering*, December, 1971. Copyright © 1971, by McGraw-Hill, Inc., New York.

expenses being compounded from the time of expenditure to plant start-up may be similarly neglected in most cases without serious problem for most practical situations in the chemical process industries. Thus, DCF return on investment on this simplified basis is that value of r for which the following equation applies

$$\sum_{n=0}^{n = \text{project life}} = C_n\left(\frac{e^r - 1}{r}\right) e^{-rn} + \left(\begin{array}{c}\text{plant salvage value} \\ \text{plus working capital}\end{array}\right) e^{-rn}$$

$$- \left(\begin{array}{c}\text{total plant investment,} \\ \text{including land value}\end{array}\right) = 0 \qquad (6\text{-}9)$$

Example

A trial and error solution is required to find the value of r that satisfies Equation (6-9) using the terms in the equation or discount factors given in Tables 6-1, 6-2, and 6-4. An example of such a calculation is as follows.

PROBLEM. You have estimated that the total capital requirement for a new project is $25 million, including $2.5 million working capital. The project would be allowed a 10-year (straight line) depreciation period, but would actually be an efficient operation for 20 years. At the end of that time there would be no salvage value. The estimated after-tax (40%) profits are $0.25, $0.50, $0.75, $1.00, $1.25, and $1.50 MM/yr for the first six years, and then the pretax rate is constant for the life of the project. Assume continuous compounding, and using the tables, what is the project's discounted cash flow rate of return?

ANSWER. It is first necessary to assume a time zero and then calculate the cash flow for each year of the project's life. It is best to enter this information in a table, as shown in Table 6-3, to simplify the record keeping of the multiple step, trial and error calculation that will be required. For simplicity let's assume that time zero occurs when the plant is ready to start operating, and that all of the capital expense has occurred instantaneously at that time. This will then be entered in the table as a negative $25 million, with a discount factor of 1.00. Cash flows can next be tabulated for the first 10 years as depreciation plus after tax profit. The depreciation is (total capital − working capital)/allowed depreciation period = $(25 − 2.5)/10 = \$2.25$ million. Adding this to the specified after tax profit gives the first 10 years' cash flow.

For the last 10 years there is no longer any depreciation, so after-tax profits will appear larger, but the cash flow will be smaller. With the after-tax profit in year 10 = $1.50 MM, and the tax rate 40%, the pretax profits must have been $0.6 \times P = 1.50$, or $2.50 MM. Since $2.25 MM depreciation "expenses" had reduced this amount, the pretax, predepreciation profit must be for years 11–20, $2.25 + 2.50 = \$4.75$ MM. The after-tax profit would then be $4.75 \times 0.6 = \$2.85$ MM, which is also the cash flow for this period.

Finally, at the end of year 20 (again for simplicity assume it to occur in an instant) the working capital ($2.50 MM) and salvage value ($0.00 MM) would be recovered as one final cash flow entry.

Table 6-3.
Working Table for DCF Calculations.

		10% Interest			12% Interest	
Year	Cash Flow, $ MM	10% Discount Factor	From Table	Present Value, $ MM	12% Discount Factor	Present Value, $ MM
0	−25.00	1.0		−25.0000	1.0	−25.0000
0–1	2.50	0.9516	6-2	2.3790	0.9423	2.3558
1–2	2.75	0.8611	6-2	2.3680	0.8358	2.2985
2–3	3.00	0.7791	6-2	2.3373	0.7413	2.2239
3–4	3.25	0.7050	6-2	2.2913	0.6574	2.1366
4–5	3.50	0.6379	6-2	2.2327	0.5831	2.0409
5–10	3.75 × 5 =	0.7869	6-4		0.7520	
	18.75	×0.6065	6-1		×0.5488	
		=0.4773		8.0495	=0.4127	7.7381
10–20	2.85 × 10 =	0.6321	6-4		0.5823	
	28.5	×0.3679	6-1		×0.3012	
		=0.2326		6.6277	=0.1754	4.9986
20	2.50	0.1353	6-1	0.3383	0.0907	0.2268
Present value (Total)				$2.5228 MM		$−0.9808 MM

$$(\Delta = 3.5036)$$

$$\text{Extrapolation: } \frac{2.5228}{3.5036} = 0.72 \times 2 = 1.54$$

$$\text{DCF} = 11.54\%$$

With the cash flows now established each one can be discounted to its present value at an assumed interest rate. When the correct interest rate has been established the sum of the cash flows will be zero. Thus, a trial and error series of guesses and discount calculations must be made on interest rates until a zero present value of cash flow is bracketed, and then the DCF can be extrapolated.

For this calculation a rough approximation of (early) steady-state cash flow divided by the total capital is $3.75/25 = 15\%$, so an initial DCF guess of 10% interest will probably be low, but may be in the correct range. Using this interest rate, Table 6-2 could be used if desired for the discount factor for every year (except for year 20 with the working capital and salvage return), but it is perhaps easiest to only use it for the first five years. Thus, for the first year, in Table 6-2, along the 0–1 year line, under the 10% column the discount factor is found to be 0.9516. When this number is multiplied by the first year's cash flow, $2.50 MM, the present value (at time zero) becomes $2.3790 MM. This step is repeated for each of the first five years.

For the periods 5–10 and 10–20 years a more rapid calculation can be made using Table 6-4. This table provides the discount factor for (in this case) the 5-

Table 6-4
Later period yearly discount factors

These continuous-interest discount factors give present worths for cash flows that occur uniformly over a period of T years after the reference point.

$100rT^*$	0	1	2	3	4	5	6	7	8	9
0	1.0000	0.9950	0.9901	0.9851	0.9803	0.9754	0.9706	0.9658	0.9610	0.9563
10	0.9516	0.9470	0.9423	0.9377	0.9332	0.9286	0.9241	0.9196	0.9152	0.9107
20	0.9063	0.9020	0.8976	0.8933	0.8891	0.8848	0.8806	0.8764	0.8722	0.8681
30	0.8639	0.8598	0.8558	0.8517	0.8477	0.8438	0.8398	0.8359	0.8319	0.8281
40	0.8242	0.8204	0.8166	0.8128	0.8090	0.8053	0.8016	0.7979	0.7942	0.7906
50	0.7869	0.7833	0.7798	0.7762	0.7727	0.7692	0.7657	0.7622	0.7588	0.7554
60	0.7520	0.7486	0.7452	0.7419	0.7386	0.7353	0.7320	0.7288	0.7256	0.7224
70	0.7192	0.7160	0.7128	0.7097	0.7066	0.7035	0.7004	0.6974	0.6944	0.6913
80	0.6883	0.6854	0.6824	0.6795	0.6765	0.6736	0.6707	0.6679	0.6650	0.6622
90	0.6594	0.6566	0.6537	0.6510	0.6483	0.6455	0.6428	0.6401	0.6374	0.6348
100	0.6321	0.6295	0.6269	0.6243	0.6217	0.6191	0.6166	0.6140	0.6115	0.6090
110	0.6065	0.6040	0.6016	0.5991	0.5967	0.5942	0.5918	0.5894	0.5871	0.5847
120	0.5823	0.5800	0.5777	0.5754	0.5731	0.5708	0.5685	0.5663	0.5641	0.5618
130	0.5596	0.5574	0.5552	0.5530	0.5509	0.5487	0.5466	0.5444	0.5424	0.5402
140	0.5381	0.5361	0.5340	0.5320	0.5299	0.5279	0.5259	0.5239	0.5219	0.5199
150	0.5179	0.5160	0.5140	0.5121	0.5102	0.5082	0.5064	0.5044	0.5026	0.5007
160	0.4988	0.4970	0.4952	0.4933	0.4915	0.4897	0.4879	0.4861	0.4843	0.4825
170	0.4808	0.4790	0.4773	0.4756	0.4739	0.4721	0.4704	0.4687	0.4671	0.4654
180	0.4637	0.4621	0.4605	0.4588	0.4571	0.4555	0.4540	0.4523	0.4508	0.4491
190	0.4476	0.4460	0.4445	0.4429	0.4414	0.4399	0.4383	0.4368	0.4354	0.4338
200	0.4323	0.4308	0.4294	0.4279	0.4265	0.4250	0.4236	0.4221	0.4207	0.4193
210	0.4179	0.4165	0.4151	0.4137	0.4123	0.4109	0.4096	0.4082	0.4069	0.4055
220	0.4042	0.4029	0.4015	0.4002	0.3989	0.3976	0.3963	0.3950	0.3937	0.3925
230	0.3912	0.3899	0.3887	0.3874	0.3862	0.3849	0.3837	0.3825	0.3813	0.3801
240	0.3789	0.3777	0.3765	0.3753	0.3741	0.3729	0.3718	0.3706	0.3695	0.3683
250	0.3672	0.3660	0.3649	0.3638	0.3627	0.3615	0.3604	0.3593	0.3582	0.3571
260	0.3560	0.3550	0.3539	0.3528	0.3517	0.3507	0.3496	0.3486	0.3476	0.3465
270	0.3455	0.3445	0.3434	0.3424	0.3414	0.3404	0.3393	0.3384	0.3374	0.3364
280	0.3354	0.3344	0.3335	0.3325	0.3315	0.3306	0.3296	0.3287	0.3277	0.3268
290	0.3259	0.3249	0.3240	0.3231	0.3221	0.3212	0.3203	0.3194	0.3185	0.3176
300	0.3167	0.3158	0.3150	0.3141	0.3132	0.3123	0.3115	0.3106	0.3098	0.3089
310	0.3080	0.3072	0.3064	0.3055	0.3047	0.3039	0.3030	0.3022	0.3014	0.3006
320	0.2998	0.2990	0.2982	0.2974	0.2966	0.2958	0.2950	0.2942	0.2934	0.2926
330	0.2919	0.2911	0.2903	0.2896	0.2888	0.2880	0.2873	0.2865	0.2858	0.2850
340	0.2843	0.2836	0.2828	0.2821	0.2814	0.2807	0.2799	0.2792	0.2785	0.2778
350	0.2771	0.2764	0.2757	0.2750	0.2743	0.2736	0.2729	0.2722	0.2715	0.2709
360	0.2702	0.2695	0.2688	0.2682	0.2675	0.2669	0.2662	0.2655	0.2649	0.2642

Table 6-4 (Continued)

100rn*	0	1	2	3	4	5	6	7	8	9
370	0.2636	0.2629	0.2623	0.2617	0.2610	0.2604	0.2598	0.2591	0.2585	0.2579
380	0.2573	0.2567	0.2560	0.2554	0.2548	0.2542	0.2536	0.2530	0.2524	0.2518
390	0.2512	0.2506	0.2500	0.2495	0.2489	0.2483	0.2477	0.2471	0.2466	0.2460
400	0.2454	0.2449	0.2443	0.2437	0.2432	0.2426	0.2421	0.2415	0.2410	0.2404
410	0.2399	0.2393	0.2388	0.2382	0.2377	0.2372	0.2366	0.2361	0.2356	0.2350
420	0.2345	0.2340	0.2335	0.2330	0.2325	0.2319	0.2314	0.2309	0.2304	0.2299
430	0.2294	0.2289	0.2284	0.2279	0.2274	0.2269	0.2264	0.2259	0.2255	0.2250
440	0.2245	0.2240	0.2235	0.2230	0.2226	0.2221	0.2216	0.2212	0.2207	0.2202
450	0.2198	0.2193	0.2188	0.2184	0.2179	0.2175	0.2170	0.2166	0.2161	0.2157
460	0.2152	0.2148	0.2143	0.2139	0.2134	0.2130	0.2126	0.2121	0.2117	0.2113
470	0.2108	0.2104	0.2100	0.2096	0.2091	0.2087	0.2083	0.2079	0.2074	0.2070
480	0.2066	0.2062	0.2058	0.2054	0.2050	0.2046	0.2042	0.2038	0.2034	0.2030
490	0.2026	0.2022	0.2018	0.2014	0.2010	0.2006	0.2002	0.1998	0.1994	0.1990
500	0.1987	0.1949	0.1912	0.1877	0.1843	0.1811	0.1779	0.1749	0.1719	0.1690
600	0.1663	0.1636	0.1610	0.1584	0.1560	0.1536	0.1513	0.1491	0.1469	0.1448
700	0.1427	0.1407	0.1388	0.1369	0.1351	0.1333	0.1315	0.1298	0.1282	0.1265
800	0.1250	0.1234	0.1219	0.1206	0.1190	0.1176	0.1163	0.1149	0.1136	0.1123
900	0.1111	0.1099	0.1087	0.1075	0.1064	0.1053	0.1043	0.1031	0.1020	0.1010
1000	0.1000	0.0990	0.0980	0.0971	0.0962	0.0952	0.0943	0.0935	0.0926	0.0917
1100	0.0909	0.0901	0.0893	0.0885	0.0877	0.0869	0.0862	0.0855	0.0847	0.0840
1200	0.0833	0.0826	0.0820	0.0813	0.0806	0.0800	0.0794	0.0787	0.0781	0.0775
1300	0.0769	0.0763	0.0758	0.0752	0.0746	0.0741	0.0735	0.0730	0.0725	0.0719
1400	0.0714	0.0709	0.0704	0.0699	0.0694	0.0690	0.0685	0.0680	0.0676	0.0671
1500	0.0667	0.0662	0.0658	0.0654	0.0649	0.0645	0.0641	0.0637	0.0633	0.0629
1600	0.0625	0.0621	0.0617	0.0613	0.0610	0.0606	0.0602	0.0599	0.0595	0.0592
1700	0.0588	0.0585	0.0581	0.0578	0.0575	0.0571	0.0568	0.0565	0.0562	0.0559
1800	0.0556	0.0552	0.0549	0.0546	0.0543	0.0541	0.0538	0.0535	0.0532	0.0529
1900	0.0526	0.0524	0.0521	0.0518	0.0515	0.0513	0.0510	0.0508	0.0505	0.0502
2000	0.0500	0.0497	0.0495	0.0492	0.0490	0.0487	0.0485	0.0483	0.0481	0.0478

*r = nominal interest compounded continuously; percent/100; $T = n$ = number of years in time period.
Source: Leibson & Trischman 1971. Excerpted by special permission from *Chemical Engineering*, December, 1971. Copyright © 1971, by McGraw-Hill, Inc., New York.

and 10-year intervals. For the 5-year interval in Table 6-3, $T = 5$ and $r = 0.10$, so the "0" column on the 50 line shows a discount factor of 0.7869 for that interval. To bring that period back to time zero, Table 6-1 for year 5 and 0.10 interest (50 line) gives a second discount factor of 0.6065. The multiple of these two discount factors is $0.7869 \times 0.6065 = 0.4773$ which is the average discount factor for the 5-year interval between years 5 and 10. Multiplying this number by the cash flow for the period ($3.75 MM) times the number of years (5) gives the present worth for the entire 5-year period. In a similar manner the 10–20 year period can be discounted for a 10-year interval (Table 6-3), and then discounted again to bring it to time zero (i.e., $0.6321 \times 0.3679 \times 10 \times 2.85 = \$6.6277$ MM). In both of these intervals a year by year tabulation could have been made from Table 6-2 to give the identical answer, but using Tables 6-4 and 6-1 is faster.

The final cash flow is the working capital and salvage value return discounted from Table 6-1. All of the present worths' are then added, yielding in this case a fairly large positive value ($2.52 MM), indicating that the interest rate guess was too low. A second tabulation was consequently made assuming a 12% interest rate. This gave a negative number, so an extrapolation between 10 and 12% indicated an 11.5% DCF for the project. A third trial at 11% interest would have given a more accurate estimate, but this accuracy is seldom warranted because of all of the other uncertainties involved.

PRESENT WORTH (OR PRESENT VALUE)

Most companies have a criterion for the return on investment (ROI) expected to be realized on all new projects. The present worth (PW) or present value (PV) method of comparing investment opportunities takes this into account by discounting the annual cash flows calculated for a new project to their present value using the normal discount factors and this company desired ROI interest rate. Thus, a direct project evaluation is possible without a trial and error calculation. With this method the project alternative having the greatest present worth is the one selected, assuming that risk and other factors are equal.

Present worth is most valid for the comparison of projects that have similar lives or cover comparable time spans. The interest rate used for discounting the annual cash flows of the project represents the minimum desired return on investment capital to the company. Investments made in future years of the venture are also discounted back to their present value using the same rate factor, just as in DCF calculations. The present worth method is particularly useful in comparing small with large projects, since an actual dollar value is calculated for the excess return over the assumed minimum acceptable return rate. The same holds true for comparing projects with large DCF values since again, actual dollar returns are shown.

In the previous DCF example the present worth of the project at 10% interest would be $2.52 million. This number could be used to compare with other projects, but just as with the DCF analysis, this number or an 11.5 DCF return are extremely low values, and considering all of the uncertainties and risks with any new project, for most companies neither would be acceptable, or if they were, the decision to proceed might well be based upon other considerations.

SIMPLIFIED CALCULATION METHODS

It has been noted in the previous section that DCF can be calculated by formula or charts in a trial and error manner. This procedure is laborious and time consuming for multiple calculations, but does have the advantage of providing a feel for the importance of all of the cash flow input into the analysis, and the rapidity with which time changes (decreases) the value and importance of cash flow. At the same time present worth values are automatically determined by these calculations.

There are many ways by which the DCF calculations can be made easier, such as for simplified problems by graphical presentations. An example of such a graph is shown in Figure 6-1. In this case it was assumed that the initial investment (C_0) was made at time zero, and that all future cash flows (C_n) were constant over the project life (n years). Cash flow was assumed to be received at one time at the end of each year, and compounded on an annual basis. Salvage and working capital recovery were ignored. With the original investment and cash flow known, the DCF can be read from the graph for any project life. This becomes essentially an annuity calculation.

If financial calculations are to be made very often, however, because of the very low cost, accuracy, and speed of small hand-held calculators (or any computer), essentially all DCF calculations are now made with them. Every engineer is strongly urged to obtain a hand-held calculator or computer with financial calculation programs, and preferably one that can be used at work. It is essentially the only practical means of making frequent or repetitive DCF calculations with any reasonable speed and consistency. The same calculators or software programs can also solve present worth and many other compound interest calculations.

DATA PRESENTATION; PRO FORMA ANALYSIS

Since there are several different methods of calculating the discount factor involved with compounding both the interest and cash flow (yearly increments; more frequent intervals such as quarterly, monthly, daily, etc.; and continuously), one must specify which method is being used in the project financial analysis and comparison presentations. The difference is comparatively small, and usually far less than the inaccuracy of estimating the capital, operating cost, and sales value of the product, but nevertheless it should be noted. Likewise, each of the major discretionary variables, such as the method and period of depreciation, income tax percentage, life of the project, sales revenue, operating cost, amount and period of capital expenditure, working capital, salvage value, and so on, should also be listed when the data are reported.

The easiest way to do this is to prepare a chart with columns listing these variables for each year over the life of the project. The cash flow totals for each

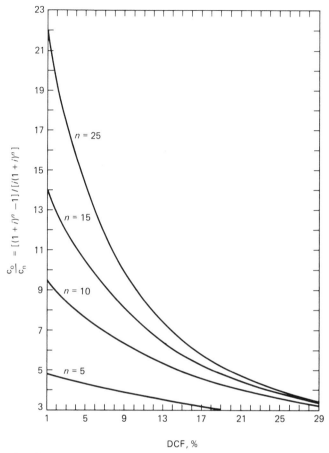

Source: Horowitz 1980. Excerpted by special permission from *Chemical Engineering* May 19, 1980. Copyright © 1980, by McGraw-Hill, Inc., New York.

Figure 6-1. Graphical solution for simplified DCF calculation (constant cash flow, compounded annually, no salvage or working capital return).

year will then be the input for the DCF calculations, and as with an ROI presentation, any reader can adjust the variables as they desire. An example of a presentation of this type is shown in Table 6-5. It is part of what would be called a *pro forma* analyses, or a detailed economic presentation of a project. Unless the economic analysis is summarized in such a manner one can be assured that most financial departments, and thus the management whom they advise, will not take it seriously until it is redone on this basis. Usually when the financial department reviews such estimates they will increase the DCF based upon their more precise knowledge of depreciation and income tax, but decrease it

Table 6-5.
Abbreviated Example of a Pro Forma Analysis.

Proposed plant to produce _____ tons per year of _____

	Years								
	1	2	3	4	5	6	7	8	9 · · ·
Capital spending									
Working capital									
Salvage value anticipated									
Sales volume, t/yr									
Sales price, $/ton									
Sales, $/yr									
Production cost, $/yr									
Pretax profit, $									
Income tax (--%),[a] $									
Net profit after tax, $									
Depreciation, $[b]									
Cash flow, $									
Interest expense, $[c]									
Present worth at --% interest[d]									
Discounted cash flow, DCF[e] (synonomous with internal rate of return, IRR—one entry only)									

[a]List assumed state and federal tax rates.
[b]List depreciation method and IRS-allowable life of the plant.
[c]List amount of borrowed funds and interest rate.
[d]Only list this if of interest.
[e]Note compounding basis; total capital or company funds only; and all other pertinent factors.

by adding additional contingency to the capital cost, and assuming a lower sales realization (the sales price less freight and discount allowances, etc.). It is seldom that the engineer is involved with the financial department's review, so one must anticipate needing extra profitability in the engineer's estimate to meet the company's minimum DCF requirements.

MINIMUM ACCEPTABLE DCF

Most companies establish a lower limit and a desired DCF for new capital projects to be accepted. The risk of any new project is always the most important consideration, since sales price and quantity are always a calculated gamble, and historically the industry as a whole only operates at about 80% of capacity because of limited sales. However, assuming the best, one of the factors used in considering the DCF limit is an evaluation of the *cost of debt* for the corporation or the income that could be made from purely financial and relatively safe and secure investments such as purchasing short-term bonds or certificates of deposits (CDs in a bank). As an example of such an analysis for the minimum acceptable DCF, assume that $100 could be used to purchase bonds earning 10% interest, or be invested in a new plant with a 10-year life earning the same income. Assume that the working capital is $10, (part of the $100), that there would be no plant salvage value, and that the actual total income tax rate is 25%. Also assume that the return of the original investment through depreciation is roughly the same as selling the bonds (to return the capital) 10 years hence.

The bonds would yield $10 per year gross income, less $2.5 taxes, or $7.5 after tax income. The plant investment would provide ($100 − $10 working capital)/10 = $9/yr depreciation, which would reduce taxes by 9 × 0.25 = $2.25/yr, making the equivalent after-tax income (to the bonds) 7.50 − 2.25 = $5.25/yr. The plant's cash flow would be $9 + 5.25 = $14.25 for 9 years, and 24.25 on the tenth year (with the return of working capital). On an annual interest compounding basis this would yield a DCF of 8.05%. If we wanted the new plant investment to earn its share of the stockholder's dividend payment, assume that one-third of the profit goes to the shareholders, so $(\frac{1}{3})(5.25 + x) = x; x = 2.625. The total profit then needs to be $7.875 per year, and the DCF = 11.66%.

By an analogous calculation if the company borrowed the $100 at 11% interest, the equivalent DCF's that should be earned to just equal this amount of interest on an after-tax basis would be 9.11 and 13.15%, respectively. It is seen that these numbers would change with the current interest rate, the income tax rate, depreciation schedules, and obviously many other items, but in general this estimated range of minimum DCF of 8–13% is about correct for the values that might be competitive with low-risk alternative investments or be equal to the cost of borrowed money, and thus should be about the lowest rate that a project's DCF should return.

Inflation

Also, there are two other basic considerations to take into account. The first is inflation. Generally in DCF calculations this is ignored as a first approximation since it is very hard to predict, and it usually can be assumed that the value of the product roughly rises with inflation, as does profits. The income numbers some years hence may not be the same as originally calculated, but on a constant dollar basis the results and DCF should normally be approximately the same. This is not always true, however, so it is another uncertainty that makes management desire a higher initial DCF, and sometimes a correction for anticipated inflation should be made if it is suspected that profits may not rise commensurately with inflation. The far greater likelihood, however, is that in the above minimum DCF calculation inflation would disproportionately reduce the potential earnings from other (financial-type) investments or reduce the cost of borrowed money. Realistically, therefore, the acceptable minimums for plant investments usually should be lowered somewhat in proportion to the anticipated inflation rate.

Basic Businesses

The second consideration in establishing the minimum DCF is the broad balancing of risk, rate of return, and the value a project may have for the company. Profitability could be considerably reduced for the basic or core businesses of the company, or with projects with high potential for future growth. As an example, a company that is strong in the field of chlorinated hydorcarbons certainly should be willing to accept any reasonable DCF from needed new capacity, or much more efficient plants, to make chlorine-caustic. This is an integral basic building block operation with very little risk and technical or marketing uncertainty. Considering the very low effective DCF that most chemical companies' existing operations now produce (i.e., 2-8%), any reasonable DCF number (i.e., >10%) should be very attractive for such a basic, important operation.

In a similar manner, if a company has emphasized research in a new field such as biotechnology, lower DCF values might be quite appropriate for the first plants to help "get a foot in the door" or to speed developing production and marketing know-how.

For most projects and most companies, however, considering the risk of any new project, and/or the desire of the company to improve its financial position, a DCF of 15% is usually considered the absolute minimum, and +20% a target value. This range is the cutoff objective for most chemical companies, with the exact limit being set by the uncertainty of the project and the availability of other attractive projects.

Considerable skill must be employed by management in evaluating DCF rate of return numbers, since even though it is an excellent measurement tool for comparing projects, it does not: (1) consider project size or the total dollars involved; (2) it does not consider project risk or uncertainty. A number of statistical probability additions have been suggested to augment DCF analysis (i.e., "Decision Trees," etc.) but it is very rare that enough data exist to make the risk estimations any better than management's basic knowledge, skill and, judgment; and (3) many mitigating factors enter into the project selection, such as previously discussed. Any management that relies too heavily on minimum DCF standards, or judges projects solely on the highest DCF numbers will surely pass up many excellent projects, will (probably unwisely) favor one group or division over another, and will probably be deluding itself on their accuracy and reliability.

SENSITIVITY ANALYSIS

Once a DCF rate has been calculated for a project, and it appears to show an attractive economic potential, a number of additional calculations should be made before the project is presented to management (Neal 1982). A typical method of presentation would be as shown in Figure 6-2, along with perhaps simplified work sheets or pro forma analyses of the type shown in Table 6-5. These "sensitivity analyses" attempt to provide some data on the question, "What if the initial estimate was wrong on each or several of the major variables?" Since this examination is done with quick checks of less accuracy than the original estimate, comparatively small curves, as in Figure 6-2, provide a somewhat more appropriate presentation of the influence of changes in these variables. If the curves indicate a great DCF sensitivity for a certain variable, and it is one that could in its lower values make the project unattractive, then more precise data would be required to provide greater accuracy for that variable, and management must be informed of this possibility. Sometimes the sensitivity analysis will show that the plant size that has been selected is too small to provide an adequate profit, and the sales department will have to redetermine whether they can sell the output of a more economically sized plant.

The desirability of such a large number of DCF estimates for the sensitivity analysis of a project underscores the strong need for using a hand-held calculator or computer in the calculations. The estimation and tabulation of capital and operating costs is quite laborious in itself, but to recalculate the DCF values by trial and error each time would make this an extremely time consuming job, plus would often not provide the accuracy desired for comparison purposes. There are also many software programs for computers that are not only set up for DCF calculations, but also allow a complete array of sensitivity analyses to be made.

The calculation of after-tax earnings for each new case in a sensitivity analyses is usually accomplished by shortcut methods. For instance, if the plant size is to be considered larger or smaller than the base case, then the new capital costs are usually obtained by using exponent factors unless more accurate data (actual price quotations, etc.) are available. If the sizing exponent is not known, use the value 0.6 or 0.64 for first approximations. Thus, if a plant 20% larger than the $10 million base case is to be considered, the new cost may be estimated at $10 \times (1.2)^{0.64} = \$11.2$ million.

In a similar manner manufacturing costs may be adjusted by changing only those variables involved: labor may remain constant, or raw materials, utilities, and operating labor changed in direct proportion to the production rate; the labor related costs by the previously discovered total factor, such as 1.60 times operating labor; capital related charges by its percentage of capital (e.g., 26% of capital); sales related cost by its factor, such as 20% of sales; working capital by its fraction of manufacturing cost (e.g., 10–35%) or total plant cost (e.g., 10–20%); and so on. For example, in the above 20% plant size increase case, the cost of raw materials and utilities might increase by 20%, the capital related manufacturing costs by 26%, or perhaps $0.26 \times \$1.2$ MM $= \$312,000/\text{yr}$, the sales related costs by 20%, the working capital by perhaps 15% of the plant

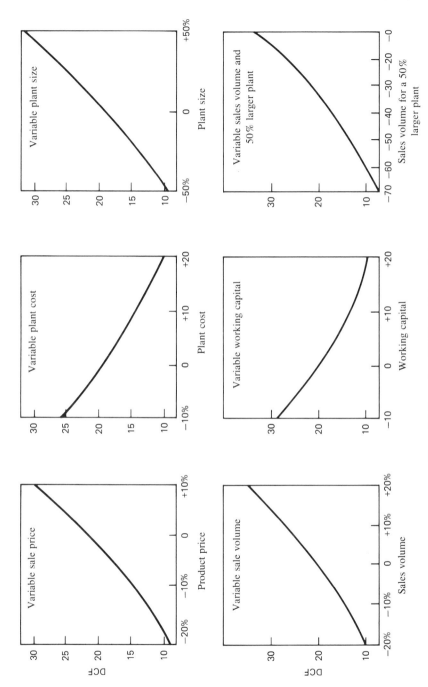

Figure 6-2. Examples of DCF sensitivity analysis.

cost, or perhaps $11.2 MM $\times$ 0.15 = $1.68 MM, and the depreciation by 12% (perhaps to become $1.12 MM/yr). Care must be taken to consider all of the items that change when making the sensitivity analysis, but shortcut methods such as the above should reduce the time and effort required in preliminary analyses. Of course, much more detail and firm data would be required for more definitive analyses and many more of the individual cost components would change.

EFFECT OF BORROWED CAPITAL; LEVERAGE

The leverage on profitability resulting from borrowing part of the capital required for a project is frequently an essential part of new major capital expenditures. However, even before being discussed in more detail, it should be cautioned that generally the design engineer should not consider leverage in his evaluations. Management can and will factor it into their projections, and thus more fully appreciate the risks involved as well as the gain with this maneuvering of the company's cash flow and debt position.

Most companies have the capability of borrowing some, or even a large fraction of the funds required for any new project. Consequently, the company money invested in a project is almost always less than the total, and if the project realizes an adequate profit, the return on the company's money is leveraged, or greatly increased. Furthermore, since interest paid on a loan is tax deductible, and inflation reduces the value of the funds repaid, not all of the loan's cost is actually a drain on the project (Barna 1984).

For instance, consider a $10,000,000 capital cost project (including $1,000,000 working capital) with several different potential amounts borrowed at 10% interest and depreciation, and $1,000,000 after-tax profit for the no-debt case. Any debt will be repaid at the end of the 10-year project life. The results are shown in Table 6-6.

From this simplified example it is seen that the more capital borrowed, the greater the profitability on the company funds, but since bankers generally dislike greater than about a 0.5 debt to equity ratio for a company, the middle case is probably somewhat more typical of the average project, although each company's current debt and policy on this matter varies widely.

In addition to increasing the ROI and DCF, borrowing naturally allows many more projects to be implemented, providing greater growth. Thus, it has become a somewhat necessary and constructive tool for most progressive managements. Remember, however, that the company is usually at risk for the entire amount of capital involved in the project, so the process engineer should generally ignore debt in his evaluations unless directly told not to. The profitability on a nondebt basis should be the basic decision making consideration and then management can analyze the entire corporate position before deciding upon debt.

Table 6-6.
Example of Leverage upon DCF and ROI.

	Entirely Company Capital	*$3,333,000 Borrowed*	*$5,000,000 Borrowed*
Company capital, $	10,000,000	6,667,000	5,000,000
Interest on loan, $/yr	0	333,000	500,000
Interest deduction from after-tax profit at 40% tax, $/yr	0	200,000	300,000
After-tax profit, $/yr	1,000,000	800,000	700,000
Cash flow, $/yr	1,900,000	1,700,000	1,600,000
Payout period on company capital, yrs	4.7	3.3	2.5
Rate of return on company capital, %	10	12.0	14.0
DCF (annual compounding) on company capital, %	14.4	20.4	27.1
Debt to equity ratio	0	0.50	1.00

An interesting example of how the use of debt can be both good and bad is seen from a recent analysis of farm profitability, as shown in Table 6-7. For the first period shown, the greater the use of debt, the greater the farmer prospered, and there was a rush to more borrowing, or high debt to equity ratios. Then interest rates increased, farm prices dropped, and land values (their principal equity) dramatically decreased. This resulted in greatly decreased profits, and caused a major farm crisis for most of the country. Those farmers with reasonable debt to equity ratios were not seriously effected. A comparable

Table 6-7.
Results of Alternative Farm Financial Strategies over the Periods
1972–1977 and 1979–1983

Initial Debt to Equity Ratio, %	*Results After 5 Years of Operation*	
	Change in Net Worth	*Average Annual Rate of Return on Equity*
	1972–1977	
25	+13%	+20%
75	+20%	+54%
	1979–1983	
25	+3%	+2%
75	−35%	−9%

Table 6-8.
Effect of Leverage with Decreasing Income. (Example as Table 6-6, but with Variable Income.)

	Entirely Company Capital	$3,333,000 Borrowed	$5,000,000 Borrowed
Company capital, $	10,000,000	6,667,000	5,000,000
a. Profit on sales, one-half normal[a]	500,000	300,000	200,000
After-tax profit, $/yr			
ROI, % (on company funds)	5	4.5	4.0
DCF, % (on company funds)	6.64	7.29	8.55
b. Profit on sales, zero[a]			
After tax profit, $/yr	0	−333,000	−500,000
ROI, % (on company funds)	0	−5.0	−10.0
DCF, % (on company funds)	0	0	0

a. For the entirely company capital case.

situation can exist within the chemical and process industries at any time as indicated in the example of Table 6-8.

PRODUCTION RATE; BREAK-EVEN POINT

A second form of sensitivity analysis is required to consider the economics of operating a plant at less than full capacity. During the start-up period (which may last for years in exceptional cases), other periods of operating difficulty, and with decreased sales the plant will only operate at partial capacity. The industry average operating percentage for all products in the mid-1980s was somewhat below 80%, and in some cases it takes many years to build up to even this sales capacity. Predicting the effect of reduced operation leads to a very important secondary point of information, the break-even capacity for the plant. This is the amount of production that it takes to just pay all of the bills, and neither make a profit nor take a loss. Cash flow break-even is when all costs, excluding depreciation, are just equal to the sales realization. With sales below this number there is an actual out-of-pocket loss for the operation.

Such calculations require holding all facets of operating cost constant (as if at full production) except for raw materials, utilities, and sales expenses. These items theoretically vary in direct proportion to the production rate, but in point of fact utilities and sales expense may not change much, and in such cases should be separately analyzed. If sales are made with company staff, the number of salesmen may not decrease with decreasing sales (and might even increase). Likewise, standby utility charges may cut into the decreased useage savings, and reduced production may decrease the efficiency of raw material conversion. On the other hand, some of the other manufacturing cost items should also decrease with reduced production, such as shared or even directly reduced

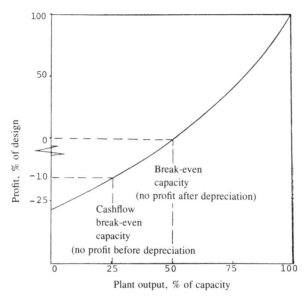

Figure 6-3. Example of a project break-even curve.

manpower. The net result of such calculations are usually shown in a break-even graph like Figure 6-3.

COMPARISON OF PROJECT EVALUATION METHODS

Several project evaluation methods have now been discussed: discounted cash flow (DCF or IRR), present worth (PW or PV), rate of return (ROI), payback period, and capitalized cost. Three of these utilize the time value of money, and all are useful. Each method allows a comparison of projects: DCF and PW based upon their present value, assuming for PW a desired return (interest) rate and discounting all of the cash flows with it. This avoids trial and error calculations, but provides two numbers for the comparison, the assumed interest rate and the net present worth. In some cases it is the simpler method to work with, and of course, it is obtained as the intermediate total for each DCF trial at an assumed interest rate. DCF calculates the equivalent effective interest rate of the project. Capitalized cost is an alternative method for considering the time value of money for the comparison of different equipment purchases.

It has also been suggested that simple rate of return or return on investment (ROI) and payback period calculations be performed on every project no matter how complex, since these are very basic and easily understood project measurements. For simple and small projects they may be all that is required.

Finally, there are numerous other economic comparison methods, each with some individual advantages or virtues, and many excellent project evaluation books (Allen 1980; Clark and Lorenzoni 1985; Holland et al. 1974; Kurtz 1984; Newman 1983; Riggs 1982; United Nations 1972) and articles (Berkoff et al 1986; Carlson 1984; Pinches 1984; Powell 1985; Shinnar and Dressler 1986; Ward 1986) frequently are published. However, the above evaluation methods have become the major ones in common use, so the other more complex or statistical (based upon uncertainty considerations) procedures will not be considered here. Once the simple ROI, payback, and DCF concept of the time value of money are understood, any other method can be quickly learned if needed at a later date.

REFERENCES

Allen, Derek H. 1980. *A Guide to Economic Evaluation of Projects.* Rugby, Warks, England: Institute of Chemical Engineering.

Barna, Bruce A. 1984. Leverage, risk and project economics. *Chemtech* (May):295–297.

Berkoff, Charles E. et al. 1986. The process profile. *Chemtech* (Sept.):552–559.

Carlson, Rodney O. 1984. Avoiding the game in project evaluation. *Chemical Engineering Progress* (Nov.):11–13.

Clark, Forrest D., and A. B. Lorenzoni 1985. *Applied Cost Engineering.* Marcel Dekker, New York.

Holland, F. A., F. A. Watson, and J. K. Wilkinson. 1974. *Introduction to Process Economics.* John Wiley & Sons, London.

Horwitz, Benjamin A. 1980. The mathematics of discounted cash flow analysis. *Chemical Engineering* (May 19):169–174.

Kurtz, Max. 1984. *Handbook of Engineering Economics.* McGraw-Hill, New York.

Leibson, Irving, and Charles A. Trischman, Jr. 1971. When and how to apply discounted cash flow and present worth. *Chemical Engineering* (Dec.).

Neal, Gordon W. 1982. Project evaluation. *Chemical Engineering* (Sept. 6; Nov. 1; Dec. 22).

Newman, D. G. 1983. *Engineering Economic Analysis.* Engineering Press, San Jose, CA.

Pinches, George E. 1984. Effective use of capital budgeting techniques. *Chemical Engineering Progress* (Nov.):11–13.

Powell, Terence E. 1985. Project evaluation. *Chemical Engineering* (Nov. 11):187–194.

Riggs, J. L. 1982. *Essentials of Engineering Economics.* McGraw-Hill, New York.

Ruben, Allen G. 1984. Choose process units via present worth index. *Chemical Engineering* (Oct. 15).

Shinnar, Reuel, and Ofer Dressler 1986. Return on investment; myth and reality. *Chemtech* (Jan.):30–41.

United Nations Publication. 1972. *Guidelines for Project Evaluation.*

Ward, Thomas J. 1986. Profitability analysis. *AIChE Plant Design and Cost Engineering* 1:22–32.

7

ECONOMY OF THE CHEMICAL INDUSTRY

The chemical industry has been a major component of the world's manufacturing activities since the early days of the industrial revolution. The Leblanc process for making soda ash (Na_2CO_3, initially used to aid in the bleaching of cloth for the emerging textile industry) from salt and sulfuric acid in 1791 was truly the start of the large-scale synthetic production of chemicals. Sulfuric acid was being produced commercially before then, but in many small facilities, and it was not until later that it became a tonnage product with the advent of the Chamber process. These two commodities became essential to the then-infant industrialization of the world's economy, and the chemical industry has been a large, vital, and prosperous force ever since.

INDUSTRY STATISTICS

The production of chemicals has become so widespread, and practiced by so many other industries that it is hard to define exactly which companies are part of the "chemical industry." For this purpose there are two broad categories considered by the U. S. Commerce Department in gathering industrial statistics that shall be used in this book.

Chemical Process Industries (CPI)

This grouping of industries includes chemicals and allied products where chemical production, processing, or processing equipment is employed; pulp, paper, and paperboard; the processing of petroleum and natural gas; rubber and plastic products; products of stone, clay, and glass; primary nonferrous metals; sugar refining, wet corn milling, and similar processing of foods and beverages; textile dyeing and finishing; leather tanning; dry cell and storage batteries; semiconductor materials; carbon and graphite products; and hard-surface floor coverings. It is a very large segment of U. S. industry, with the value of its shipments in 1986 and 1988 (est.) being 16.6 and 17.7%, respectively, of the total U. S. gross national product. The health and vigor of this industry is quite obviously a major factor in the U. S. economy.

The Chemical Industry

This grouping of industries is a part of the chemical process industries, but only includes companies with a major chemical production: pharmaceuticals, detergents, and other sanitizing and polishing preparations, toiletries and cosmetics, paints and coatings, fertilizers and pesticides, printing inks, carbon black, adhesives, additives and catalysts, as well as industrial chemicals and synthetic materials. For 1986 and 1988 (est.) the chemical industry was only 35.4 and 33.9%, respectively, the size of chemical process industries, and 5.9 and 6.0%, respectively, of the total U. S. gross national product. This is still a reasonably large industrial group with about $230 billion in sales, 580 thousand production workers, and disproportionately large in such areas as maintaining a positive balance of trade for the United States in the world's markets. Some of the statistical data for these industry groupings and the total U. S. production are given in Table 7-1.

Industry Data

Largest Companies. Very few chemical manufacturing companies now produce only chemicals, so company size can be measured by either the company's total business or the size of its chemical operations alone. Using the later criteria, Table 7-2 lists the 50 largest companies for 1986. Table 7-3 lists the 15 largest of these companies on the basis of their total sales (of all commodities) for the same period. It is seen that the top 50 companies in chemical sales only averaged 53.7% of their total sales in chemicals, and that they consisted of a wide range (11) of other industrial groups than chemicals and petroleum. It is also seen that 10 of the country's 50 largest chemical producers are in the petroleum business, and about half of the country's largest companies produce at least some chemicals.

Financial Analyses. Some additional financial data for a selected group of the larger chemical companies with a relatively high percentage of chemical sales in 1986 are given in Table 7-4. Figures 7-1 to 7-10 present much of the same data, but as industry averages and in a graphical form. It is seen that as an industry the profit margin determined either as after-tax income as a percent of sales or of stockholder equity is relatively low. About one-third of their total assets have been obtained by borrowing (a debt to equity ratio of 0.5), and their yearly sales value is about equal to their total (depreciated) assets. Their plants are about 50% depreciated, and in the mid-1980s only ran at about 77% of capacity. The industry manpower requirements (per unit of investment or sales)

Table 7-1
General Data on the U.S. Economy, Chemical and Chemical Process Industries

	1986	1987	1988 est.
The U. S. Economy			
Gross national product adjusted for inflation (billions of 1982 dollars)	$3,679.0	$3,814.8	$3,882.3
Consumption expenditures (billions of 1982 dollars)	$2,412.8	$2,498.2	$2,528.8
Industrial production index (1977 average = 100)	125.0	129.4	131.9
Corporate cash flow, after taxes (billion dollars)	$377.5	$369.5	$358.4
Merchandise trade deficit (billion dollars)	$145.3	$152.0	$134.6
The Chemical Process Industries			
Value of shipments (billion dollars)	$611.8	$625.1	$687.2
Production index (1977 average = 100)	126.7	131.6	135.5
Average operating rate (percentage of capacity)	86.7%	85.3	83.5
Capital spending (billion dollars)	$43.7	$45.1	$52.0
The Chemical Industry			
Value of shipments (billion dollars)	$216.8	$214.6	$233.2
Production index (1977 average = 100)	133.5	139.7	144.4
Average operating rate (percentage of capacity)	76.9%	81.1	80.2
Producers' price index (1967 average = 100)	299.6	336.2	351.7
Capital spending (billion dollars)	$16.9	$16.6	$19.3
Production workers (thousands)	571.0	568.0	580.0
Workers' average earnings (dollars/hour)	$11.96	$12.38	$12.87
Chemical exports (billion dollars)	$22.9	$26.0	$29.3
Chemical imports (billion dollars)	$15.5	$16.7	$18.2

Source: *Chemical Week* 1988; 1987. Excerpted by special permission from *Chemical Week*, January 6–13, 1988 and January 7–14, 1987. Copyright © 1987 and 1988, by McGraw-Hill, Inc., New York.

Table 7-2
50 Largest Chemical Producers in 1986 (Based upon Their Chemical Sales Only)

Rank 1986	Chemical sales 1986, $ millions	Chemical sales as % of total sales	Industry classification	Chemical operating profits, $ millions	Chemical operating profits as % of total operating profits	Identifiable chemical assets, $ millions	Chemical assets as % of total assets
1 Du Pont	$11,839	43.6%	Diversified	$1764	52.1%	$8739	36.2%
2 Dow Chemical	8,863	79.8	Basic chemicals	978	76.1	8139	66.5
3 Exxon	6,079	8.1	Petroleum	817	9.9	5508	7.9
4 Union Carbide	5,001	78.8	Basic chemicals	748	94.6	5047	74.3
5 Atlantic Richfield	4,915	32.8	Petroleum	398	64.7	2085	9.7
6 Monsanto	4,701	68.3	Basic chemicals	896	141.1	3860	46.7
7 BASF Wyandotte	3,600	100.0	Basic chemicals	na	na	na	na
8 Shell Oil	3,292	19.6	Petroleum	648	35.1	3447	13.1
9 Amoco	2,928	14.5	Petroleum	467	23.0	2545	11.0
10 Celanese	2,891	100.0	Basic chemicals	258	100.0	na	na
11 Allied-Signal	2,819	23.9	Diversified	346	32.9	2161	19.2
12 W. R. Grace	2,492	66.9	Specialty chemicals	285	1187.5	1633	53.1
13 Mobil	2,391	4.8	Petroleum	205	4.5	2103	5.7
14 Eastman Kodak	2,379	20.6	Photo equipment	227	31.4	2266	17.6
15 Chevron	2,378	9.1	Petroleum	147	20.6	3207	9.3
16 General Electric	2,331	6.3	Diversified	424	9.8	3602	10.4
17 Occidental Petroleum	2,088	13.0	Petroleum	176	36.2	2393	13.7
18 Rohm & Haas	2,067	100.0	Basic chemicals	280	100.0	1470	100.0
19 American Cyanamid	1,821	47.7	Basic chemicals	135	38.3	1275	49.6
20 Air Products	1,798	90.7	Basic chemicals	363	129.4	2173	80.3
21 Mobay	1,710	100.0	Basic chemicals	150	100.0	na	na
22 Hercules	1,709	65.4	Basic chemicals	101	49.5	1464	71.6

23 Phillips Petroleum	1,698	17.0	Petroleum	299	30.3	1053	8.5
24 Borden	1,638	32.7	Dairy products	176	35.3	1399	40.3
25 Ciba-Gelgy	1,490	61.3	Specialty chemicals	na	na	na	na
26 Ashland Oil	1,477	20.3	Petroleum	71	15.0	527	13.9
27 B. F. Goodrich	1,458	57.1	Rubber products	134	79.9	1199	82.5
28 FMC	1,399	46.6	Machinery	224	71.6	1344	50.0
29 American Hoechst	1,331	77.8	Basic chemicals	na	na	na	na
30 Texaco	1,279	4.0	Petroleum	91	3.9	1114	3.2
31 Ethyl	1,272	57.0	Basic chemicals	247	69.0	829	48.4
32 Olin	1,114	65.3	Basic chemicals	80	50.3	840	59.6
33 National Distillers	1,087	62.8	Alcoholic beverages	124	67.0	1385	47.7
34 Dow Corning	1,085	100.0	Specialty chemicals	111	100.0	na	na
35 International Minerals	1,077	87.1	Agrochemicals	35	57.1	2042	84.1
36 National Starch	1,064	100.0	Specialty chemicals	140	100.0	743	100.0
37 Unocal Corp.	1,056	12.6	Petroleum	46	26.1	722	7.1
38 Borg-Warner	1,043	30.9	Auto equipment	153	43.8	562	23.6
39 Lubrizol	889	91.0	Specialty chemicals	140	108.6	589	67.1
40 PPG Industries	843	18.0	Glass products	103	14.8	1185	25.5
41 Aluminum Co. of America	840	18.0	Nonferrous metals	na	na	na	na
42 Pennwalt	777	70.1	Basic chemicals	93	78.8	649	73.8
43 Cabot	769	58.7	Specialty metals	138	69.3	494	32.7
44 CF Industries	766	100.0	Agrochemicals	−45	def	842	100.0
45 Reichhold Chemicals	766	100.0	Basic chemicals	53	100.0	458	100.0
46 Aristech	751	100.0	Basic chemicals	98	100.0	na	na
47 Naico Chemical	736	100.0	Specialty chemicals	108	100.0	586	100.0
48 NL Industries	734	57.2	Petroleum services	na	na	na	na
49 Eli Lilly	699	18.8	Drugs	53	5.8	766	16.7
50 Engelhard	666	29.1	Specialty metals	77	80.5	607	58.2
Average		53.7			56.6		44.8

Source: *Chemical and Engineering News* 1987e. Reprinted with permission from Chemical and Engineering News, June 8, 1987 and April 13, 1987. © 1987, American Chemical Society; na. = not available; − or def = deficit.

Table 7-3
15 Largest U. S. Companies Producing Chemicals in 1986.

Company	Rank in			Sales, $ Millions	Net Income, $ Millions	Overall Rate of Return as % of		
	Size	Income	Chem. Prod.			Sales	Assets	Stock Holders Equity
Exxon	2	1	3	69,888	5,360	7.7	7.7	16.7
Mobil	5	8	15	44,866	1,407	3.1	3.6	9.2
General Electric	6	5	14	35,211	2,492	7.1	7.2	15.5
Texaco	8	17	31	31,613	725	2.3	2.1	5.3
E. I. Du Pont	9	6	1	27,148	1,538	5.7	5.8	11.5
Chevron	10	18	11	24,351	715	2.9	2.1	4.6
Amoco	13	15	9	18,281	747	4.1	3.2	6.6
Shell Oil	15	12	7	16,833	883	5.3	3.4	6.2
Proctor & Gamble	18	19	—	15,439	709	4.6	5.4	11.9
Occidental Petro.	19	96	22	15,344	181	1.2	1.0	3.4
Atlantic Richfield	20	25	6	14,586	615	4.2	2.9	11.7
USX	22	476	32	14,000	(8,833)	—	—	—
Allied Signal	25	27	17	11,794	605	5.1	5.4	15.6
Eastman Kodak	26	50	13	11,550	374	3.2	2.9	5.9
Dow Chemical	27	16	2	11,113	732	6.6	6.0	14.2
					Average	4.2	3.9	9.3

Source: *Fortune* 1987. Courtesy of Fortune, April 27, 1987, © 1987 Time Inc. All rights reserved.
( ) = loss.

$ Billions

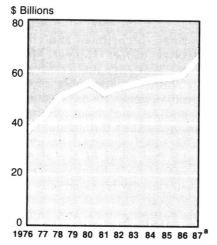

1976 77 78 79 80 81 82 83 84 85 86 87[a]

Source: Storck 1988. Reprinted with permission from *Chemical and Engineering News*, February 22, 1988. © 1988, American Chemical Society.

Figure 7-1. Annual sales.[a]

$ Billions

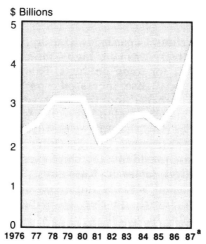

1976 77 78 79 80 81 82 83 84 85 86 87[a]

Source: Storck 1988. Reprinted with permission from *Chemical and Engineering News*, February 22, 1988. © 1988, American Chemical Society.

Figure 7-2. Annual earnings.[a]

[a]30 major chemical companies

Table 7-4.
1986 Financial Analysis for Various of the Larger Chemical Companies.

Chemical Sales, % of Total	Rel. Size in Chem.		Year	Profit Margin[a]	Return on Investment[b]	Debt as % Debt Plus Equity	% Total Sales Abroad[c]	% Total Assets Abroad[d]	Sales per Employee[e]	Dividends As % of Net Income	Capital Spending[f] As % of Sales	As % of Net Plant	R & D As % of Sales	Sales As % of Assets	Net Plant As % of Gross Plant
90.7	20	Air Products	1986	6.9%	3.3%	38.9%	19.5%	25.0%	$119.4	32.9%	18.5%	20.0%	3.1%	73.3%	53.2%
			1985	7.8	3.8	30.9	17.3	18.8	97.8	26.6	21.4	22.0	2.8	70.5	56.1
47.7	19	American Cyanamid	1986	5.3	4.4	25.4	33.6	24.2	110.7	43.6	6.9	20.7	7.3	104.1	47.4
			1985	3.4	2.7	25.8	29.6	21.9	97.1	76.0	5.7	18.0	7.1	103.9	44.3
58.7	43	Cabot	1986	5.6	4.1	41.9	38.4	16.9	198.5	35.4	5.5	11.3	2.6	86.8	53.0
			1985	5.1	4.1	37.9	28.8	15.7	182.8	41.4	10.1	22.7	2.3	88.3	56.3
79.8	2	Dow Chemical	1986	6.7	4.3	39.7	53.5	49.4	216.6	49.1	8.0	16.6	5.4	90.8	42.1
			1985	0.5	0.4	40.0	54.8	46.6	216.9	587.9	7.0	15.7	4.7	97.5	43.2
57.0	31	Ethyl	1986	11.3	10.7	30.2	19.6	8.7	150.4	24.0	8.5	22.5	3.0	92.1	49.3
			1985	7.6	7.5	33.6	28.3	9.2	147.3	31.7	5.8	16.1	3.0	99.4	50.6
57.1	27	B. F. Goodrich	1986	0.5	0.6	34.4	19.1	14.3	214.6	289.9	5.6	16.9	2.2	140.3	58.3
			1985	0.3	0.3	41.7	15.4	15.1	122.2	374.5	6.3	19.4	2.0	141.6	55.1
66.9	12	W. R. Grace	1986	1.2	0.9	45.6	32.3	16.9	90.0	263.0	5.3	12.6	2.5	90.9	48.3
			1985	2.5	1.9	38.8	25.4	16.4	58.3	111.7	6.7	13.4	1.8	95.8	57.1
65.4	22	Hercules	1986	8.7	6.8	24.3	21.1	15.4	104.1	41.4	9.8	22.3	2.7	89.7	51.4
			1985	5.1	4.4	26.5	25.0	14.9	101.7	64.9	9.1	23.4	2.9	97.3	50.1
87.1	35	International Minerals	1986	def	def	48.1	13.2	9.7	84.7	def	8.0	10.3	na	48.5	57.7
			1985	7.3	4.4	23.9	17.3	8.3	187.0	59.5	7.1	10.3	2.8	83.1	52.3
91.0	39	Lubrizol	1986	8.0	7.1	8.4	51.6	31.8	203.4	59.3	4.1	14.0	5.2	111.3	45.0
			1985	6.7	5.7	12.4	47.7	29.5	173.7	78.1	4.4	13.6	4.8	105.7	48.3

68.3	6	Monsanto	1986	6.3	30.1	32.6	21.2	133.1	46.0	7.6	17.9	8.7	83.2	46.0
			1985	2.7	38.0	28.5	18.8	120.3	102.7	9.6	21.3	8.1	76.0	44.4
62.8	33	National Distillers	1986	3.2	37.6	na	na	na	136.1	5.9	7.3	na	59.6	71.9
			1985	3.4	23.2	na	na	na	96.9	3.3	8.5	na	113.3	60.0
65.3	32	Ollin	1986	4.4	29.7	6.1	8.6	129.3	45.6	7.5	17.8	3.3	110.5	37.6
			1985	2.3	34.0	7.2	10.3	117.5	86.9	8.8	21.5	3.0	109.6	39.3
18.0	40	Pennwalt	1986	4.7	33.6	23.5	21.1	115.4	68.0	4.9	12.5	4.1	111.1	50.7
			1985	1.1	30.1	19.8	18.3	102.4	319.6	8.9	20.7	3.9	105.7	53.3
70.1	42	PPG Industries	1986	6.8	32.8	25.5	27.4	128.4	35.3	7.6	13.4	4.3	101.0	58.2
			1985	7.0	35.4	21.5	23.7	115.9	36.4	10.4	18.2	4.0	106.4	59.3
100.0	18	Rohm & Haas	1986	6.7	21.7	38.7	28.4	171.4	39.1	8.7	27.6	6.4	112.2	39.8
			1985	6.9	20.7	32.5	26.6	173.2	34.8	7.8	27.4	6.0	118.3	38.7
78.8	4	Union Carbide	1986	2.1	75.3	28.6	33.2	124.2	110.8	8.4	12.0	2.4	82.5	51.2
			1985	2.3	30.3	29.2	28.9	98.4	113.4	7.2	11.2	3.1	85.1	54.2
40.1	61	Witco	1986	4.8	17.0	16.7	15.8	169.4	35.9	4.4	16.3	1.6	165.3	47.8
			1985	3.9	24.7	14.0	13.5	na	38.4	4.9	19.7	na	178.8	39.3
66.9		Median 1986		5.5%	33.2%	25.5%	21.1%	$129.3	44.5%	7.6%	16.5%	3.1%	91.5%	50.0%
		Median 1985		3.7%	30.6%	25.2%	18.6%	$118.9	77.1%	7.2%	16.8%	3.1%	101.7%	52.8%

Source: *Chemical and Engineering News* 1987e. Reprinted with permission from Chemical and Engineering News, June 8, 1987. © 1987. American Chemical Society.

Notes: Net income is from continuing operations, excluding extraordinary and nonrecurring items where possible.

a. Net income as a percentage of sales.
b. Net income as a percentage of current assets plus gross plant.
c. Consolidated sales only.
d. Foreign identifiable assets as a percentage of total assets.
e. Thousands of dollars.
f. Actual spending on construction of new facilities and purchase of new equipment and land in consolidated businesses.
na = not available.
def = deficit.

115

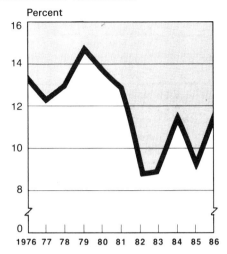

Source: Storck 1988. Reprinted with permission from *Chemical and Engineering News*, February 22, 1988. © 1988, American Chemical Society.

Figure 7-3. Return on equity.[a]

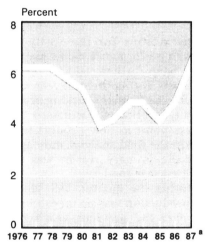

Source: Storck 1988. Reprinted with permission from *Chemical and Engineering News*, February 22, 1988. © 1988, American Chemical Society.

Figure 7-4. Annual profit margins.[a]

(After-tax earnings as % of sales.)

[a]30 major chemical companies

are quite low. Capital spending has only averaged slightly higher than the industry's depreciation rate.

Industry Products. The chemical process industries make an amazingly large variety of products, with the 50 largest basic chemicals on a tonnage basis, and the major plastics and fibers listed in Tables 7-5 and 7-6. The total output for most commodities has been slowly rising as shown in Figure 7-5, with modest annual growth rates. It is seen that the production of inorganic chemicals has risen about 1% per year since 1976, even though declining about 1.5% during the mid-1980s. Plastics have increased 5.5%, synthetic fibers 1.3%, and all organics 2.9% during the mid-1970s to mid-1980s.

All of these factors show the chemical industry to be very large, with generally slowly rising sales, and steady or declining profits until 1986. Its reinvestment in new capital equipment has been modest (Figures 7-7 and 7-8), its environmental spending (Figure 7-11) heavy, and its research and development (R & D) spending (Figures 7-4 and 7-5; Tables 7-9 and 7-4) not very aggres-

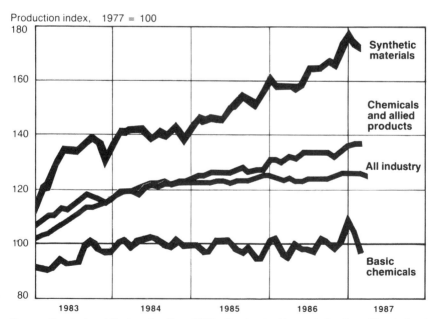

Source: *Chemical and Engineering News* 1987*e*. Reprinted with permission from *Chemical and Engineering News*, June 8, 1987. © 1987, American Chemical Society.

Figure 7-5. Relative yearly sales.

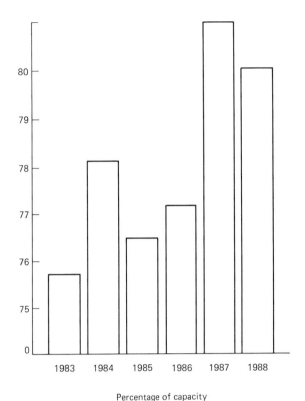

Percentage of capacity

Source: *Chemical Week* 1988. Excerpted by special permission from *Chemical Week*, January 6–13, 1988. Copyright © 1988, by McGraw-Hill, Inc., New York.

Figure 7-6. U. S. chemical industry average operating rate.

sive, at least compared to its foreign competition. It has become a large and diversified, but mature industry that still has an excellent positive trade balance (Figures 7-8, 7-9, and 7-10) (export sales minus import sales), but is increasingly having difficulty meeting worldwide competition. It would appear to be in the decision period of its history between allowing a conservative, nontechnical or long-range thinking management letting it become slowly more obsolete and noncompetitive (like the steel industry), or continuing to be a more aggressive industry investing in new innovations, processes, and equipment as a large net exporter. This decision point will be examined in more detail in the following sections.

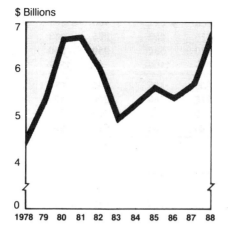

Source: Reisch 1987b. Reprinted with permission from *Chemical and Engineering News*, December 21, 1987. © 1987, American Chemical Society.

Figure 7-7. Capital spending,[a] $.

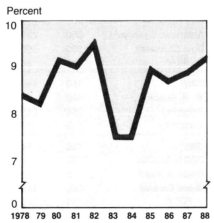

Source: Reisch 1987b. Reprinted with permission from *Chemical and Engineering News*, December 21, 1987. © 1987, American Chemical Society.

Figure 7-8. Capital spending,[a] % of sales.

[a]12 major chemical companies

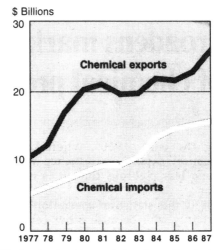

Source: Anderson 1987. Reprinted with permission from *Chemical and Engineering News*, December 4, 1987. © 1987, American Chemical Society.

Figure 7-9. U. S. chemical exports, imports, $.

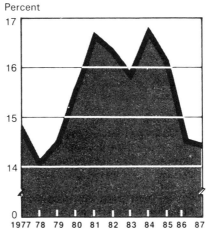

Source: Anderson 1987. Reprinted with permission from *Chemical and Engineering News*, December 4, 1987. © 1987, American Chemical Society.

Figure 7-10. U. S. chemical exports, % of world total.

PROFITABILITY

The U. S. chemical industry was founded, and continued all through its early years on the basis of: (a) a strong and innovative technical capability, and (b) a technical and venturesome management that was willing to invest in sound new research and development opportunities. Companies such as Dow Chemical, Monsanto, and du Pont have continuously had some of the strongest and most brilliant research and development and engineering groups in the world, and their management has in the past funded and brought into production one major new development after another. All of this made the U.S. chemical industry in its pre-1960 days a very high-profit, dynamically growing industry that generally dominated world chemical production, if not by export quantities alone, by licensing its processes and serving as the world's role model for most chemical production.

Since the mid-1950s, however, this has slowly changed. Figures 7-12 and 7-13 show the decline in the rate of return, based upon constant dollar values (corrected for inflation) and a pseudo-DCF measurement. The numbers from 1980 through 1986 indicate a slight further decline to an industry average pseudo-DCF of only about 4.3%. It is seen that the return on investment was very high following WWII because of the technical and managerial excellence noted above, and since there was only limited competition. This began to change during the 1950s as new production was built, largely with U. S. know-how and foreign aid, in Europe and Japan. These plants were new and employed the most modern technology, providing the initial competition for the U. S. producers. Some profits fell, but they still stayed comparatively high. This foreign competitive pressure, however, has steadily increased and grown more vigorous over the intervening years.

This brought on the next surge of competition which was from other U. S. companies, predominately the lower profit oil industry in the 1960s and 1970s, who had surplus funds to invest, but were generally less well (and nontechnically) managed. Their new plants were often very large, and resulted in many commodities having excess production capacity over the market demand, which reduced industry operating rates and profits. This situation initiated a profound change in the makeup of the chemical industry, for as seen in the current top 50 chemical producers (Table 7-5), the chemical companies started to diversify into other fields, and companies in many other fields became chemical producers. In most cases the newcomers' research and development capability was limited, and the management did not know the chemical business or technology too well. Fewer new research and development projects were conceived, and far fewer were commercialized, which further reduced the overall industry rate of return.

In the mid-1970s the OPEC oil cartel was formed, and energy prices increased as much as 20-fold over a relatively short period of time. This was costly for

Table 7-5
Top 50 Chemicals Produced in the United States

Rank			Billions of lb		Common Units[b]			Average Annual Growth		
1986	1985		1986	1985	1986	1985	1985–86	1984–85	1981–86	1976–86
1	1	Sulfuric acid	73.64	79.30	36,822 tt	39,651 tt	−7.1%	−5.1%	−2.0%	0.5%
2	2	Nitrogen	48.62	47.46	671 bcf	655 bcf	2.4	−0.9	6.5	8.8
3	4	Oxygen	33.03	32.53	399 bcf	393 bcf	1.5	2.1	−1.5	0.4
4	6	Ethylene	32.81	29.85	32,811 mp	29,847 mp	9.9	−4.9	2.2	3.9
5	5	Lime	30.34	31.60	15,172 tt	15,800 tt	−4.0	−1.2	−4.3	−2.3
6	3	Ammonia	28.01	34.64	14,005 tt	17,319 tt	−19.1	3.8	−6.0	−1.8
7	7	Sodium hydroxide	22.01	21.79	11,007 tt	10,893 tt	1.0	−0.2	1.1	0.5
8	9	Chlorine	20.98	20.79	10,489 tt	10,395 tt	0.9	−2.8	−0.5	0.1
9	8	Phosphoric acid	18.41	21.04	9,206 tt	10,518 tt	−12.5	−7.5	−1.6	1.5
10	11	Propylene	17.34	14.89	17,343 mp	14,887 mp	16.5	−4.3	5.2	5.6
11	10	Sodium carbonate	17.20	17.19	8,600 tt	8,597 tt	0.0	1.0	0.8	5.1
12	15	Ethylene dichloride	14.53	12.10	14,529 mp	12,101 mp	20.1	13.0	7.8	6.1
13	12	Nitric acid	13.12	14.73	6,562 tt	7,364 tt	−10.9	−4.7	−6.3	−1.7
14	14	Urea	12.06	13.36	6,029 tt	6,678 tt	−9.7	−10.2	−5.6	4.4
15	13	Ammonium nitrate	11.11	13.55	5,556 tt	6,776 tt	−18.0	−5.2	−8.9	−2.5
16	17	Benzene	10.23	9.39	1,389 mg	1,275 mg	8.9	−3.3	1.2	−0.3
17	20	Ethylbenzene	8.92	7.39	8,915 mp	7,386 mp	20.7	−2.3	2.7	4.4
18	18	Carbon dioxide	8.50	9.25	4,252 tt	4,623 tt	−8.0	1.3	2.3	8.0
19	16	Vinyl chloride	8.42	9.46	8,415 mp	9,463 mp	−11.1	55.5	4.1	4.0
20	19	Styrene	7.84	7.62	7,838 mp	7,622 mp	2.8	−1.1	3.3	2.2
21	21	Terephthalic acid	7.68	6.49	7,684 mp	6,490 mp	18.4	9.8	4.3	0.6
22	27	Methanol	7.33	5.00	7,327 mp	5,003 mp	46.5	−38.9	−3.1	1.6
23	22	Hydrochloric acid	5.97	5.61	2,983 tt	2,807 tt	6.3	2.7	3.0	1.6
24	24	Ethylene oxide	5.94	5.43	5,943 mp	5,430 mp	9.4	−4.7	2.9	3.6
25	22	Formaldehyde (37% basis)	5.89	5.61	5,885 mp	5,606 mp	5.0	−3.6	0.6	0.8
26	26	Toluene[j]	5.82	5.07	802 mg	699 mg	14.7	−4.0	−1.2	−2.2

27	25	Xylene	5.55	5.31	771 mg	738 mg	4.5	-2.6	-13.6	0.7
28	30	Ethylene glycol	4.76	4.18	4,759 mp	4,178 mp	13.9	2.8	-13.4	3.6
29	28	p-Xylene	4.67	4.78	4,669 mp	4,779 mp	-2.3	0.6	12.1	4.8
30	29	Ammonium sulfate	4.17	4.19	2,086 tt	2,093 tt	-0.3	-0.9	1.3	0.4
31	31	Cumene	3.70	3.35	3,695 mp	3,345 mp	10.5	2.2	-10.9	3.1
32	32	Acetic acid	2.93	2.90	2,931 mp	2,897 mp	1.2	-0.9	10.6	1.8
33	34	Phenol	2.92	2.78	2,921 mp	2,777 mp	5.2	2.5	-3.9	3.3
34	39	Butadiene	2.59	2.34	2,593 mp	2,340 mp	10.8	-2.8	-4.6	-3.0
35	35	Carbon black	2.59	2.57	2,585 mp	2,571 mp	0.5	-1.1	-11.1	-1.6
36	33	Potash (K_2O basis)	2.58	2.84	1,169 tmt	1,288 tmt	-9.2	-11.5	-17.6	-6.0
37	36	Aluminum sulfate	2.52	2.54	1,258 tt	1,268 tt	-0.8	-0.6	12.3	0.5
38	37	Propylene oxide	2.48	2.40	2,480 mp	2,400 mp	3.3	6.1	34.9	3.1
39	38	Acrylonitrile	2.31	2.35	2,314 mp	2,346 mp	-1.4	3.0	5.7	4.3
40	40	Vinyl acetate	2.25	2.11	2,249 mp	2,112 mp	6.5	3.0	4.3	4.3
41	41	Methyl *tert*-butyl ether	2.24	1.89	2,237 mp	1,891 mp	18.3	24.1	37.4	nm
42	44	Cyclohexane	2.07	1.66	2,071 mp	1,657 mp	25.0	2.6	-16.9	-0.5
43	42	Acetone	1.94	1.79	1,936 mp	1,788 mp	8.3	-2.0	-4.0	0.4
44	43	Titanium dioxide	1.83	1.72	917 tt	860 tt	6.6	3.8	3.0	2.5
45	48	Sodium silicate	1.57	1.44	786 tt	721 tt	9.0	0.4	-3.9	0.5
46	46	Calcium chloride	1.56	1.88	780 tt	940 tt	-17.0	-3.2	-10.4	-4.2
47	45	Sodium sulfate	1.55	1.65	775 tt	827 tt	-6.3	-6.9	-5.2	-4.5
48	47	Adipic acid	1.52	1.45	1,522 mp	1,453 mp	4.7	2.0	4.5	0.0
49	50	Isopropyl alcohol	1.28	1.24	1,275 mp	1,235 mp	3.2	-5.2	-11.4	-4.1
50	49	Sodium tripolyphosphate	1.27	1.25	634 tt	625 tt	1.4	-1.8	-7.4	-1.3
		Total organics	188.00	172.16			9.2%	1.8%	-1.9%	2.9%
		Total inorganics	350.60	369.56			-5.1%	-1.5%	-2.0%	0.8%
		Grand total	538.60	541.72			-0.6%	-0.5%	-2.0%	1.5%

Source: *Chemical Engineering News* 1987e. Reprinted with permission from Chemical and Engineering News, June 8, 1987. © 1987, American Chemical Society.
b. tt = thousands of tons, bcf = billions of cubic feet, mp = millions of pounds, mg = millions of gallons, tmt = thousands of metric tons.

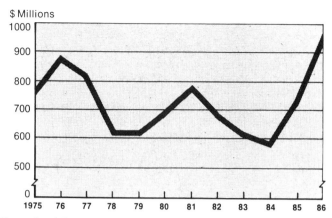

Source: *Chemical and Engineering News* 1987*a*. Reprinted with permission from *Chemical and Engineering News*, February 16, 1987. © 1987, American Chemical Society.

Figure 7-11. Chemical firms spending on pollution abatement.

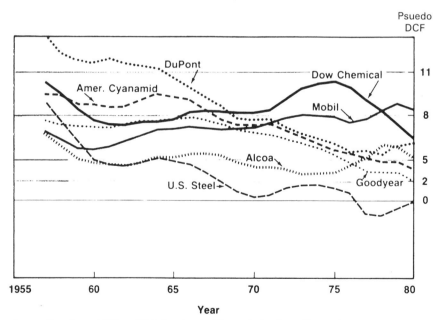

Source: Dressler and Shinar 1986. Reprinted with permission from *Chemtech*, January, 1986, pp. 30–41. © 1986, American Chemical Society.

Figure 7-12. Estimated Psudo DCF returns for various companies.

Table 7-6
Total Polymers Production in the United States

	Billions of lb		Common Units					1985–86	Average Annual Growth		
	1986	1985	1986	1985	1984	1981	1976		1984–85	1981–86	1976–86
PLASTICS			millions of lb								
Thermosetting resins	5.83	5.63	5,834	5,631	5,549	5,010	3,633	3.6%	1.5%	3.1%	4.9%
Phenol, other tar acid resins	2.72	2.62	2,721	2,621	2,502	2,333	1,340	3.8	4.8	3.1	7.3
Urea resins	1.27	1.21	1,271	1,210	1,199	1,165	821	5.0	0.9	1.8	4.5
Polyesters (unsaturated)	1.27	1.22	1,271	1,223	1,232	997	1,042	3.9	−0.7	5.0	2.0
Epoxies (unmodified)	0.40	0.39	398	385	406	336	244	3.4	−5.2	3.4	5.0
Melamine resins	0.17	0.19	173	192	210	179	186	−9.9	−8.6	−0.7	−0.7
Thermoplastic resins	33.57	31.53	33,569	31,525	30,036	25,671	19,481	6.5%	5.0%	5.5%	5.6%
Low-density polyethylene	8.89	8.89	8,888	8,889	8,413	7,693	5,813	0.0	5.7	2.9	4.3
PVC and copolymers	7.26	6.77	7,256	6,772	6,760	5,707	4,716	7.1	0.2	4.9	4.4
High-density polyethylene	7.17	6.67	7,171	6,671	6,085	4,695	3,125	7.5	9.6	8.8	8.7
Polystyrene	4.44	4.05	4,442	4,054	3,838	3,621	3,195	9.6	5.6	4.2	3.4
Polypropylene	5.81	5.14	5,812	5,139	4,940	3,955	2,632	13.1	4.0	8.0	8.2
Total	39.40	37.16	39,403	37,156	35,585	30,681	23,114	6.0%	4.4%	5.1%	5.5%
SYNTHETIC FIBERS			millions of lb								
Cellulosics	0.62	0.56	619	558	589	770	791	10.9%	−5.3%	−4.3%	−2.4%
Rayon	0.40	0.35	404	353	390	509	493	14.4	−9.5	−4.5	−2.0
Acetate	0.22	0.21	215	205	199	261	298	4.9	3.0	−3.8	−3.2
Noncellulosics	7.82	7.56	7,815	7,564	7,473	7,982	6,615	3.3%	1.2%	−0.4%	1.7%
Polyester	3.30	3.34	3,304	3,341	3,392	4,173	3,341	−1.1	−1.5	−4.6	−0.1
Nylon	2.52	2.34	2,515	2,343	2,412	2,333	2,076	7.3	−2.9	1.5	1.9
Olefin	1.38	1.25	1,380	1,249	998	785	577	10.5	25.2	11.9	9.1
Acrylic	0.62	0.63	616	631	671	691	621	− 2.4	−6.0	−2.3	−0.1
Total	8.43	8.12	8,434	8,122	8,062	8,752	7,406	3.8%	0.7%	−0.7%	1.3%
SYNTHETIC RUBBER			thousands of metric tons								
Styrene-butadiene	1.75	1.62	792	735	958	1,032	1,333	7.8	−23.3	−5.2	−5.1
Polybutadiene	0.72	0.73	325	330	359	342	352	−1.5	−8.1	−1.0	−0.8
Ethylene-propylene	0.51	0.47	230	215	215	178	130	7.0	0.0	5.3	5.9
Nitrile	0.13	0.12	58	53	67	66	73	9.4	−20.9	−2.6	−2.3
Other	1.28	1.11	582	505	556	404	416	15.2	−9.2	7.6	3.4
Total	4.38	4.05	1,987	1,838	2,155	2,022	2,304	8.1%	−14.7%	−0.3%	−1.5%

Source: *Chemical and Engineering News* 1987e. Reprinted with permission from *Chemical and Engineering News*, June 8, 1987. © 1987 American Chemical Society.

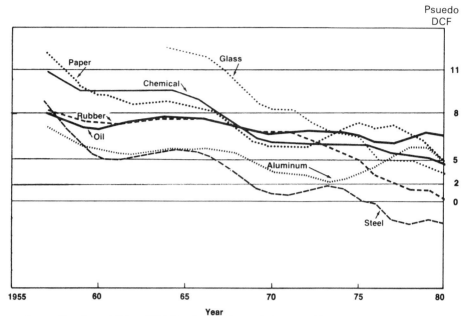

Psuedo
DCF

Source: Dressler and Shinar 1986. Reprinted with permission from *Chemtech*, January, 1986, pp. 30–41. © 1986, American Chemical Society.

Figure 7-13. Estimated Psuedo DCF returns for selected industries.

the energy intensive chemical industry, since the original design of most chemical processes was based upon very inexpensive energy. It resulted in a massive shift in the economic balance between operating cost (energy), and capital cost (heat exchangers, etc.). The resultant need for more energy efficient new processes and equipment is still present. Shortly after the energy profit squeeze was first felt, the U. S. environmentalists also began a massive crusade, first with air and water pollution legislation, and more recently focusing on hazardous materials and hazardous wastes. This has resulted in governmental laws and ''enforcement'' supervision that appears to be far more punitive in its effect than helpful in attempting to solve problems. Other countries have become equally concerned about the environment, but in general their governments have worked with industry in a cooperating rather than a punishing mode which has been both more effective and far less damaging to their industries.

As a final and more recent factor reducing the rate of return and the competitiveness of the U. S. chemical industry certain chemicals began being sold on the world market at relatively low costs because of either: (a) state subsidies for their production, or (b) very low cost raw materials. An example of the former are the government controlled companies producing potash (potassium chloride) in Israel, Russia, and Canada. All three countries would like to sell their product at the highest price possible (i.e., the ''world price''), but particularly in the case of Israel, they must sell it at any price to keep this large segment of their industrial economy employed, so they undercut price if they must, and U. S. private industry either must meet the reduced price or sell what they can (a reduced amount) at the higher price. Both actions hurt profitability. Russia can control the price either for political purposes or to obtain much needed foreign currency, and Canada has very cheap potash, but can also control the price as a matter of policy.

A more serious problem which is only now beginning to have an important impact, and which will become much more dominant in the future, is chemical production from areas where there are very inexpensive natural resources. A good example of this is the large number of basic chemicals that can be made from natural gas and cheap energy. When crude oil is produced, natural gas is always a coproduct and sometimes the major product. In remote or undeveloped areas this gas must be flared (burned) merely to get rid of it so that the oil can be produced. For some time less developed areas like Alaska, Mexico, and Venezuela have been producing ammonia from this gas. The cost of the gas is only that of the gathering equipment, perhaps $0.05 to $0.50 per Mscf, while competitive plants in the United States have to pay $1.50–3.30 per Mscf (in 1987). This gives the low cost countries a $45–100 per ton of ammonia manufacturing cost advantage on a $70–140 per ton commodity (again, in 1987), which is far greater than the freight cost to ship the ammonia to the United States. This means that U. S. production must be displaced or sold at very low margins to be competitive. Ammonia is now also supplied by the Middle East, Russia, Canada, and other low priced gas countries.

This same trend is steadily growing with a wide variety of basic petrochemicals such as vinyl chloride, ethylene, methanol, and many others which have a similar low cost gas or energy base. Saudi Arabia has rapidly built plants to become a world leader with such products, and a number of other oil rich countries are adding to the flow (Block 1987). These countries are also doing more complete refining of their crude oil, which provides a further diversification to their upgraded fuel and chemical production through intermediates such as benzene, naphtha, and separate hydrocarbon cuts. The United States will definitely have to adjust its operations to consider these chemical sources in the future.

A final (and lesser) consideration, but also important for the economics of the chemical industry in the future, is the cost of utilities. Despite the abundance of intrinsically cheap utility fuels, oil, gas, coal and lignite, in the United States the problems with acid rain, sulfur dioxide, and NO_x emissions, and perhaps more pressure in the future on the ''greenhouse'' effect caused by CO_2, have resulted in U. S. electric prices rising far faster than those of our worldwide competitors. The obvious answer for most of the industrialized world is nuclear power (see Figure 7-14), but the U. S. antinuclear lobbying groups have caused the government to make building and operating nuclear power plants in the United States extremely time consuming and expensive, which has stopped

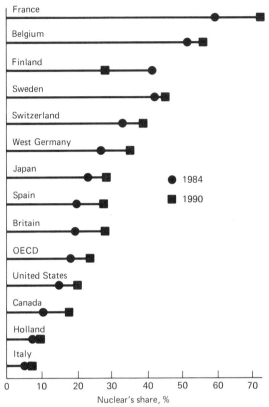

Source: Nuclear Energy Agency; Adapted from The *Economist*, May 24, 1986. Copyright 1986. *The Economist*. Reprinted with permission from the *New York Times* Syndication Sales Corporation.

Figure 7-14. Nuclear power as a percent of each country's total electric generation

almost all new construction. Because of this it has been predicted that by the year 2000 electric rates in the United States will be about double that of much of the rest of the world, and the U. S. chemical industry costs will be further increased.

Natural gas has a similar problem but for different reasons. First, there is considerable competition for this premium fuel, and even though it was plentiful in the mid-1980s, it should be in shorter supply within 5–10 years. Also, the natural gas distributing companies are private, but they are heavily regulated by the government, pressured by environmentalists, and basically work on a "cost plus" basis. There is little incentive for efficiency of operation, and many times no competition, so within reason the more it costs the gas companies to operate, the more profit they make. The regulatory agencies generally are not skilled in the concepts of processing technology or efficiency, so gas prices are now high in the United States by worldwide standards, and will be going higher.

All of these factors have contributed toward placing the U. S. chemical industry in the low profit and highly competitive position that it is now facing. We shall examine what the industry is doing about these problems in the next section.

PRESENT AND FUTURE CHANGES

As discussed above, by the mid-1980s the chemical industry found itself to be a mature industry with a quite low rate of return on its investments. The U. S. market for most of its products was growing at a very slow rate, some export markets (such as heavy-debtor nations) (*Chemical and Engineering News* 1987 *c*) were disappearing, and there was increasingly severe foreign competition. Many foreign companies had equal or better processing technology, equal or newer plants, and generally lower operating costs. Foreign competitors also generally had their government's assistance or at least general support, while in the United States the government often took a more adversary position against industry in environmental, monopoly, and many matters relating to trade, financing, and competition. This has resulted in the top management of many companies spending most of their time, and often being selected because of their skill in dealing with the government, handling legal matters, and satisfying the financial community with short-term profits and growth. All of these factors would tend to indicate that the U. S. chemical industry should continue to decline and eventually follow the exact pattern of the steel industry. What actually is happening to counteract this pattern, to change it, or perhaps to eventually succumb to it?

Cutting Costs

As the above situation became more clear to management in the mid-1980s, essentially every chemical process industries company went through a major cost cutting and efficiency program. A strong catalyst for this movement, however, and one of the reasons that it occurred so universally and rapidly, actually was from quite a different motivation. A small oil company, Messa Petroleum, and its president, T. Boone Pickens, came to the conclusion that most of the major oil companies were poorly managed and making such low returns on their investments that their stockholders were not being adequately compensated. Also, the total value of all of their stock was often less than the value of the assets of the companies. He believed that because of this his company could either take over companies more than 10 times his size, or at least greatly profit in the attempt by buying some of their stock cheaply (i.e., 5–15% of the total) at its original price, offering much more for the rest of the stock to purchase the company, and if need be, selling the original stock after his takeover attempt had failed. He did exactly that with a number of companies in succession, and in every case it worked perfectly. Most of the larger oil companies immediately gave in and rushed into the arms of a ''white knight'' (another company) to avoid him, while a few encumbered themselves with massive debt and sold off highly profitable divisions to make themselves less attractive. In either case they or their white knights repurchased his stock at an attractive price.

Since that time other ''raiders'' have emulated Picken's technique in many industrial and commercial fields, generally with success. This naturally created some panic among many in management (since in a takeover or merger they would probably loose their jobs), and most companies immediately instituted major cost cutting and efficiency programs. This efficiency effort has been very beneficial to the industry, but most raiding now appears to have been taken to the point of being relatively harmful to the industry's long-term well-being. In T. Boone Picken's case he appeared to be fully prepared to operate any company that he made a bid for, and with his innovative ideas, deep knowledge of the industry, and ability to promote efficiency, he probably would have done so better than the existing management. Many of the other raiders, however, are either strictly financial people or groups, or nondevelopment oriented companies, and when they gain control of a company they (at least partly) liquidate, since the value of the divisions is usually greater than the company as a whole, or operate it strictly for the short-term gain. Research and development and new capital investment on new or improved products or equipment have little part in their plans. With the extraordinary skill that will be required to make the chemical industry remain vital and dominant, there would appear to be little or no possibility that the financial groups or other raiders can maintain this vigor. Once strong technical companies were acquired, like Stauffer Chemical, Mallinckrodt, and Borg Warner, they would appear to be in a very weakened position. Nevertheless, the efficiency moves that the raiders indirectly helped to initiate for the rest of the industry have been very important to its strengthened future by resulting in a much more diligent management and a reduced, more efficient operation.

Staff Reductions. The principal cost cutting technique used during this period has been through staff reduction, also called ''downsizing,'' which has often resulted in massive employee layoffs. (See Figure 7-15.) Sometimes companies instead have practiced the ''golden handshake'' technique, where bonuses were given for early retirement. Headquarter staffs were often also severely trimmed to reduce the overhead costs (general and administrative, G & A), and as a result the productivity for the CPI rose (Figure 7-15) and the unit labor costs fell, as seen in Figure 7-16. This in turn increased profits

% year-to-year change

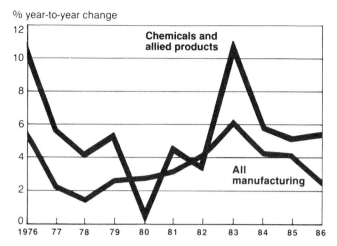

Note: Productivity is output per workhour, calculated by dividing indexes for production by indexes for workhours of production employees. Unit labor costs are labor costs per unit of output, calculated by dividing indexes for wages by indexes for output per workhour. **Sources:** Department of Labor, Federal Reserve Board, C&EN calculations

Source: Storck 1987. Reprinted with permission from *Chemical and Engineering News*, March 9, 1987. © 1987, American Chemical Society.

Figure 7-15. Chemical productivity.

% year-to-year change

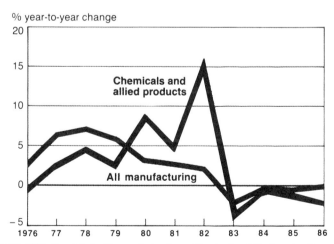

Source: Storck 1987. Reprinted with permission from *Chemical and Engineering News*, March 9, 1987. © 1987, American Chemical Society.

Figure 7-16. Unit labor costs.

considerably, which was greeted so enthusiastically by the financial community that the stock prices (Figure 7-17) and price per earnings (dividend) ratios rose to all time highs in mid-1987.

An example of such staff reductions is Monsanto, who in 1985 announced a major transition program to its employees. It was aimed at controlling costs and increasing productivity by decentralizing staff groups by moving them under line responsibility, and streamlining management. They also initiated a program of voluntary early retirement, and of the 3,880 employees eligible, 2,340 accepted. The du Pont company also went from a 141,000-person work force in 1982 to 112,000 in 1987. Exxon reduced its employees by 17%. Its U. S. work force of 40,000 was cut by 7,000; 6,200 through voluntary separation under a special early retirement or termination inducement package, and another 800 from layoffs. In the 1970s, there were more than 2,000 people at headquarters; in 1987 there were 325. At the same time, Exxon consolidated several regional operating organizations, and eliminated various headquarters divisions.

There is little doubt that reduced manpower has been beneficial for most companies, especially where bureaucratic-type overhead could be reduced (some feel that most U. S. companies are still overstaffed by 15–20% in their administrative offices), or obvious worker related costs such as union "feather bedding" or jurisdictional restrictions could be eliminated. However, in many cases engineering, technical, or research and development staffs were the first to be let go, and beyond the point of upgrading or removing nonproductive groups, this could be very damaging to the industry's long-term position. Even though all companies saw their profits rise from staff reduction, for only those who did it carefully and thoughtfully was it a large step toward more efficient and competitive future production. For those who did it more randomly, or focused on technical groups, many will have to quickly undo it, or probably suffer severe future problems.

Such was the case of a large fertilizer complex who were advised by management consultants that they had too many engineers. Most of the engineers were then fired, and almost immediately operating problems began to accumulate to the point that production and profits were severely reduced. When new engineers were hired the plant began to run well again, but in many cases equipment deterioration had occurred, and customers considered the company unreliable and with product quality problems. This case involved a management with unusual naivete or incompetence, but even though the example is exaggerated, the general point is a valid one for the current staff reduction programs.

Divestitures

A second efficiency step that often accompanied staff reduction was the sale or closing down of businesses that were losing money, not making an adequate profit, or were felt to not fit in with the new direction of the company. This

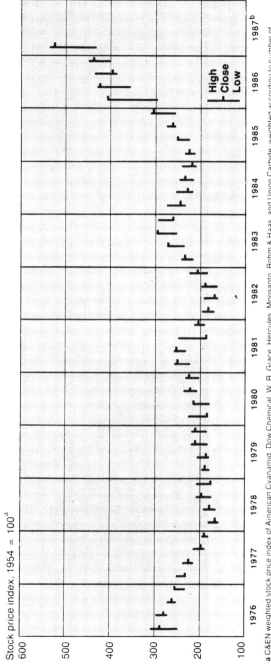

Figure 7-17. Prices of chemical stocks.

a C&EN weighted stock price index of American Cyanamid, Dow Chemical, W. R. Grace, Hercules, Monsanto, Rohm & Haas, and Union Carbide; weighted according to number of shares outstanding for each company. b First quarter

Source: *Chemical and Engineering News* 1987*e*. Reprinted with permission from *Chemical and Engineering News*, June 8, 1987. © 1987, American Chemical Society.

practice has always been an option of management, but in the mid-80s it became almost an obsession, and very large numbers of divisions from many companies were divested. In general there was an active interest from companies desiring to purchase the unwanted groups, since often that manufacturing facility or division might fit better with another company where there would be related, final product, or raw material production facilities to make a more integrated product line. Also, other companies, such as foreign operators, were actively looking for an entry into the United States or that product line, or to obtain that know-how, or more modern facilities. For all of these reasons and many others there usually have been customers even for divested plants or divisions that have not made a profit, and there has been lively bidding on the more attractive operations.

Leveraged Buy-outs. When corporate customers were not available, however, or would not offer a reasonable price, the parent company had one further divestiture profit opportunity, the "leveraged buy-out," usually through an "ESOP" (employee stock ownership plan). This type of purchase contract was originally designed to sell an operation to the employees, and the tax laws were changed to provide an incentive for financial groups to loan the funds, such as the lender only having to pay taxes on half of the principal loaned, and some of the debt repayment to be considered as an operating cost. Only a small amount of investor cash was required (i.e., usually 0 to 10%), so the purchase was highly "leveraged." Since a currently operated business was involved, the purchase price could be set fairly high, making it attractive to the seller, and the financial group could examine prior and anticipated profits to determine if enough cash flow (after tax profits plus depreciation) could be generated to pay the interest and repay the debt. If this wasn't adequate, the new owners could consider selling off some of the assets to see if they could then make the required profit. Usually a purchase price and loan agreement could be worked out that was attractive to everyone.

As leveraged buy-outs (LBO) became more common they were dominated by financial groups who became the majority owners, although some employees had to be included to satisfy the government's tax requirements. Generally the LBO companies have a dedicated and knowledgeable management, and operate with a very low overhead (McGuire 1987). This often gives them a reasonable profit potential, but since all of their energies are directed toward interest and debt repayment, they must sell their output, no matter what the demand, and cost cutting and sales discounts often result. This makes them difficult competitors, but also weakens them. In other cases they have "gone public" (i.e., sold stock to the general public) in order to raise money; some for the company's operation, perhaps more for the LBO organizer's profit (Goldhaum and Lune 1987). It is too early to know the percentages, but with an economic downturn many would fail (Loomis 1987), while with increased product demand and prices most should slowly become "normal" companies, developing new products and processes, and reinvesting in enough research and development to become stable competitors.

An example of a large financial group's leveraged buy-out is the Sterling Group, Inc. They formed Cain Chemical, Inc. to operate seven LBO petrochemical plants, as shown in Table 7-7. Cain Chemical emerged as a billion dollar company with 1,500 employees, and was almost wholly owned by the financial group (Sterling), management, and employees. Each of the plants produce bulk, highly competitive, slow growth, and usually oversupplied commodities with a history of up and down, and generally low profits. None of the original companies were well integrated from raw materials through to finished products, while Cain Chemical consumes about 80% of its ethylene in its own derivatives plants. It became a relatively well-balanced, integrated olefins business, supported by a low-overhead management, but with limited technical service capability. The divesting companies had fairly modern plants, with some expansion capability, which gave Cain a reasonable (but certainly not guaranteed) chance for success. Sterling sold the operation to the Occidental Petroleum Corp. in 1988 after a turn in the economy caused the market demand for their products to rise dramatically (Reisch 1988).

Mergers and Acquisitions

One of the major avenues for corporate growth has always been through purchasing other companies, or by mergers or acquisitions. Du Pont and Allied became dominant chemical producers in the 1920s by acquiring a number of large basic operations. In a similar manner in the 1970s and 1980s some of the current major chemical manufacturers also obtained their present size by mergers and acquisitions. As noted previously, in the mid-1980s acquisitions (just as divestitures) became almost epidemic, as indicated in Table 7-8 (*Chemical Week* 1987*a*; Goldbaum 1987, etc.). Typical examples were:

Some groups were most active in divesting, such as Union Carbide who was struggling under the financial uncertainty of the Bopal disaster (poisonous gas escape, killing thousands in India) and the heavy debt to fight off GAF Corp.'s take over attempt. They sold their consumer products business to First Brands, a leveraged-buyout (LBO) company for $800 million; the Eveready Battery business to Ralston Purina for $1.4 billion; and their agrichemicals business to France's Rhone-Poulenc for $575 million. Also, in a $340.5 million lease-back deal Carbide sold its headquarters complex at Danbury, Conn. Several companies that were involved in previous major acquisitions were later acquired, such as Chesebrough-Pond's, which in 1985 took over Stauffer Chemical, and was acquired itself in 1986 by Unilever, the British-Dutch consumer products giant for $3.1 billion. Unilever then sold the Stauffer operations (in 1987) to ICI for

Table 7-7
Cain Chemical Company Acquisitions

Divesting Company	Facility			
	Location	Operation	Product	Capacity
Du Pont	Chocolate Bayou, Alvin, TX	Ethylene cracker	Ethylene Propylene Butadiene Benzene	1 billion lb 650 million lb 135 million lb 90 million gal
	Mategora, TX (Bay City)	High density poly- ethylene	HDPE	460 million lb
	Orange, TX	High density poly- ethylene	HDPE	(205) 200 million lb
	Victoria, TX	High density poly- ethylene	HDPE	250 million lb
	Ponca City, OK	High density poly- ethylene	HDPE	Pilot plant
Corpus Christie Petrochemical	Corpus Christi, TX	Ethylene cracker	Ethylene Propylene Butadiene Benzene	1.4 billion lb 560 million lb 200 million lb[a] 60 million gal
ICI America	Bayport, TX	Ethylene glycol	Ethylene glycol	520 million lb
		Ethylene oxide	Glycol ethers, acetates and amines	150 million lb (450)
PPG Industries (Interest in PPG/Du Pont plant)	Beaumont, TX	Ethylene glycol	Ethylene glycol	620 million lb (570)

Summary:	% of U. S. Production	U. S. Rank		
	7	6	Ethylene	2.4 billion lb
	7	4	Benzene	150 million
	13	2	Ethylene glycol	gal 1.09 million lb
			Propylene	1.2 billion lb
			Butadiene	300 million lb
			Benzene	150 million
			Glycol and derivatives	gal 900 million

[a]Being expanded and updated. ( ) alternate estimate
Source: *Chemical Week* 1987c; Excerpted by special permission from *Chemical Week*, June 10, 1987. Copyright © 1987, by McGraw-Hill Inc., New York.

Table 7-8
Chemical Industry Acquisitions

Year	No. of Transactions	Value, $ Million
1980	79	947
1981	89	4,197
1982	80	2,307
1983	71	1,446
1984	117	2,820
1985	139	12,298
1986	700	33,000 (Chemical Processing 1987)
1987 (est.)	—	47,000 (L.A. Times 1987)

Source: Nordhoy 1987. Chemical Engineering Progress, February, 1987, 9–13, Reproduced by permission of the American Institute of Chemical Engineers.

$1.7 billion (Reisch 1987), who in turn sold Stauffer's specialty chemicals division to Akzo America for $625 million, and the basic chemicals groups to Rhone-Poulenc for $522 million. Other business areas were later sold.

Occidental Petroleum's chemical division purchased various commodity-type chemicals in the chlorine and vinyl chloride field, (Brockington 1987), including Firestone's PVC operations, Diamond Shamrock's chemical plants, Shell Oil's vinyl chloride monomer business, Du Pont's Corpus Christi chlor-alkali production, and Tenneco's polyvinyl chloride operations.

Most mergers, however, were in specialty chemicals and pharmaceuticals, such as in 1986 International Minerals buying Mallinckrodt for $675 million from Avon who had acquired it only four years earlier. Eli Lilly bought Hybritech for $300 million, and Bristol-Meyers paid $294 million for Genetic Systems. Key Pharmaceuticals, with its new drug delivery systems, became a part of Schering-Plough in an $800 million acquisition. Revlon sold its USV Pharmaceutical and Armour Pharmaceutical operations to the Rorer Group for $690 million. Britain's Boots also increased its stake in the U. S. pharmaceutical market with its $555 million purchase of Baxter Travenol's Flint Laboratories, as did Du Pont by the purchase of another Baxter Travenol unit, American Critical Care. In specialties and biotech, Eli Lilly purchased Hybri-tech for $300 million, and Bristol Meyers purchased Genetic Systems for $294 million.

These limited examples of major corporate restructurings were typical of what was happening to the U. S. chemical industry during the 1980s with companies selling large and small units which no longer meshed with their strategies, or to raise capital, and which other companies were eager to buy. Non-U. S. companies did a great deal of the purchasing in order to strengthen their U. S. holdings and improve their global marketing strategies (Chemical and Engineering News 1987b). Table 7-9 shows the beginning of this foreign purchase trend, which grew much stronger in the following years because of

Table 7-9
Number of U. S. Versus Foreign Acquirers, Chemicals and Allied Products

Acquisitions	1981	1982	1983	1984	1985	Total 1981–85	Cost, Billion $ 1986	1987
U. S. abroad	12	34	48	59	83	236	—	—
Foreign in United States	18	43	25	76	121	283	22.9	26[a]
U. S. excess (deficit)	(6)	(9)	23	(17)	(38)	(47)	—	—

Source: Nordhoy 1987. Chemical Engineering Progress, February 1987, 9–13. Reproduced by permission of the American Institute of Chemical Engineers.
[a]About 70% of all 1987 CPI acquisitions (Kiesche 1988).

this divestiture "fad," and the sharply falling dollar (value compared to foreign currencies). Gaining access to valuable technologies also proved to be a key incentive to acquisitions, especially in the most popular areas of acquisition, downstream products, specialty chemicals, pharmaceuticals and other health care items, and fabricated plastics and rubber.

There are obviously many good reasons for acquisitions and divestitures, and just as obviously many of them have been of considerable advantage to both the acquiring and acquired company. When a strong chemical company acquires another company it is probable that they will build upon the capability and expertise acquired, add more capital and research and development support, and both groups will prosper. Unfortunately, however, many (and perhaps even most) acquisitions do not work out so well, and are not that beneficial to the acquiring or acquired company, or the industry as a whole. Acquisitions have at least partly become popular because it is much easier for management to spend their time searching for and acquiring other companies than constructively building or improving their own. Acquisitions provide immediate entry into fields with considerable gain in sales and income which are very visible to the financial community, while the possible poor performance of their basic operations, debt, or stockholder dilution incurred from the purchase or merger are far less noticeable.

It has become almost a rule with few exceptions that new discoveries, innovations, or improvements are first commercialized in the United States by smaller, better managed companies, and then when the technology is more visible and widely accepted, larger companies acquire the smaller ones to obtain the new technology, or are themselves willing to build the "second generation" plants. This is comparatively safe and effective for the large company managers, but has the following disadvantages: (a) generally only smaller or less expensive new developments can be executed this way because of the limited resources of the smaller companies, and (b) foreign competitors can introduce the new technology when the small U. S. companies do, or they can recognize its virtues sooner, and be far ahead of the large U. S. companies. An opposite problem occurs when technology is developed by the larger U. S. companies that are so conservative and concerned with mergers that the developments are not utilized themselves, but licensed and left for foreign or smaller U. S. companies to commercialize, if at all.

A striking example of this is in the U. S. steel industry where company laboratories, along with university cooperative effort, developed over the years many major potential cost improving or higher profit technical innovations such as the basic oxygen injected furnace, electric arc furnaces, continuous casting and rolling, hydrogen reduction of ore, super alloys, and so on. However, no large steel company would commercialize any of these major developments (Ross 1987), so foreign companies did. This resulted in greatly increased competition, and when the big U. S. companies finally did start to slowly modernize it was much too little and perhaps too late (Szekely 1987). Steel company management also preferred to go the acquisition route rather than invest in new technology, with several weak companies merging with several others to form equally weak new companies. USX (U.S. Steel) appears to have merged itself partly out of the steel business rather than attempt to solve its industry's problems. On the other hand, many of the smaller steel companies who did modernize (mainly at first with small electric furnaces, continuous casting and reduced labor, production, and management overhead costs) have been very successful (*Fortune* 1987).

Whether the chemical industry will follow the major steel companies' nondevelopment pattern as well as drain much of their energies playing merger musical chair, or will be truly strengthened by the recent efficiency moves and mergers is not yet known. However, the substitution of mergers for sound new technical development appears to be quite common. As an example, one large firm in the field of electro- and chlorine chemistry was among the leaders in developing the revolutionary new dimensionally stabilized electrodes and polymer (cell) membranes. In fairly typical fashion, however, when the firm built a new chlorine-caustic plant it did not use this technology and stayed with improved, but old designs. The firm had also developed better vinyl chloride technology, but did not modernize or build a new plant. Later the firm acquired companies with both operations, stating the purchase of more modern plants as one of its main reasons for the acquisitions. The mergers will help the firm to catch up, but do not give much confidence that the firm will be ready with the next generation of improvements, or can even currently be competitive with more progressive companies. Mergers would appear to be a poor substitute for internal development if management pursues strictly one path or the other, or has no intention of capitalizing on new technical developments.

As a second factor, acquisition can often be a more risky route in reshaping a company than is first envisioned (Berney 1987). Frequently mismatches of operational methods (or "cultures") (Lefkoe 1987), losses of key personnel, a zeal for change, or a conservative, inattentive, or inept new management can destroy the anticipated profitability all too soon. A typical, hypothetical history of mergers is shown in Figure 7-18. In it a profitable company is acquired in year 0. After one year the new owner, confident now that he understands the business, reorganizes and installs a new general manager. Profits promptly begin to fall, but eventually bottom out around year 3, and remain low. Meanwhile the acquirer changes general managers about every 18 months. Finally, after five years and three general managers, one of two things occur: (1) the division is divested or even abandoned, or (2) after a second reorganization, profitability gradually recovers to an acceptable level, and the unit is retained as a division of the acquiring company. This later case only happened in from 13 to 83% (average about 50%) of the U. S. mergers during the 1970s according to a Harvard Business School study (Cory 1987). In a later study (*Fortune* 1987) it was noted that the profitability of most companies deteriorated significantly, and that 90% of mergers never live up to expectations.

A final example of potential problems with mergers or "raiding" is where the vigor of a dynamic company may be destroyed by the merger. An example of this may be the Borg-Warner's plastics division (Flanigan 1987). Building on technology learned in World War II when it made synthetic rubber insulation for radar wiring, it had developed into a technological and industry leader. It

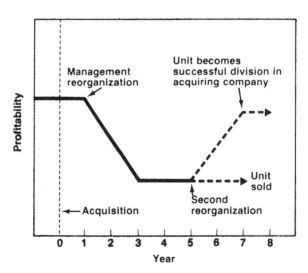

Source: Kline 1986. Reprinted with permission from *Chemtech*, August 1986, pp. 478–483. © 1986, American Chemical Society.

Figure 7-18. Typical performance curve of mergered companies.

dominated the market for ABS plastics (acrylonitrile–butadiene–styrene) used for telephones, automobile bumpers, etc., was the world's largest producer, and had 30% of the market. It was a highly profitable business, accounting for less than one-third of Borg-Warner's $3.4 billion in revenue in 1986, but contributing about 60% of the total profit. The ABS resins accounted for $754 million of the plastics group's 1986 sales of $1.042 billion.

The company had also successfully developed a low-cost plastic that is capable of withstanding extreme heat or cold, and therefore was a candidate to replace steel in large parts of automobile bodies. If successful in preliminary tests in Detroit, the development could mean a big step forward for American plastics technology and for Borg-Warner. However, such breakthroughs don't come easily or cheaply. To get to this development, Borg-Warner started thinking about plastic cars in the 1960s, and built 10 prototypes. Years of expensive research chemistry stood behind the new product and an investment of $91 million in 1986, with much more ahead. The company would spend $200 million on a new plant if the car–plastic test was successful.

The question is can it do as much after being taken over in 1987, especially considering the $4 billion debt incurred to avoid the merger? When the debt burden is so high companies find ways to cut capital spending that are not immediately visible. The penalty for reduced spending today shows up in an uncompetitive business five years from now. The take-over action has probably resulted in such high fees, costs, and debt burden, plus probably an inferior (financial organization) management, that the company may never again be the dynamic force that it once was. This may remain the case even with GE's later purchase of the plastics division from the financial group.

In summary, the extensive merger activity of the mid-1980s would appear to have potentially offered some benefit to the U. S. CPI, but it has resulted in much more foreign control and financial company management, and all too often appears to have been done as the path of least resistance by management rather than face up to the uncertaintes of commercializing new technology and the difficulty of sound long-term cost improvement and growth policies. Merging is simple, easy to analyze, sometimes spectacular, and the results are instantly placed on the company books. Obviously it can do some companies a lot of good, but for the industry as a whole it would appear to provide far more publicity and short-term value than new developments, or basic and long-term strengthening against low profits or foreign competition. With the extensive new foreign ownership (23% in 1986; 26% in 1987) (Kiesche 1988; Reisch 1988) that has resulted the United States has lost ownership in a fair percentage of its chemical industry, and this segment has become the most research, development and growth (new capital addition) oriented part of the U. S. CPI. Perhaps this development will spur on the remaining U. S. companies (*Chemical Engineering and News* 1987g; *Chemical Week* 1987b; Allen 1988).

Strengthening Existing Production

Without question, the soundest method of improving costs and competitiveness is through the strengthening of existing business. This can be done in many ways, such as modernizing, developing new low-cost processes or product innovations, shifting to new and cheaper feedstocks, reducing energy costs, providing cheaper distribution, or acquiring the businesses of weaker suppliers, either to operate or to shut down and reduce industry capacity. The lowest cost producers should be able to make a profit even with heavy competition, but of course it requires efficient operation, reasonably priced raw materials, excellent technical capability, and the support of management to provide low overhead and the needed capital. It also requires that the needed new investments be economically justified, and this raises the question of whether management will accept a modest rate of return on a basic commodity rather than the very high rates that most companies aspire for on new investments. This is a general problem facing U. S. industry, in that almost all companies feel that they must obtain very high rates of return (which usually do not happen) on new capital spending even to the detriment of basic or core businesses where a secure, but lower return would still be considerably better than their present average.

A number of companies have or are using the merger route to strengthen their basic businesses by becoming larger, more modern, and integrating both backward to raw materials and forward to finished products. We have all ready mentioned Cain Chemical in the olefins business and Occidental Chemical building dominance in chlorine and vinyl chloride fields through mergers (Storck 1988). Obviously mergers can build size, which may improve strength and efficiency, but just as with the weak steel companies example, it will not provide benefits into the future unless there is also continued internal development, plant and product improvement, operating efficiency, management support, and wise selection among alternatives in this endeavor. This type of production strengthening implies that weaker operators drop out to leave probably only a handful of competitors for each product, and that capacity be adjusted to the amount that can be sold with high operating rates competitively in the United States and wherever else the pricing is favorable.

The United States has always excelled in this type of technical improvement and product and process innovation, but as with the steel companies, CPI management has become very reluctant to commercialize new innovations or improvements. It is primarily in this area of management courage to utilize new technology (even though the U. S. CPI did spend an estimated $15.6 billion in 1987 on research and development, with about 57% of this funding coming from the chemical industry—see Figure 4-4) that the U. S. CPI will either retain its position or continue to steadily decline.

Differentiation and Segmentation. Another method of improving profitability is to distinguish the company's total offering, product service, and image from those of its competition by finding a unique position through unusual product features, multiple grades tailored to specific uses, superior distribution, good technical service, and rapid response to new needs (Portnoy 1988). The company looks for new customer needs, and prices its products on their value in actual use. They attempt by continuous innovation and the maintenance of close customer ties to achieve higher sales and profits.

Alternatively, a company may find specific products, market segments, or special situations. Many low-volume fine chemicals and specialized grades of specialty and tonnage chemicals are examples of such product "niches" (see Table 7-10). An example of such a market area is water treating chemicals, where municipal and small facilities require a sales force and distribution network quite different from those of the industrial market. Regional segmentation is also common in chemical distribution, where a distributor may elect to serve only one area by specializing in that area's needs. Similar results can be obtained by innovative research and development, higher product purity and uniformity, customer service, and rapid response to customers' needs. These attributes generally allow pricing based upon the customer's need more so than on the basic material sold.

Move into Higher Value-Added Products

Just as with mergers, there is a real stampede in the CPI to diversify, or even change the entire business, into higher value-added products. The chemical industry has led the way to develop new materials and products in the past, and

Table 7-10
Specialty Chemical Niches, 1987

Category	Recent Average Annual Increase, %	Niche	Recent Average Annual Increase, %
Adhesives	7	Structural adhesives	10
Cosmetic additives	4	Ultraviolet absorbers	8
Diagnostics	10	Toxicology	15
Electronics	11	Polymer thick films	25
		Galium arsenide	30
		Specialty gases	20
Industrial coatings	5	High solids	8
		Powder	11
Specialty surfactants	5	Amphoterics	8
Synthetic lubricants	5	High-water-based hydraulics	10
Water management	4	Institutional	9
		Metalworking	8
		Food processing	8

Source: Portnoy 1988. Exerpted by special permission from *Chemical Week*, March 2, 1988. Copyright © 1988, by McGraw-Hill, Inc., New York.

certainly in the future many new products will also produce high profits and large-volume business. Consequently, this rush to value-added products is very sound, and hopefully will further strengthen the chemical industry for the future. However, the current rush into these fields would appear for many companies to be another case of management doing what is popular rather than what is sound, and merely following a "me too" path.

Some of the areas that are being investigated are (Kline 1986):

- Specialty chemicals, the oldest and best known downstream products, comprising some 50 categories, from dyestuffs to photoresists, and being sold to over 30 different industries;
- Pharmaceuticals, agricultural chemicals, and other life science products, also a familiar field to many old-line chemical companies but now in the forefront with biotechnology. This and the next two fields are considered to be "high technology";
- Advanced materials, a newer area, including composites, ceramics, specialty metals, surface-modified materials, super conductivity, and other products with new and special properties; and
- Instruments, systems, services, and devices based on chemical technology—for example, du Pont's automated clinical analyzer. Also, included is technical service, licensing, consulting, management, contract maintenance, and so on.

Entry into these new fields is not easy. All four are already served by established suppliers, companies such as Ciba-Geigy and Rohm & Haas in specialty chemicals; the drug and pesticide companies in the life sciences; Corning, Hercules, Norton, 3M, and aerospace companies in advanced materials; and electronic and mechanical goods companies in instruments, systems, and devices. A few examples of the merger entry route into these areas has previously been discussed; there were literally hundreds during the 1980s (Nordhoy 1987).

Several companies have already stumbled badly in these areas, such as National Lead (NL Industries) which changed so that much of its chemical business was specialty chemicals in the oil field area, just as oil prices dropped precipitously and the industry stayed depressed for many years. Others are entering fields where there is obviously great potential, but good profits are probably many, many years away, and very large research and development and market entry investments will be required. Based upon prior experience with similar highly popular fields (solar energy, fuel cells, insecticides, etc.), few companies have adequate staying power to remain with these new technologies until they are profitable.

Specialty chemicals also represent somewhat of a mixed opportunity. The classic specialty chemical company is now seeing quite heavy competition and reduced opportunity for growth, as shown in Tables 7-11 and 7-12. There is also considerable consolidation among these companies with 1986 alone seeing 47 mergers. However, the old definition that specialty chemicals are sold on the basis of what they do, not what they are, generally low-volume chemicals with a degree of "uniqueness," is changing from "niche," "effect," or "performance" chemicals to more generally, "highly profitable chemicals."

As an example of this trend, Dow Chemical targeted a change in 1978 to have 50% of its business in 1987 be in specialty chemicals. Dow began realigning business, and by 1986 the specialties contributed 49% of sales revenue and 54% of operating income. Included in Dow's rather liberal definition of specialty chemicals are pharmaceuticals, electronics chemicals, high-performance ABS and polycarbonate resins (for automotive uses), polyolefin and polystyrene polymers (for food packaging), styrofoam and chlorinated polyethylene products (for the construction industry), mining chemicals, adhesives and sealants, chemicals for the oil and gas industries, specialized

Table 7-11
Current Growth Prospects for Specialties, 1986

Slow Growers	Medium Growers	Fast Growers
Ag chemicals (3%)	Specialty polymers	Advanced polymer
Cosmetic chemicals (3%)	(8%)	composites (13%)
Foundry chemicals (3%)	Specialty adhesives	Electronic chemicals
Industrial chemicals	and sealants (7%)	(12%)
(3%)	Specialty surfactants	High-performance
Industrial institutional	(6%)	ceramics (11%)
chemicals (3%)	Plastic additives (5%)	Diagnostics chemicals
Petroleum additives (3%)	Reagents (5%)	(10%)
Metal finishing chemicals	Refinery chemicals	
(3%)	(5%)	
Paint and coating	Biocides (4%)	
additives (3%)	Catalysts (4%)	
Printing ink additives	Food additives (4%)	
(3%)	Oil field chemicals	
Textile chemicals (3%)	(4%)	
Explosives (2%)	Paper chemicals (4%)	
Mining chemicals	Photo chemicals	
(−3%)	(4%)	
	Rubber processing	
	chemicals (4%)	
	Specialty lubricants	
	(4%)	
	Water management	
	chemicals (4%)	

Source: Horiszny 1987. Courtesy of *Chemical Business*.

Table 7-12
Estimated Future U. S. Specialty Chemicals Sales Growth

Segment	Sales, Million dollars 1985[a]	1991[b]	Estimated Average Annual Growth, %/yr
Adhesives and sealants	$1,415	$2,245	8
Advanced polymer composites	1,370	4,090	20
Agricultural chemicals	5,840	6,975	3
Biocides	560	710	4
Catalysts	1,520	1,925	4
Cosmetics	555	665	3
Diagnostic chemicals	1,815	3,216	10
Electronic chemicals	3,930	8,625	14
Explosives	430	485	2
Food additives	1,300	1,645	4
Foundry chemicals	135	160	3
Fuel additives	230	290	4
Industrial coatings	4,245	5,070	3
Industrial and institutional cleaners	3,080	3,680	3
Metal finishing chemicals	610	730	3
Mining chemicals	340	285	−3
Oil field chemicals	4,285	5,420	4
Paint and coating additives	895	1,070	3
Paper chemicals	365	460	4
Petroleum additives	1,595	1,905	3
Photographic chemicals	1,165	1,475	4
Plastic additives	1,765	2,365	5
Printing inks	185	220	3
Reagents	410	550	5
Refinery chemicals	285	380	5
Rubber processing chemicals	510	645	4
Specialty lubricants	630	795	4
Specialty polymers	2,115	3,545	9
Specialty surfactants	475	675	6
Textile chemicals	1,320	1,575	3
Water treatment chemicals	1,600	2,025	3
Total	$44,975	$63,900	3

Source: Wilson 1987. Excerpted by special permission from *Chemical Week*, May 13, 1987. Copyright © 1987, by McGraw-Hill, Inc., New York.
a. Actual. b. Estimated.

ceramics and epoxy resins (for aircraft and aerospace applications), powder metallurgy, and a number of household consumer specialties.

In concentrating on certain markets, Dow added special qualities to what others call commodity chemicals, thereby blurring the traditional line between specialty and commodity. They do not plan on abandoning their basic chemicals and plastics business, believing that the largest profit increase over the late 1980s and early 1990s will come in basic chemicals and plastics since restructuring may have removed much of the overcapacity that plagued the industry during the 1970s. It takes about three years to build a basic chemical plant, and in the mid-1980s there were few plans by anybody to build these large chemical plants around the world.

One of the new areas that would appear to offer very large potential rewards is biotechnology, but it can be seen already that legal (even moral), political, and sales problems on top of the technical demands make this a difficult and long-range field. Small companies have prospered on a speculative basis to date, but by the time technical, marketing, permitting (with the government), and environmental pressure group problems are solved, it is quite possible that only large companies and perhaps even consortiums with very large resources and long-term commitment will prosper. As with computer chips and pharmaceuticals, this gives foreign competitors such as Japan with its major governmental sponsorship, consortium action, and willingness to invest heavily in long-term projects before a payout is expected, or other countries where governmental permitting is much easier and faster, a considerable advantage. U. S. companies are leading in this technology and first applications race at present, but the future must be considered as quite uncertain. Hopefully the U. S. government will not be too harsh with their restrictions, costs and time delays, and the U. S. companies involved will be correspondingly more resourceful and persistent.

Other higher value-added fields also offer considerable promise, but they are in production areas that already are well supplied by a number of strong producers. There is always room for new more efficient operators and for new commercial developments, so some of the new entries will prosper. However, for many it will be the "grass being greener on the other side of the fence" situation, and it is doubtful that this is a major panacea for the U. S. chemical industry. This is especially true considering the foreign competition that is also developing in this area. It does not mean that it is not deserving of a major effort, for in many cases it represents the essence of innovation and skillful marketing, but this movement should be balanced with efficiency and excellence in the companies' basic commodities.

Foreign Trade and Production

Many CPI executives feel that U. S. companies will have to undergo a striking change toward a more international outlook (Shamel 1986; Mullen 1987). To profit in the late 1980s and 1990s, some claim that more will have to become lean, global competitors. The modern manager feels the effects of decisions made in other countries very quickly. The CPI is no longer a regional or even a national business, as chemical products are now made all over the world. The new global industry is very competitive, fast moving, and changing.

The U. S. chemical industry has for many years been a world trader, with large export sales, considerable import purchasing, and a steadily growing manufacturing presence overseas. This has considerably added to its strength and vigor, but even this situation is changing. Europe and Japan have seen a decline in the growth rates of their markets and greatly increased competition. A high percentage of their chemical production (i.e., 50% or more) is exported, so they must react. Many of their companies have also announced a change from commodities to high-technology products like biotechnology and specialty chemicals. In Italy, Montedison announced a shift from 20% specialties to 80%. Japan announced an annual investment of $200 million per year in biotechnology, the "next major frontier of the century." Most European and Japanese companies also rely heavily on new technical development, such as West Germany's Bayer, which invested one-third more for research (as a percentage of sales) than did du Pont in 1984. This kind of investment has marked Bayer for decades, and as a result, "40% of its roughly $16 billion in worldwide sales comes from products that didn't exist 15 years ago." The results are a global market, and Bayer's 1984 North American sales were larger than its West German sales. Foreign companies were also issued 46% of all U. S. patents in 1986.

Developing countries, on the other hand, have rushed into commodity or basic chemicals, and mostly for export (*Chemical and Engineering News* 1987*b*). Taiwan, for example, has moved into petrochemicals, South Korea into fertilizers, polypropylene, and so on, and South America into "oil field" chemicals, particularly plastics. Saudi Arabia now produces 4% of the world's ethylene, 3% of the world's ammonia, 6% of the world's methanol, and many other basic commodities (*Chemical Week* 1987). Most of this new production is accomplished with very low-cost raw materials, low-cost labor, and large, modern, and efficient plants.

All of these factors point toward the U. S. CPI needing to develop an even greater global outlook. There are still many new markets in both developed and developing countries that the United States can profitably operate in with intelligent and aggressive salesmanship. This may require more knowledgeable multilingual local representation, perhaps more barter (trade one commodity for another) arrangements, and so forth. It will also definitely require more foreign plants and joint venture relationships to take advantage of highly favorable raw material, labor, energy, or sales opportunities. Care must be taken, however, not to rush excessively into foreign operations for cheaper labor or raw materials when equivalent modern plants in the United States with improved innovations would result in a much sounder long-term position. It will finally require an even closer knowledge of what is happening with foreign chemical production, and an improved product quality and service. The chemical industry has indeed become global in its nature, and the U. S. companies, if they are to compete, must emphasize its most unique skills (that it at least once had): technical

innovations, quality products and service, venturesome and skillful management, and an ability to react reasonably quickly to the changing world situation.
Some U. S. chemical companies already are very skillful world traders, and
hopefully many others will be able to develop these skills.

Diversification

It was noted earlier in this section that very few of the major chemical producers
manufacture chemicals alone. Among the top 50 chemical operators in Table
7-2 there are 11 different industry categories, with the average of the 50 only
having about 54% of their business in chemicals. Many of the companies in
other industries became chemical manufacturers for any of a wide range of
reasons including increased profitability, diversification, their use of the
products, or because they had inexpensive raw materials. The chemical companies added their nonchemical products for similar reasons such as diversification
or because they controlled or needed raw materials, were in related businesses,
or in some cases, just to grow in a hurry. The du Pont white knight take-over
of Conoco was probably an example of the later cause, more than doubling its
size, and making du Pont more of an oil than a chemical company. Only 10 of
the top 50 chemical manufacturers were listed in 1986 as being pure chemical
companies, and this number has been reduced since that time. The trend for
U. S. chemical companies is strongly toward having a diversity of products,
and pure U. S. chemical producers will probably be a rarity in the future.

There is a very positive management rational for this diversification
movement. For many products a period of depressed sales for one product may
not be the same for another. This occurs within the chemical industry itself, for
example, since a period when fibers and their raw materials are selling poorly
or at depressed prices might very well be a period when some of the inorganic
commodities are selling well. For this reason most chemical companies are
reasonably well diversified with respect to different types of chemicals, but the
current trend is far more than that. Companies are also branching into nonchemical products in their desire to smooth out economic cycles of product demand
and profits. For instance, when du Pont's oil business has experienced profit
declines, its chemicals have carried the company, and vice versa.

Diversification, however, has many other advantages such as providing an
opportunity to enhance profits through the control of business areas throughout
their entire cycle. Petrochemical companies are large consumers of hydrocarbon
feed stocks, so the acquisitions of an oil company would not only provide
another business area for profits, but also guarantee feed stocks at a reasonable
price. Several chemical companies have done this. Other chemical companies
have diversified into mining, such as soda ash (Na_2CO_3) production by Allied,
FMC, Stauffer, and Tenneco; potash by PPG and International Minerals; and
agriculture by Quaker Oats (corn cobs to produce furfural). There are many

such examples of raw materials processing in a different industry to produce chemicals.

A similar branching out can result from companies that produce intermediate materials to carry them one step or more to the ultimate consumer. Examples were Union Carbide's production of carbon for electrodes, and electrolyte chemicals, and then combining them into the final consumer product, batteries in their former Eveready division. In a similar manner they marketed some of their glycols directly to the consumer as Prestone antifreeze liquids. Until the early 1980s not too many companies did such marketing directly with the public other than Union Carbide as just noted, and du Pont with paints and lacquers. However, this tendency is being strongly reversed with the industries' new interest in higher value-added products, and many more companies are diversifying in this manner.

A general case study of this combined with innovative marketing and excellent management is Monsanto's current production of acrylic fibers (Reisch 1987c). This activity had made up one-fourth of its chemical revenues, and was in trouble. The company had more capacity than business, having build megaplants with high capacity and little flexibility, so that while there were increasing economies of scale, far eastern competitors were cutting into their sales. In addition, Monsanto was relying heavily on commodity fibers sales, half the volume being derived from low-margin exports.

The company was forced to change its strategy, and businesses totaling $2 billion in sales have been dropped since 1979. In 1986 Monsanto was a $4.7 billion chemicals company with returns three times what they were in 1979. They have attempted to emphasize increased profitability more than sales alone (*Chemical and Engineering News* 1987*f*). The changes include more distribution points for their fibers, and major investments in flatbed knitting machines that can manufacture many styles more quickly than older methods. When a market survey indicated customers were unhappy with the amount of wear they were receiving from their socks, Monsanto increased its ''Acrilan'' socks' lifetime by 50% in less than 18 months, and its market share climbed 50%. The company also initiated a ''Wear-Dated'' apparel warranty program that has proven to be very popular. Finally, Monsanto has now invested in ''the plant of the '90s''. It has electronic networks linking Monsanto to its customers, ensuring the quick response time so crucial in competing with foreign suppliers.

This type of diversification and management response to changes, competition, and opportunities is obviously very beneficial to the individual companies and the entire industry, but it is not without its potential problems. The entire chemical industry owes its past success to technical innovation, and managements' ability to commercialize the resultant new processes and developments. But this is a very difficult job, since even with extensive pilot plant testing, comprehensive market research, and detailed staff reports there are always major uncertaintes and some failures. Also, there is invariably competition among different projects for the company's limited capital resources. Consequently, for a company to truly succeed and prosper over the long period, management must have outstanding knowledge and experience to make these difficult technical decisions, which means expertise in the fields involved. If a company diversifies into fields too different from its general area of corporate skill and knowledge, the chances for success become very poor and many of the new ventures will ultimately fail or falter.

As noted previously, 85–95% of most companies' top management time is taken up with financial matters, government relations, management or legal problems, and public relations or employee concerns, all of which are generally similar from one company to another. Many chief executives are selected on the basis of these skills, and are felt to be readily transferable from one industry or business area to another. Of course these functions are important and can greatly contribute to a company's short-term success, but without the deep technical knowledge of the industry involved, such an executive has to be very conservative and in the majority of cases will essentially stop all new development projects, and thus the long-term growth and vigor of the company.

As an example of this, there is little doubt that the executives of the steel industry were excellent administrators, and probably performed well by "Harvard Business School" standards. However, their conservatism, lack of perception of the industry changing around them, and refusal to sponsor any new technical developments essentially destroyed their industry (Ross 1987). Of course, foreign government subsidized imports, out-of-control union pensions, wages, and work rules, competition with plastics and other metals, and a slowed U. S. economy and minirecessions caused a great deal of the problem the industry now faces, but basically a nontechnical, noninnovative, very bureaucratic and high-overhead management must take most of the blame.

The lesson to be learned by the CPI in its diversification program is that new ventures in other industries should be as closely related in some aspects of the business as possible so that there will still be management expertise, or the management of the new companies must be allowed to operate on an independent "cost center" basis so that they can make the important decisions and have their own capital investment budgets. Otherwise, as an extreme example, with an oil producing division competing for capital with a basic chemical division, no matter how attractive a new project for the later group may look, the oil group can always claim the possibility of finding a new Prudho Bay oil field, and a conservative or nonthinking management will make everything (except perhaps the thrill of mergers) stop in favor of oil drilling. On an independent profit center basis both divisions would have their own business objectives, profits, and capital decisions to be responsible for, and thus presumably would be better managed. This is the only possible way that basic commodities, new technology, or new venture groups have a chance of growing or prospering in a large organization, but it is very hard for most managements to let them operate semiindependently with their own divisional objectives for long-term programs.

In summary, diversification is rapidly giving the chemical industry a totally different appearance than it has ever had before. It should help it to be more profitable and active, and if done well should assist in chemical production remaining the innovative major industry that it has always been.

REFERENCES

Allen, Deborah, and John Rutledge. 1988. We should love the trade deficit. *Fortune* (Feb. 29):125.
Anderson, Earl V. 1987. U.S. surpluss in chemical trade. *Chemical and Engineering News* (Dec. 4):29–31.
Berney, Karen. 1987. Just merged. *Nations Business* (Dec.):30–36.
Block, Paul M. 1987. Saudi petrochemicals. *Chemical Week* (Dec. 9):30–36.
Brockington, Langdon C. 1987. OxyChem's investment in PVC pays off. *Chemical Week* (Sept. 2):30–33.
Chemical and Engineering News. 1987*a.* Companies spend more to curb pollution. (Feb. 16):9.
Chemical and Engineering News. 1987*b.* Third world leads in chemical growth. (March 16).
Chemical and Engineering News. 1987*c.* U. S. chemical trade hurt by developing country debt. (April 13):25–26.
Chemical and Engineering News. 1987*d.* Cain chemical (May 11):4.
Chemical and Engineering News. 1987*e.* Facts and figures. (June 8):24–76; (April 13):22–24.
Chemical and Engineering News. 1987*f.* Monsanto stresses profitability over sales. (Aug. 10).
Chemical and Engineering News. 1987*g.* Foreign ownership of U.S. patents still rising. (Aug. 17).
Chemical and Engineering News. 1987*h.* Foreign investment in U.S. chemicals soars. (Oct. 5):11.
Chemical Processing. 1987. Acquisitions in chemicals. (Aug. 12):12.
Chemical Week. 1987*a.* Mergers and acquisitions. (Jan. 7–14):51–53.
Chemical Week. 1987*b.* Foreign takeovers: Boost or bane. (April 29):19.
Chemical Week. 1987*c.* Cain chemical. (June 10):7–9.
Chemical Week. 1987*d.* Saudi petrochemicals. (Dec. 9):30–36.
Chemical Week. 1988. Forecast, 1988. (Jan. 6–13):34–48; 1987 (Jan. 7–14):32.
Cory, Peter. 1987. Diversity spells failure. *Fresno Bee* (May 7):D-1,3.
Dressler, O., and R. Shinar. 1986. Return on investments. *Chemtech* (Jan.):30–41.
Flanigan, James. 1987. Parasites may drain health of U. S. industries. *L.A. Times* (April 21).
Fortune. 1987. 500 largest companies. (April 27):355–414.
Goldbaum, Ellen. 1988. 1987 calms the CPI acquisition pace. *Chemical Week* (Jan. 20):61–62.
Goldbaum, Ellen, and Meyer G. Lurie 1987. Another LBO goes public. *Chemical Week* (June 24):8.
Horiszny, Jean. 1987. Specialty chemicals. *Chemical Business* (April):10–13.
Kiefer, David M. 1987. World chemical outlook. *Chemical and Engineering News* (Dec. 14):25–48.
Kiesche, Elizabeth S. 1988. Non-U.S. bids-benefit or risk? *Chemical Week* (March 16):10–13.
Kline, Charles H. 1986. Reshaping the chemical industries. *Chemtech* (Aug.):478–483.
L.A. Times. 1987. Mergers fall But value rises. (Oct. 19).
Lefkoe, Morty. 1987. Why so many mergers fail. *Fortune* (July 20):113–114.
—— 1988. The new J.P. Morgans. *Fortune* (Feb. 29):44–58.
Loomis, Carol J. 1987. LBO's are taking their lumps. *Fortune* (Dec. 7):63–68.

McGuire, Jane. 1987. Life among the LBO's. *Chemical Business* (June):38–39.

Mullen, Theo. 1987. How to thrive in a global market. *Chemical Week* (Feb. 25):16–17.

Nordhoy, Frode. 1987. Acquisitions. *Chemical Engineering Progress* (Feb.):9–13.

Portnoy, Kristine. 1988. Realizing specialties' potential. *Chemical Week* (March 2):9–12.

Reisch, Marc. 1987a. Chemicals key to ICI's Stauffer plans. *Chemical and Engineering News* (July 6):17–18.

—— 1987b. Chemical capital spending to jump 20% in 1988.*Chemical and Engineering News* (Dec. 21):9–11.

—— 1987c. Monsanto stresses profitability over sales. (Aug. 10):17; *Chem. Mkt. Reptr.* (Apr. 20):3.

—— 1988. Occidental acquires Cain Chemical. *Chemical and Engineering News* (Apr. 25):4.

Ross, Irwin. 1987. Is steel's revival for real. *Fortune* (Oct. 26):94–99.

Savage, Peter 1988. Cain Chemical builds a name in basic chemicals. *Chemical Week* (March 23):22–26.

Schwartz, Irvin. 1987. Cain: Countin on an ethylen upturn. *Chemical Business* (Aug.):18–21.

Shamel, R. E. 1986. Strategies for the global chemical industry. *Chemical Engineering Progress* (Aug.):8–12.

Storck, William J. 1988. Chemical earnings. *Chemical and Engineering News* (Feb. 22):10–14.

—— 1988. Occidental Petroleum is still building by acquisition. *Chemical and Engineering News* (May 2):19–22.

—— 1987. Chemical industry productivity continued to rise in 1986, *Chemical and Engineering News* (March 9):9–10.

Szekely, Julian. 1987. Can advanced technology save the U.S. steel industry? *Scientific American* (July):34–41.

Wilson, Linda J. 1987. Specialty chemicals. *Chemical Week* (May 13):12.

—— 1988. Foreign investment in U.S. chemical industry continues steady climb. *Chemical and Engineering News* (April 25):7–10.

8

ACCOUNTING AND BUDGETS

We live in a world in which budgets of all types have a profound influence on our lives. Contact with budgets ranges from somewhat informal personal budgets for our income and time, to the direct or indirect influence of governmental budgets, to the strict control exercised by our employer's business or corporate budgets. In a budget the funds that have been established for specific purposes very much determine what activities will be pursued and the extent that they will be accomplished. This pattern is definitely the case in the chemical process industries, varying from such simple and basic functions as whether there are funds to hire additional engineers, the amount of money allocated for specific projects, and the extent that the company will invest in new projects and make capital expenditure based on recommendations you have made after performing cost estimates and economic analyses.

Budgets basically address the fact that there are limited resources available for almost all activities, and they present a plan on how these resources may be distributed in order to obtain the maximum benefit. Budgets are also utilized for the allocation of time, again ranging from your own personal activities to very complex planning assignments on major projects. In this chapter simple budgets will be considered for capital and cost allocation, and in a following chapter project management budgets will be considered for time and manpower programming.

The basis for the formulation and control of all budgets is through accounting records, subdividing the expenditures into categories and then compiling the actual amount spent to determine how the costs compare with the original planning of the initial budget. Since accounting records control the budgets both in their conception and execution, this subject will be discussed first.

BASIC ACCOUNTING PRINCIPLES

Accounting departments perform a critical and indispensable role in every business organization. They accumulate and systematize the records of expenses made by all operations within the facility, keep track of monies due or payable, receive or make payments, monitor all transactions required on debt or tax payment, pension and health insurance monies, and so on, and in general keep track of all of the money input, output, and commitments to and from the corporation. Their basic job is to record the company's business transactions in a systematic manner.

The average chemical engineer interfaces with the accounting department only in the budgeting and control function, and on occasion with input and feedback on the monitoring of plant construction and operating performance. If suitable information is available it allows the engineer to better perform these functions, control his assignments, and with luck to compare previous economic estimates with actual performance. This later function, however, is often exceedingly difficult to do because of the complexities and interaction of the many facets of an operation, but if complete analysis is not available, at least a partial analysis can often be obtained so that an engineer may have enough feedback on some of the actual project costs to improve his estimates for the future. Accounting also can provide a file of prior information on other projects to use in the estimates. For every engineer who aspires to understand his company's business and to progress into management a general working knowledge of accounting, its methods, and its output are essential.

To better understand this accounting function, and in order to work with accountants to obtain the maximum benefits from their department, the following fundamentals on accounting methods are presented: There are many excellent books, articles, and courses given on this subject when greater detail is required (Clark and Iorenzoni 1985; AMR 1971; Ernst and Whinney 1986; etc.).

Journals

The outline of Table 8-1 indicates some of the basic functions utilized in general accounting procedures. It starts with a business transaction, whether it be a purchase, sale, operating expense, debt transaction, or any other of the financial activities performed within the company. The transaction is entered into an appropriate journal, in prior times and for very small companies at present, manually. These were separate books or pages kept on each major subdivision of the company's operation. At present, however, the transaction is entered into a computer, along with code numbers that allow it to be organized into the appropriate sections of the accounting records, which is the equivalent to it having been entered into journals. With the aid of the computer, however, one entry with the appropriate coding will allow the software program to not only classify it and enter it as part of totals within that class, (such as accounts payable, accounts receivable, capital, numerous subdivisions of operating costs, etc.), but also if appropriate, prepare for payment through automatic check writing, invoicing customers for sales, inventory and reordering records, and so on. For small companies where the computer capability isn't quite so elabo-

Table 8-1
Partial Outline of Business Accounting Functions

1. Company transaction
2. Enter in appropriate journals (actually enter in computer and distribute appropriately)
   a. Purchase journal, with a sequence of documents:
      (1) Purchase order
      (2) Delivery receipt
      (3) Billing
         (a) Accounts payable
      (4) Payment made
   b. Accounts payable journal
      (1) Salaries, wages
      (2) All other bills (utilities, taxes, fees, services, and so on)
      (3) Other charges and debt
   c. Sales journal, with a sequence of documents:
      (1) Purchase order
      (2) Delivery invoice
      (3) Billing
         (a) Accounts receivable
      (4) Payment received
   d. Accounts receivable or cash journal
      (1) All funds due and received
   e. Other journals
3. Post in the general ledger
4. Handle tax, debt, and other obligations
5. Prepare financial statements
   a. Balance sheets
   b. Income statements
   c. Flow of funds statement
   d. Operating cost statements

rate, some of these functions must be done separately or by hand, but nevertheless one entry into even a simple computer allows the business transaction to be not only entered into the journals but tabulated into various other segments of the accounting records.

The accounting records for each transaction can be quite demanding in order to ensure that each entry is valid and properly completed. With a purchase, for instance, a purchase order (PO) signed by an authorized manager must first be obtained and then filled out with adequate detail by the purchasing agent. When the equipment is delivered a delivery receipt, signed by an authorized company representative joins the file. Later a billing is received, checked against the PO and receipt, and entered in the accounts payable ledger. Finally, payment is made, and the entire group of documents are filed together for future reference or auditing. Such a procedure provides a verification that the correct equipment was properly authorized for purchase, ordered, received, and paid for. The records can also be very useful for reordering and providing the basis for cost and maintenance analysis. A similar procedure is used for product sales.

Ledgers

Once again referring to the older manual entries, after a transaction had been posted into the appropriate journal it was then reentered into an overall or general ledger, again with the appropriate coding. For very small operations the ledger might utilize columns on the same pages that would be equivalent to separate journal entries for larger operations. With computer entries an equivalent ledger printout from the machine can be directly obtained with all of the appropriate journal categories and subdivisions that are required.

Basis

Each company may select its own accounting method (basis) and accounting period (it need not be from January 1 through December 31 of each year), and these can have a significant impact on the records of a business organization. The methods are:

Cash receipts and dispursements is the simplest and most commonly used accounting procedure. Under this system, income is entered in the company books (for tax purposes) at the time it is actually received, rather than when the sale was made or the income was earned. Likewise, deductions or expenses are shown on the books when the payment is actually made.

Accrual accounting, conversely, enters the sales in the books when it was made, regardless of when payment is received, and shows expenses when the liability was incurred regardless of when the purchase payment is made. The accrual method for purchases and sales is required by the government when it is necessary to maintain an extensive inventory, since inventories have a major influence on income when the production, purchase, or sale of merchandise is a sizable part of the business. Thus, any company that produces or sells merchandise must use the accrual method for at least its purchases and sales; other items of income and expense could be accounted for on a cash basis if desired.

Hybrid methods of accounting are a combination of accounting methods, acceptable by the Internal Revenue Service if the combination properly reflects income and is consistently applied. For example, a small manufacturing business might deduct operating expenses when payments are made, but inventories would be included, and gross profit from sales would be determined by accruing sales and purchases. In general, the method used in computing taxable income must ''clearly reflect income,'' and if it does not, the government can substitute a method which does. Of course, for the government's own books it does not do this.

FINANCIAL STATEMENTS

Financial statements of all types can be prepared from the information placed in ledgers. For the company as a whole this would mean balance sheets and income statements for the entire operation, and subtabulations would be made for the individual groups or divisions within the company. Even for a single plant, for instance, separate operating statements would be sent to the research, maintenance, production and sales departments so that each of these groups can be separately monitored and compare their actual expenditures against their own budgets. A well-run accounting department would have immediate information available for each of the divisions within the company to check the meaning or details on any charges, and they in turn could request of the operating division explanations about budget deviations. For the overall organization the balance sheet would list all of the physical assets, liabilities, and equity of the corporation, while the operating statements would itemize the detailed breakdown of all expenses and the income that was received.

Assets

Balance sheets will be discussed again in the chapter on corporate annual reports, but in general they show the assets of a corporation in two categories: (1) current or liquid assets such as cash, stock, bonds, or other securities, accounts receivable or funds owed to the company, inventories, and any other accounts that might be reasonably quickly converted into cash; and (2) fixed assets, or all the machinery, equipment, and property the company owns. The initial cost or purchase price is first listed, followed by the accumulated depreciation, leaving a net fixed asset value. The concept of depreciation has been previously discussed as a hypothetical decrease in the value of assets each year, based upon various Internal Revenue Service (IRS) rules. The government allows specific deductions as operating costs for tax purposes from the initial value of each of the fixed assets (except land) as a certain percentage of the original cost each year in recognition that funds should be set aside for the ultimate replacement of that asset, or perhaps assuming that the equipment is wearing out or becoming obsolete and is no longer worth as much as it originally was. This hypothetical decrease in value (depreciation) is written off as an operating expense each year, which results in an apparently reduced income, which in turn permits a reduced tax payment.

Liabilities

The word liabilities refers to all of the debt owed by the corporation. It also has several components, which are broadly divided into two categories. The first is current liabilities, which represents "cash" demands against the company, including accounts payable for equipment or purchases the company has made and not yet paid the vendors, several categories of short-term debt that are due within the year, and that portion of long-term debt that is also due within the year. It may include accrued charges, such as the current portion of funds that might have been previously spent or due, but are only partially payed, or charged within the current year in order to expense them over a longer period. Examples are major research and development projects, taxes, and so on. The second portion of liabilities is primarily long-term debt, and consists of the funds the company has borrowed to be paid back many years hence. It includes bonds and debentures, either: (1) convertible, which means that they may change from pure debt with definite interest payments into stock (company ownership) at a certain time or at the discretion of the borrower, (2) subordinated debentures which mean that the debt is somewhat like a second trust deed where bond holders that are not subordinated have first call on the companies assets in case of a default, and (3) bonds and long-term debt with a fixed interest rate, which are secured by the assets of the corporation. Preferred stock has a specified dividend rate, so it is similar to debt as far as "interest" payments are concerned, but it is usually not listed as debt.

Equities

The final term that is used in accounting relationships is equities. Equity in general represent the net worth of the corporation, and commonly it is called stockholders equity. It can also be called proprietorship or capital, and it is obtained by the relationship that

$$\text{assets} - \text{liabilities} = \text{stockholders' equity}$$

It is thus seen to be the net, or free-and-clear depreciated property and assets of the corporation, and it often becomes the primary measure of the total net worth of the company. It is probably the most conservative and reliable indicator of a company's value, even though the original purchase price, before depreciation of the land and equipment, plus the current assets, represent the actual total capital investment.

COST ACCOUNTING

Figures 8-1 and 8-2 give a rough indication of the relationships between various departments and their budgets in a large manufacturing plant, which are generally monitored by financial or asset accounting, cost accounting, and managerial accounting. For most engineers the industrial accountant most frequently dealt with is the cost accountant, whose main concern is monitoring the costs incurred in manufacturing a product.

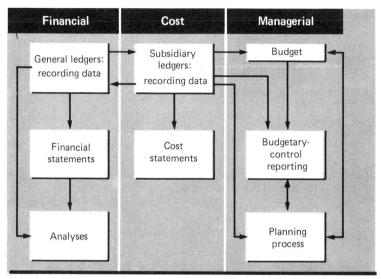

Source: Correia 1980. Excerpted by special permission from *Chemical Engineering*, March 24, 1980. Copyright © 1980, by McGraw-Hill, Inc., New York.

Figure 8-1. Industrial accounting types.

Cost accounting specifically refers to the gathering, classification, recording, summarizing, and reporting of manufacturing expenses and other data, including the recording of the amount of production, the inventory situation, and the purchase of raw materials. The data are compiled into reports both to management and to the individual departments within the company for them to review, and if required they can prepare various analyses and deviation reports. The review function may sometimes be considered as managerial accounting, through which the various division heads, group leaders, and so on are given information to supervise the operations and to plan and control. It provides all levels of management with the information needed for detailed budget making, production pricing, long-range forecasting, and for making the necessary efficiency moves that may be required and indicated.

Plant Operating Statements

A somewhat typical example of a plant operating statement for a chemical manufacturing plant is shown in Table 8-2. As much detail or as many categories as desired may be listed with this statement, but usually they are limited to the more meaningful items that represent a significant dollar value. Also, statements are often broken down into groupings such as: items that vary directly with production, like raw materials and utilities; "controllable" items such as

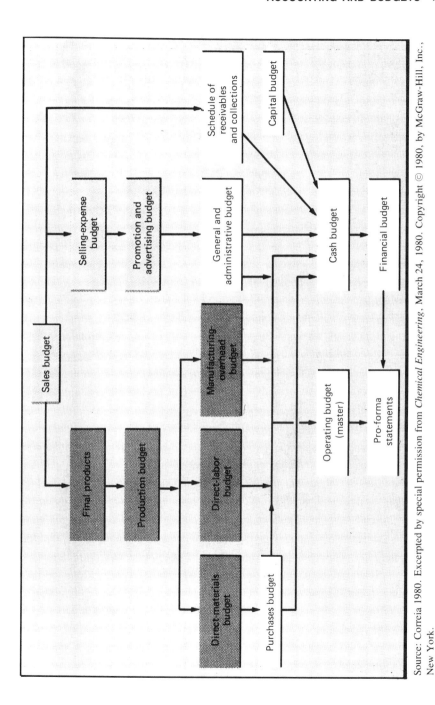

Figure 8-2. Industrial budget and accounting relationships.

Source: Correia 1980. Excerpted by special permission from *Chemical Engineering*, March 24, 1980. Copyright © 1980, by McGraw-Hill, Inc., New York.

Table 8-2
Partial Example of a Monthly Production Department Operating Statement

Plant_____   Period_____

	Actual Expenses				Budget			
	Current Period		Year to Date		Current Period		Year to Date	
Item	Amount	$/ton	Amount	$/ton	Amount	$/ton	Amount	$/ton
Production, tons								
Net value to plant, $								
Controlable Costs:								
Raw materials: A								
B								
Additives, catalysts:								
A								
B								
Packaging materials:								
A								
B								
Freight in, out								
Operating labor								
Overtime								
Other labor								
Group insurance								
Retirement plan								
Workman's compensation								
Payroll taxes								
Other payroll expenses								
Warehouse, stores expense								
Maintenance labor								
Maintenance material								
Contact maintenance								
Gas								
Electricity								
Water								
Telephone								
Office supplies								
Dues, subscriptions, books								
Professional services								
Laboratory expense								
Travel								
Environmental								
Miscellaneous								
Subtotal								
Fixed Costs:								
Depreciation								
Taxes								
Insurance								
G & A								
Management salaries								
Sales salaries								
Sales expense								
Interest, rent, royalties								
R & D								
Subtotal								
Total								

travel, and so on; and fixed charges that do not change appreciably with the plant operating output, and are not directly within the control of the plant staff, such as depreciation and corporate general and administrative (G & A) overhead charges. The budgetary control and analysis of spending in each of these groups is quite different, and decisions on efficiency improvements such as new plants, changes in the operation, or manufacture or purchase raw materials analyses must be made with the differences and new budgets compared to the old ones, so that fixed costs and overhead do not become ignored or improperly distributed.

For these reasons, cost accounting and cost allocation among products must be a shared responsibility between accountants, engineers, and management, with initial budgets and proposed changes made jointly, and major reviews or proposed changes carefully analyzed for all of the direct, indirect, canceled, and ongoing charges (Strauss 1987). For instance, in a proposed production change, which items will remain as real costs (G & A, contractual charges, etc.), and which as ''book'' charges only, such as depreciation on the discontinued equipment. Likewise, shared or arbitrarily allocated charges between different operations or products within the plant must be carefully examined and considered when changes are proposed to safeguard improvements that would genuinely increase efficiency, but might not appear attractive because of previously allocated charges. In the reverse direction, it is often easy to assume that general and administrative, plant security, facility maintenance, research and development and other fixed costs might not increase with a new operation or a plant expansion, which if true would greatly improve its economics. This may be the case, but usually most of these fixed costs will increase, at least with time, and the new operation should share in their burden.

Standard Costs

In developing budgets and cost accounting methods most facilities will have first established a system of standard costs. This will have included records of the cost of plant-wide services: management, clerical, accounting, research and development and so on. A general allocation formula will then be determined for each product (if it is a multiproduct facility), such as based upon the FOB sales income of each product, the total tonnage produced, or some more complex relationship. In a similar manner common services such as steam, electricity and compressed air will probably be charged on a usage basis and the cost per unit will have been established. Also, individual operations' basic costs will have been established for all categories of labor and associated charges, maintenance, local taxes, insurance, raw materials, and so on.

Materials, Overhead, Labor The considerations noted above in cost accounting and cost distribution to products have become of ever-increasing concern (Chowdhury 1987; Dudick 1987; Lunde 1984; Worthy 1987) to industry because of their importance in accurately determining each product's costs when a large number of products are produced in a given plant, and when competition is severe. Only with an accurate and fair assignment of costs can the profitability of each product be determined, and a reliable assessment made of the best competitive sales price, when to expand or discontinue production of certain items, whether to buy or produce raw materials or intermediates, close, modernize, or expand other plants making the same product, and so on.

In assigning costs, materials, labor and utilities, and some services can usually be fairly accurately allocated to a product, as noted above. However, overhead charges are much more difficult to assign, especially with the ever-growing tendency to substitute equipment and controls for labor, and to share services. Improper allocation of overhead (which includes depreciation, research and development, testing, security, management, etc.) can frequently result in as much as a 50% error in assigning costs and calculating profitability. The error is most frequently encountered in large-volume, basic chemicals and in batch operations. It is easy to visualize that such inaccuracy can result in very wrong decisions on products to discontinue or price competitively. This problem appears to be surprisingly prevalent in current industrial operations (Worthy 1987) because of obsolete accounting–management practices, such as using labor costs as the assignment basis in some cases. Other problems arise from the improper sharing of inventory costs, and in considering the total, individual cost of marketing and distribution, such as even on a customer-to-customer basis. Of course, too broad a capital distribution charge, as previously noted, can also be particularly in error. All of these factors have resulted in the more progressive companies reorganizing their cost systems to a more competitive environment by more detailed and accurate breakdowns, and more complex computer assisted cost assignment.

Plant Management

A second and somewhat independent aspect of accounting in the chemical process industries is involved with sales, inventory, and operating cost feedback in the management of plants. Improved performance is often obtained through computer assisted programs such as PMS or CIM (plant management system; computer integrated manufacturing), or MRPII (manufacturing resource planning) (Chowdhury 1987). The former stresses process control, while the later emphasizes better plant scheduling strategies, particularly with batch processing plants.

The PMS software considers economic information of sales demands and prices, inventories, raw material prices and availability, energy costs, operating costs versus product output, and so on. By using such detailed data, the input costs and effort are considerable, but the plant manager can often much more efficiently manage the plant output, inventory, yields, and profitability. The profit increases can be considerable.

Control Layers.   The PMS program basically integrates four "control layers" in a plant. The first concerns the normal plant control instruments and functions at their optimum settings; the second is the management or supervision of an entire process unit at its most favorable operation. The third layer is plant management control of a complete plant, and the final layer is information control, which includes raw material, product and energy costs, product sales and inventory, business data, and so on. Information such as product quality, production costs and optimum operation from several view points all may become immediately available to management from the program.

An alternative planning scheduling and inventory control method is MRPII. When off-the-shelf programs are modified for CPI batch process plants they have proven very useful not only for inventory control, but also for production planning if sales projections are fed into the system as one of its more than a dozen data bases (e.g. bill of materials, raw material ordering policy, production scheduling, materials requirement planning, plant floor control, purchasing, cost management, order management, etc.) Extremely demanding input frequency and accuracy are required for the system to function properly.

More detailed information on cost control methods and procedures may be found in numerous articles on the subject, and in texts, such as Dudick (1987) when needed in an engineers' career.

BUDGETS

Budgets are used in every aspect of the chemical industry, and each engineer's degree of contact with the budgets depends greatly upon his position in the organization. As an entry level engineer all wage increases, projects, equipment availability, and the activities of the corporation that he sees depends upon budgets. Once an employee has been with an organization for some time, or has risen in the company, he generally becomes an integral part of the budget formulation process, as well as being governed by it. At higher levels one's advancement record is partly determined by one's skill in budgeting and accomplishing the desired results within the budget. Conversely, it often appears for those doing government contracting, success depends upon *not* completing the work, and carefully overrunning the costs (budgets), but believably, so that the government will extend the contracts or grant new ones and feel that they are advancing the project.

Operating Budgets

Preparing budgets is another form of cost estimating, but one that usually is based largely upon the individual groups and the accounting department's records. Overhead factors are generally available to be applied to actual salary and labor charges, as are other standard costs, as noted above. Business tax, local fees, utility rates and depreciation estimates are also jointly determined with the accounting department. The manufacturing overhead budget is complex because it contains both variable and fixed costs, includes many cost elements (some of which are relatively small), and represents costs for which different groups or managers are responsible.

Other aspects of manufacturing cost have been covered in Chapter 4, but will be briefly highlighted here. The amount of manpower required is often as simple as knowing the size of the group available, but in other cases it will have to be estimated based upon previous work, or as best possible from a knowledge of what is desired to be accomplished, and perhaps manning charts. Often this is done by breaking the project or work into as many component parts and types as can be visualized, and then estimating the manpower and staffing for each unit. An estimate of the materials, supplies, utilities, and so on that will be required to manufacture the desired amount of product generally can be obtained from heat and material balances, adjusted for actual plant efficiency factors. On the manufacturing cost and budget sheet shown in Table 8-2, many of the cost elements are also of considerable importance to other groups within the organization. The raw materials component is important to the purchasing department, which must make sure that the desired quantities will be available when needed. The treasurer-controller uses the dollar amounts of the total budget for input to the cash budget to ensure that the money will be available when the materials are delivered and wages and other costs are due. The maintenance equipment and supplies need to be ordered by the purchasing or maintenance departments, and so on (see Table 4-6).

Inventory Budgeting

Several different conventions may be considered in regard to the value attached to raw materials in inventory. Finished product inventory is always budgeted at cost, but since the initial purchase price may change, the value of the raw material inventory will also change. The amount of both inventories may be determined by such considerations as storage capacity, raw material availability, processing time, shipping schedules, sales demand, production capacity, and most importantly, the pattern used in withdrawing the inventory. Once the desired amount is determined, the following conventions may be used to establish the raw material inventory value and to consider for product inventory tax treatment:

1. The current average or current value method. The average price of all the inventory on hand at the time of delivery or use is employed in this method. In other words, it is the actual cost of the inventory. It may be adjusted, if appropriate, for its current value, if this is lower than the purchase price.

2. The first-in, first-out (FIFO) method. This method assumes that the oldest material is always used first, and the inventory is valued at the most recent cost.

3. The last-in, first-out (LIFO) method. With this method, the oldest prices are used for the entire inventory value. In a period of rising prices this results in lower priced inventories, higher costs of sales and, consequently, lower taxable earnings. Many businesses adopt LIFO in order to reduce taxes and thus improve cash flow.

    The dollar value LIFO inventory method is the most common LIFO technique used. It measures inventory quantities in terms of dollars, rather than physical units, and this is done for a pool of items, rather than for specific items.

4. The "just-in-time" method of sizing the inventory. In this procedure, raw material inventory is kept to an absolute minimum, assuming that new supplies can be received as needed. This reduces the working capital requirement, but if a company sells to "just-in-time" companies (as is usually the case with automotive, etc. customers), then much larger product inventories are required since the customers' inventory is so low (*Chemical Business* 1988).

5. UNICA (uniform capitalization) rules for the taxation of inventories came into effect on January 1, 1987. They are very complex, but basically require "full absorption costing" for inventory, including some overhead and administrative costs to be allocated to the goods that are being produced, and that some inventory be capitalized and not shown on the books as an expense (Gomez 1987). Engineers may need to be aware of these inventory procedures and complications, but all details will normally be handled by the accounting department.

Budget Preparation

Budgets are a key factor, as noted before, for managers to efficiently run their operations. Generally the budget preparation procedure starts at fairly low levels when each group or function within the organization is given general guidelines of what the management feels they would like to accomplish during the following year, the general priorities, and the increase or decrease in spending limitations. Then the component groups within the company assemble their data, suggest programs, and formulate detailed operational plans for accomplishment, manpower, time, and spending requirements. An engineer might be asked to participate in this budget making process by outlining what he anticipates will be required for the projects he should be working on during the coming year. The preliminary budget requests are then gathered by each line of management. Generally they are reviewed with the individual groups preparing them and are consolidated, often with considerable discussion, editing, selection, and choice between alternatives, and passed on to the next level of management. The process is repeated until finally they are presented to the top management of the corporation, and again generally with fairly detailed and elaborate reports and presentations as to why each subgroup within the organization believes that their programs should be followed. Considerable detail is given supporting these requests, including preliminary project cost estimates, manufacturing costs, and the anticipated economic analysis or DCF results. Top management then compares and evaluates the spending desires with the funds available (including increasing debt if justified) and sets out the final detailed budgets.

Once approved, budgets become a rather absolute guideline for all activities during the following year. They certainly can be changed, and all of them have some degree of flexibility, but as a first approximation management expects that they be rigorously adhered to unless they have given advance authorization for a change, or after an intermediate review period, management allows alterations to be made. This frequently happens, but it is not the prerogative of the groups within the organization to change budgets on their own, and a fairly complex review procedure is usually required for management to authorize such changes.

Budget Control

The accounting department of the company generally sends monthly reports to each group letting them know how well they are conforming to the budget (see Cost Accounting). These records show not only the direct spending but also all of the indirect expenses and the company overhead charged to them. The accounting department will clarify all questions concerning the spending, but if there are overruns the group usually must immediately explain them and lay out a corrective plan. Chronic or worrisome overruns almost always result in higher management intervention unless they can be satisfactorily explained and/ or excused.

This accounting department feedback has become a vital part of most company's operation, for with the advent of computer accounting it is not too difficult for detailed breakdowns to be given to every subdivision of the company frequently (such as on a monthly basis) and soon after the period closes. Even with maintenance work it has become common practice for the cost and repair history of most major equipment to be entered into the accountant's (or department's) computer, and be retrievable in a wide variety of analyses. This greatly assists in planning, parts inventory control, preventative maintenance, and in improved equipment selection.

Budgetary control reports based upon actual expenditures and charges thus provide the manager at each level with much of the information he needs to control his day-to-day operations. In production control reports can be found answers to such questions as raw material usage, production rates, on-stream efficiencies, total production, manpower, all categories of expenses, and the inventory of raw materials and products. In most companies monthly budget meetings are held in which any deviation from the budget is discussed and explained, and this information is then passed on to the next level of management. Such budget meetings also often serve as the focus for improvement suggestions, company news, and ''esprit de corps'' formation.

All company divisions have to plan for both local management and corporate general and administrative overhead (G & A) charges spread out to their divisions by management, service groups, research and development, and so on. These charges can be quite large, and management always has the prerogative to increase them. Such ''distributed'' charges may thus be both substantial and sometimes appear as unexpected additions. Well-run organizations have worked hard to significantly reduce these charges over the past few years by cutting back on bureaucracy and unnecessary overhead costs. This has helped to make the industry more competitive for the world markets.

Corporate policy with regard to decentralization will dictate the extent to which the overall budgets are supervised and reviewed at the plant or headquarters. In a decentralized organization the master budget and all support budgets are normally prepared and analyzed at the plant level. Although the operating managers may not be directly concerned with the overall corporate budget, they will contribute to its preparation. The operation managers acceptance of and dedication to achieving the goals set in the master budget certainly depends to some extent upon their knowledge of its preparation and the give-and-take of its components.

Variance Analysis.   Variance analysis is one of the methods used to systematize the control of an operation. By noting the variance between actual and budgeted costs it provides a number, and pinpoints where problems exist and the extent of the problem. It sometimes may help show where greater or in-depth analyses may be needed for recurring problems. It may indicate that better planning is required or that the budgets should be revised. It is the end-product of the budgeting and control system, and its value depends on whether it is used as a tool for improving operations or for finding fault. Corporate budgeting is thus the base, or beginning, of operational control.

Many excellent books (Schwartzman and Ball 1977; etc.) and courses (AMR 1971; Nevitt 1975; Symonds 1978; etc.) frequently appear on these general accounting/financial subjects.

PERSONAL BUDGETS

Many people go through life with only a vague idea of how much they should save, whether they actually can afford their lifestyle, or even if they are going to have enough money to live comfortably when they retire. A multitude of books, seminars, sales people, and television pitches are available to advise on what to do with your money, but most of them jump right into complex investing techniques. In practice, if you don't have the basics down, you're not going to have the money to get into more sophisticated investing. It is invariably necessary for you to start with a good personal budget.

To determine where you are you have to organize your papers, figure your net worth and set up expense categories. These steps will require considerable analysis on your part, and at first will not be too accurate. By organizing your documents into some sort of logical filing system you make it easier on yourself to come up with supporting evidence if you are audited by the government, and it also helps you to determine spending categories. Establishing a budget is actually a three-pronged effort that can take as little time as a month or two or as long as six months to a year. Your first need is to track your income and expenses to see where your money is going. Once you get an idea of how you are spending your money, you can allocate your spending into different headings within two general groups—fixed expenses and variable expenses.

The last step in setting your budget is living with the allocations you made for a short period and then adjusting them where needed. You'll find that fixed expenses can't be changed easily—you'd have to move to a new place where you would pay less rent or make smaller mortgage payments, for instance. And even though variable expenses are easier to rearrange, this kind of cost cutting does not make such dramatic changes. However, whether it be major structural changes or small ones, improvement usually can be obtained.

Finally, the major reason for the budgeting process is so that you can regularly and systematically save some of your income for investments. No matter how small, once the saving habit is started and maintained, your financial security is greatly increased. Most people find this to be progressively easier to accomplish with practice, and then the enjoyment and profit from investments will multiply. Your net worth is the difference between everything you own and everything you owe. If calculated once a year, as many planners recommend, the results tell you how you are progressing on the road to security. A typical budget worksheet is shown in Table 8-3.

Table 8-3
Typical Personal Budget Worksheet

INCOME

Monthly earnings
(take home) $_____

Other Income $_____

TOTAL INCOME .... $_____

MONTHLY EXPENSES

FIXED EXPENSES

House/rental payments	$_____
Car & transportation	$_____
Other loan payments	$_____
Food	$_____
Utilities	$_____
Child care	$_____
Household help	$_____
Savings & investments	$_____
Personal allowance	$_____
Other	$_____

Subtotal $_____

VARIABLE EXPENSES

	Annual cost	Monthly pro-rated cost
Tax payments	$_____	$_____
Insurance premiums	$_____	$_____
Health care expenses	$_____	$_____
Education	$_____	$_____
Household repairs	$_____	$_____
Car repairs	$_____	$_____
Clothing	$_____	$_____
Entertainment & travel	$_____	$_____
Gifts & donations	$_____	$_____
Other	$_____	$_____

Subtotal $_____

TOTAL MONTHLY EXPENSES
(fixed plus variable) .... $_____

TOTAL DISCRETIONARY DOLLARS
(total income minus expenses) .... $_____

Source: Alcott 1987. Fresno Bee, January 12, 1987, C7-8. Courtesy of *McClatchy News Service*. (McClatchy Newspapers, Inc.)

REFERENCES

Advanced Management Research. 1971. *Fundamentals of Finance and Accounting for Non-Financial Executives.* New York:118.

Alcott, Martha J. 1987. Personal budgets. *Fresno Bee* (Jan. 12):C 7-8. McClatchy News Service (McClatchy Newspapers, Inc.).

Chemical Business 1988. Just in time. (Feb.):7.

Chowdhury, J. 1987. CPI find big dividends in better plant management. *Chemical Engineering* (Nov. 28):29-33. McGraw-Hill, New York.

Clark, Forrest D., and A. B. Lorenzoni. 1985. *Applied Cost Engineering.* Marcel Dekker, New York:352.

Correia, Ernest V. 1980. Cost control begins with budgeting. *Chemical Engineering* (March 24):87-90. McGraw-Hill, New York.

Dudick, Thomas S. 1987. *Manufacturing Cost Controls.* Prentice-Hall, Inc., Englewood Cliffs, N.J.

Ernest and Whinney. 1986. Accounting and cash flow ideas. *Fertilizer Progress* (May/June):53.

Gomez, Iris. 1987. In the short term, UNICAP hurts. *Chemical Week* (June 24):15-16.

Lunde, Kenneth E. 1984. Joint product costing. *Chemical Engineering* (Dec. 10):89-92; 1985 (Jan. 7):95-100. McGraw-Hill, New York.

Nevitt, Peter K. 1975. *Project Financing.* AMR International, Inc:159.

Schwartzman, Sylvan D., and Richard E. Ball. 1977. *Elements of Financial Analysis.* Van Nostrand Reinhold, New York:133.

Strauss, Richard 1987. Use and misuse of manufacturing cost systems. *Chemical Engineering* (March 16):106; (April 27):65-70; (May 25):87-88. McGraw-Hill, New York.

Symonds, Curtis W. 1978. *Basic Financial Management.* AMACOM, New York:208.

Worthy, F. S. 1987. Accounting bores you? Wake up. *Fortune* (Oct. 12):45-52. Time, Inc.

9

CORPORATE ANNUAL REPORTS

Every employee should regularly read his company's annual reports, and other similar publications when possible. They are mailed to all stockholders, so purchasing at least a few shares of company stock will automatically have them delivered to you. A review of competitors' reports can also often be quite informative, but probably less so than technical journals until one reaches the middle-to-upper management ranks. These reports discuss what management feels is important about the business, and can give the working engineer some idea of how his work fits into the total operation. Often new directions that the company is interested in pursuing are outlined, as well as general objectives and statements concerning the company's financial health. A knowledge of these factors can usually be of general value in performing one's work, and certainly makes the engineer more responsive to the company needs. It also may occasionally open the door to otherwise unknown opportunities within the company or give a hint when one should start looking for another job.

As actual examples of this, a young research and development engineering group leader in one of a medium-sized chemical company's plants was invited to an evening social gathering when the company president visited the operation. The engineer had just read the annual report and the latest issue of *Fortune* magazine. While part of a large group chit-chatting with the president he happened to observe that if the company continued its present rate of growth, next year it would make the Fortune 500 list (the 500 largest U. S. industrial corporations) (Fortune 1987). The president had not been aware of that, was intrigued, and wanted more information, which the young engineer supplied. From that time onward the engineer was remembered by the president, and his rise in the corporation was accelerated.

In another case a section manager in a chemical plant that was part of a major corporation noted in the annual report that a large desalination plant was about to be built in a foreign country by the company's technical engineering department. The technical vice-president visited his plant's research and development laboratory frequently, so on the next visit he arranged with research and development friends to ask for an interview. This came about, and he volunteered to help start up and operate the new plant. The research and development vice-president had not yet hired a plant manager, so he checked the engineer's credentials with his boss, got his permission, and offered the engineer the job. The former section manager did an excellent job, and rather quickly was promoted to head part of the company's total operations in that country.

The above examples must be considered as quite unusual, but the point is that for all employees, reading the corporate annual reports cannot help but improve job understanding and satisfaction, and perhaps aid in job performance and advancement.

BALANCE SHEETS

All public companies are required to disclose fully "material" information (material here means any information that might affect the well-being of a company). They are not required to disclose it in a highly readable form nor does the Securities Exchange Commission (SEC) make any value judgments about what is disclosed. It's up to the reader to get what they can from the reports. They are usually written in somewhat of a jargon by accountants and lawyers, but once you figure out the code, understanding financial statements becomes more a matter of practice than of intellect.

An annual report often consists of a president's letter, a text explaining how the company has prospered during the past year and its future plans, a financial section that includes a balance sheet, income statement, often a flow of funds accounting, and a set of footnotes at the end that explains in more detail those things that didn't get sufficiently explained up front.

The balance sheet is the first indicator of the company's strength and vigor. As noted in the accounting section, for any balance sheet the total value of what is owned (assets) minus the total value of what is owed (liabilities) equals the net worth (stockholder's equity) of the company:

$$\text{assets} - \text{liabilities} = \text{stockholder's equity}$$

An abbreviated average balance sheet for 12 major chemical companies is shown in Table 9-1 (note that the 12 companies change in 1984–1985). It, like all others, must cover an exact period, such as one year, and have a precise starting and ending date, such as January 1 through December 31 of the year in question. Any accounting period and ending dates are acceptable, just so that they are clearly indicated.

Table 9-1 has fewer entries listed than is found in most annual reports, but the most important items are there, and the resulting simplicity aids in reviewing it. First, note the two subsections marked current assets and fixed assets. Current assets are essentially the working capital of the company, which at least theoretically could be fairly readily converted into cash, such as:

Table 9-1
Typical Chemical Industry Balance Sheet and Financial Data.

Assets, $ billions	1986	1985	1984	1983	1982
Current assets					
Cash and U.S. bank deposits	$ 2.52	$ 1.32	$ 1.10	$ 1.28	$ 1.04
Short-term securities	0.95	1.14	1.29	1.57	1.36
Receivables	11.69	12.33	13.12	14.01	13.26
Inventories	11.21	10.76	12.26	11.60	12.43
Other current assets	3.16	2.50	2.89	2.25	2.77
Total	29.54	28.05	30.66	30.71	30.86
Fixed assets					
Depreciable fixed assets	77.42	76.21	82.17	79.31	80.15
Land and mineral rights	2.87	2.98	3.44	3.18	35.59
Less: accumulated depreciation and depletion	41.51	40.19	41.48	38.24	35.59
Net fixed assets	38.78	39.00	44.13	44.25	44.56
Other assets, including investments and intangibles	22.76	21.80	19.27	17.20	17.54
Total assets	$91.08	$88.85	$94.06	$92.16	$92.95

Liabilities and Stockholders' Equity, $ billions	1986	1985	1984	1983	1982
Current liabilities					
Short-term debt	$ 3.27	$ 3.27	$ 1.89	$ 1.74	1.78
Accounts payable	5.76	5.90	6.75	6.79	6.28
Accrued income taxes	1.31	1.05	1.44	1.24	0.50
Current portion of long-term debt	0.85	1.03	1.27	1.11	0.91
Other current liabilities	8.87	9.40	9.69	8.91	8.53
Total	20.06	20.64	21.04	19.79	18.00
Long-term debt	20.87	19.21	18.14	19.43	22.04

Table 9-1 (Continued)

Liabilities and Stockholders' Equity, $ billions (Continued)	1986	1985	1984	1983	1982
Other noncurrent liabilities, including deferred taxes and minority interest	8.53	8.19	8.02	7.68	6.97
Total Liabilities	49.45	48.04	47.19	46.91	47.01
Stockholders' equity					
Capital stock and other capital	14.91	13.90	17.06	17.79	17.85
Retained earnings	26.72	26.92	29.82	27.47	28.10
Total stockholders' equity	41.63	40.82	46.87	45.26	45.95
Total liabilities and stockholders' equity	$91.08	$88.85	$94.06	$92.16	$92.25

Debt Ratios

$ Billions	1986	1985	1984	1983	1982	1981	1980	1979	1978	1977	1976
Industrial Chemicals and Synthetics											
Long-term debt	$20.9	$19.2	$18.1	$19.4	$22.0	$21.4	$16.8	$14.8	$13.6	$12.3	$11.8
Stockholders' equity	$41.6	$40.8	$46.5	$45.3	$45.9	$42.5	$39.8	$32.3	$29.3	$26.3	$25.0
Debt ratio[a]	33.4%	32.0%	28.0%	30.0%	32.4%	33.5%	29.6%	31.4%	31.8%	31.9%	32.0%
Chemicals and Allied Products											
Long-term debt	$45.8	$38.2	$33.8	$36.2	$38.2	$36.9	$28.3	$24.9	$23.5	$21.1	$19.4
Stockholders' equity	$100.0	$100.3	$100.3	$97.2	$95.3	$89.9	$79.5	$67.4	$60.8	$54.9	$50.8
Debt ratio[a]	31.4%	27.7%	25.2%	27.2%	28.6%	29.1%	26.3%	27.0%	27.9%	27.8%	27.6%

Table 9-1 (Continued)

Cash Flow

$ Millions, % of total	1986		1985		1984		1983		1982	
Sources of Funds										
Net income	$3,414	13.6%	$ 986	6.5%	$ 3,859	29.7%	$ 2,795	23.1%	$2,750	21.4%
Depreciation and depletion	5,117	20.4	5,010	32.8	4,710	36.2	4,601	38.0	4,341	33.8
Deferred taxes	568	2.3	−34	−0.2	507	3.9	333	2.7	623	4.8
Other internal sources	6,084	24.3	6,707	43.9	2,357	18.1	2,623	21.7	2,062	16.0
Long-term debt	8,773	35.0	2,356	15.4	1,441	11.1	1,352	11.2	2,665	20.6
Stock	1,091	4.4	264	1.7	136	1.0	396	3.3	432	3.4
Total	$25,047	100.0%	$15,289	100.0%	$13,009	100.0%	$12,100	100.0%	$12,864	100.0%
Applications of Funds										
Dividends	$ 1,992	8.0%	$ 2,049	13.4%	$ 1,998	15.4%	$ 1,868	15.4%	$ 1,791	13.9%
Capital expenditures	5,986	23.9	6,500	42.5	6,400	49.2	5,685	47.0	7,370	57.3
Additions to working capital	−567	−2.3	49	0.3	−436	−3.4	−265	−2.2	−1,402	−10.9
Reduction of long-term debt	7,669	30.6	2,555	16.7	2,529	19.4	3,249	26.9	3,592	27.9
Other applications	9,966	39.8	4,137	27.1	2,518	19.4	1,563	12.9	1,512	11.8
Total	$25,047	100.0%	$15,289	100.0%	$13,009	100.0%	$12,000	100.0	$12,864	100.0%

a. Long-term debt as a percentage of long-term debt plus stockholders' equity; Data from company annual reports.
Note: 1985–1986 Data are totals for 12 major chemical producers: American Cyanamid, Dow Chemical, Du Pont, Ethyl, W. R. Grace, Hercules, Monsanto, National Distillers, Olin, Penwalt, Rohm & Haas, and Union Carbide.
Note: 1976–1984 Data are totals for: American Cyanamid, Celanese, Dow Chemical, Du Pont, Ethyl, Hercules, Monsanto, Olin, Penwalt, Rohm & Haas, Union Carbide, and Williams Cos.
Source: Chemical and Engineering News, 1984, 1985, 1986, 1987. Reprinted with permission from *Chemicaland Engineering News*, June 8–11, 1984–87. Copyright 1984, 1985, 1986, 1987, American Chemical Society.

Cash on hand.

Marketable (or short-term) securities listed at their actual cost, not the current value.

Accounts receivable (adjusted for bad debts).

Inventories (raw materials, products, supplies, or in-process materials listed by value according to the actual, LIFO or FIFO convention).

The first two items are called liquid or quick assets because they are either cash or nearly equivalent to cash.

The fixed assets (property, plant, buildings and equipment) are listed at their original acquisition cost (direct fixed capital), and accumulated depreciation is shown separately. The net fixed assets, before depreciation, are therefore generally identifiable with the investment of the corporation in property and equipment. This number, plus the working capital (current assets) gives a first approximation of the total capital utilized by the corporation. Assets are kept on the books and shown as assets as long as they are in use, even though they may be completely depreciated. They are only removed from the balance sheet when they are discarded and physically removed from the property. At that time both fixed assets and accumulated depreciation are equally reduced. The asset value of land, buildings, mineral holdings, and so on are usually shown at their actual acquisition value, and thus are frequently tremendously undervalued, or occasionally vice versa. The numbers can be adjusted upward or downward to reflect actual market conditions if desired, but this then results in a taxable gain (profit) or a loss that year. If a gain, a commensurate tax payment is required and if a loss, it shows on the books as reduced earnings that year. Consequently, this profit or ''write down'' is generally only done under unusual circumstances.

Some other items that may be on the assets side of the balance sheet merit further explanation. Prepayments refer to payments for materials or services for which full value has not yet been received. Examples are prepayments on taxes or insurance, and down payments on construction contracts or equipment and supplies. Deferred charges are in some ways the reverse of prepayments, in that they indicate an expenditure that has not yet been made, but from which benefits will accrue over a number of years, and a growing (book) liability will increase, such as deferred taxes resulting from using accelerated depreciation (compared to straight line). As another example, with research and development on a major project, the corporation may feel that it is more reasonable (or required by the IRS) to consider the expenditure as if it were spread out over several years. The deferred charges specific to a particular expenditure then decrease annually over the full period of deferment.

Sometimes intangibles are listed, referring to assets which have no physical existence but which nevertheless have assignable value. Patent rights are an example, the value of which may be determined by the purchase cost of the rights, or by the cost of developing the patent. Goodwill is the assigned value of those qualities which bring patronage to a business. For instance, the exclusive (or nonexclusive) right to produce or distribute a product in a particular area is a goodwill item having some inherent value. The assignment of value must be reasonable, and preferably it should be based upon monetary transactions such as the cost of a franchise, license, or research and development work performed. The assessed value of intangibles must be included in the income statement when they are sold, depreciated, or cancelled.

The debit side of the balance sheet (the liabilities) consists of three major categories, current and long-term liabilities and equity. The current liabilities, just like the current assets, are identified with a potential quick demand for cash. Examples are short-term (one-year) loans, obligations, and accounts payable which represent money that is owed for materials and services that have already been ordered, provided, or delivered (raw materials, equipment, and so forth). Notes payable are the current year's portion (the part that must be repaid during that year) of longer term debt from bonds or loans. Accrued expenses payable are salaries, interest payments, insurance premiums, taxes, and other expenses which are due at that time, or at the end of the quarter or year, and remain unpaid at the date of the balance sheet.

Long-term liabilities are debts (promissory notes, debentures, bonds, bank loans, etc.) due after more than one year from the date of the balance sheet. The principal item under long-term liability usually consists of (first mortgage) bonds which guarantee that the purchasers will have top priority claims upon mortgaged assets of the corporation if for some reason the corporation cannot pay the value of the bonds at the maturity date. Many other entries may be listed, such as debentures, other types of bonds, loans, debt to subsidiaries, and so on.

The final item in a balance sheet is the stockholder's equity, which must satisfy the balance, assets − liabilities = equity. Essentially it is the "net worth" of the corporation. Capital stock, both preferred and common is listed at its originally issued, or "par" value. It should be noted that par value has no great significance, and some corporations issue stock with no stated par value. Stock is almost always sold by the corporation above par value, at an amount deemed to be equal to its current market value. The excess that is received from shareholders above the par value is called capital surplus.

The accumulated retained earnings category is the other important component of stockholders' equity. The retained earnings represent all of the profits that have cumulatively been earned over the years and saved or plowed back into the corporation. Each year's net profit (after subtracting all costs and taxes), diminished by the amount of stock dividends, contributes to the accumulated retained earnings total. This amount is not primarily cash that has been accumulated, but instead represents all of the assets and investments purchased by the corporation from these funds. Increases in retained earnings result in increasing assets and investments (fixed assets) obtained by using some of the current profits or cash flow. The accumulated retained earnings are thus a running balance which serves to satisfy, year after year, the rigid requirements assets − liabilities = equity. It may be based somewhat upon intangible and even vague elements such as goodwill and accrued or deferred charges, so consistent accounting principles are very necessary to come up with a perfectly balanced and accurate balance sheet.

In most annual reports the yearly distribution of the retained earnings or cash flow are also tabulated to show the stockholders what has happened to their profits (see Table 9-1, Cash Flow, p. 177).

INCOME STATEMENTS

Whereas balance sheets show the corporate position at one point in time, the income statement reflects the results of operations throughout one year or quarter. Income statements are generally only partially presented in annual reports, showing items of general stockholder interest such as total sales, research and development, tax payments, depreciation, perhaps profits from sale of certain assets, energy purchases, and so on. (For internal use, within the company, however, operating statements will generally be very detailed.) A hypothetical income statement is shown in Table 9-2.

Simplified statements start with the total sales revenue for the year, diminished by production cost, depreciation, selling and administrative expenses, which are normally called the total manufacturing cost. Accountants, however, often use the term "cost of sales" and list depreciation, sales, general and administrative, other expenses, and income taxes separately. The cost of sales is thus only the general expenses (or the out-of-pocket expenditures), and unfortunately you are never totally sure what items it contains (i.e., is interest included? Usually.) The revenue minus the cost of sales difference is called the gross or operating profit, and may be adjusted for interest, and other expense, plus income from interest or short-term investments such as securities held by the corporation, and gains from the sale (or book appreciation) of property. This gives a total pretax profit, unless other items not considered in the cost of sales are still unaccounted for, such as perhaps interest on bonds and other borrowed money, or in Table 9-2, research and development. The balance is finally the gross profit upon which income tax is paid. The net profit is the gross profit minus the total of all income taxes.

FLOW OF FUNDS, FOOTNOTES

Corporations are complex financial entities, and the consolidated statements found in annual reports cannot be expected to convey all of the important details of that financial complexity with only the sections noted above. For this reason corporate annual reports include a section entitled "notes to financial statements" or "footnotes" that explain the methodology and data used in formu-

Table 9-2
Example of an Income Statement.

Fiscal Year Ending	12/31/86	12/31/85
Revenues		
Net sales	$800,000	$750,000
Cost of sales		
Cost of goods	$550,000	$500,000
Depreciation of amortization	30,000	25,000
Selling, general, and administrative expenses	$120,000	$110,000
Total costs of sales and operating expenses	700,000	635,000
Gross Profit	$100,000	$115,000
R & D expenditures	$ 10,000	$ 5,000
Nonoperating income	6,500	7,000
−Interest expense	−15,000	−15,000
Income before taxes	$ 81,500	$102,000
Provision for income taxes	20,000	15,000
Net income	$ 61,500	$87,000
Common shares outstanding (in thousands)	15,000	15,000
Earnings per share	$4.10	$5.80
Prior accumulated retained earnings	$220,000	$200,000
Net income	61,500	87,000
Total	$281,500	$287,000
Less dividends on		
Preferred stock	$ 400	$ 400
Common stock	20,000	18,000
Retained earnings	$261,100	$268,600

Source: *Compressed Air Magazine* 1987.

lating the statements. In fact, such footnotes may contain a great deal of important supplemental information, and occasionally they explain otherwise unmentioned details very pertinent to the corporation's operation.

There is no specific requirement regarding the substance of the footnotes, but a typical annual report may show some or all of the items noted below, among others:

The components of income tax paid by the corporation, including a subdivision among federal, state, local, and foreign taxes.
Explanation of the effect of using different depreciation schedules for tax and book purposes.
A breakdown of plant properties.
Statistics on mineral and energy reserves.
Retirement plan statistics (often pension fund deposits can be very large, and represent complex investment requirements).
Details of long-term debts and the interest rate on various loans or bonds.
Stock options and awards; officer's salaries.
Details of divisional operations.
Product group sales analysis.

Other useful information in annual reports may include a historical financial summary covering several years, or the effects of inflation upon financial performance criteria. Some mention is always given of the earnings per share of common stock, and of the dividends distributed.

As previously mentioned, Table 9-1, p. 177, outlines the source and disposition of the total corporate annual cash flow. Among other things it provides an idea of the new investments made by the company and the source of funds. In Table 9-1 the term "other internal sources" of funds can represent any type of income, but generally implies divisional divestitures and/or equipment or property sales, or reevaluations. It also includes profits from investments, foreign currency exchange, and so on. In a similar manner, "other applications" generally means acquisitions and purchase of other businesses. It is seen that both of these areas have been unusually active during the 1980s.

RATIO ANALYSIS

One of the principal means of analyzing financial performance of corporations is by means of "ratio analysis." It facilitates the year-by-year monitoring of corporate operations and hopefully permits a more rational comparison of different companies. While net profit may be the "bottom line" of the income statement, it may not truly or clearly indicate how well the company performed. For instance, if a corporation experienced a steady increase in net profits over the recent years, does this indicate a good performance? Much of that apparent growth could be due to inflation, acquisitions, or new projects, and the profit margin could be low. The increase could also be partly due to accounting manipulations, such as the reduction of inventories, the sale of assets for a profit, or deferring income tax. Consequently, other financial comparisons, such as the various "ratios" may more clearly compare the performance. Some of the more widely used financial ratios are listed below:

Current Ratio

The current ratio is defined as

$$\frac{\text{current assets}}{\text{current liabilities}}$$

Using the data in Table 9-1, the average of 12 chemical companies for 1982 was

$$\text{current ratio} = \frac{30.86 \times 10^6}{18.00 \times 10^6} = 1.71$$

$$\text{for 1986,} = \frac{29.54}{20.06} = 1.47$$

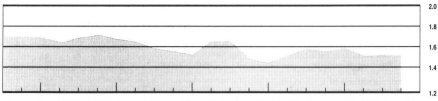

Source: *Chemical Week* 1988. Excerpted by special permission from *Chemical Week*, January 27, 1988. Copyright © 1988, by McGraw-Hill, Inc., New York.

Figure 9-1. Chemical companies' current ratio. Current assets over current liabilities.

The current ratio is a measure of the corporation's "cash" availability and its capacity to meet current obligations, withstand periods of reduced sales, and have cash for unexpected opportunities. Financial analysts often state that a safe current ratio is about 2.0, but chemical companies tend to have a lower number, probably since chemical inventories are generally somewhat smaller than needed in other industries, and the collection period on sales is slightly shorter than some other industries. Also, their large sales to other manufacturers makes the accounts receivable somewhat more secure, with fewer bad debts. The average current ratio for the industry is published quarterly in Chemical Week (see Figure 9-1). It is seen to have been declining during the 1980s as companies have endeavored to operate more efficiently, reduce working capital, and increase profits.

Acid Test Ratio

The acid test ratio, often called the quick assets ratio, is given by

$$\frac{\text{cash} + \text{marketable securities} + \text{receivables}}{\text{current liabilities}}$$

where the sum in the numerator is collectively termed quick assets. For the balance sheet in 1982,

$$\text{acid test ratio} = \frac{(1.04 + 1.36 + 13.26) \times 10^6}{18.00 \times 10^6} = 0.87$$

$$\text{for 1986} = \frac{(2.52 + 0.95 + 11.69)}{20.06} = 0.76$$

This ratio is perhaps a more valid measure of the ability of a company to quickly meet its obligations. Quick assets (without inventories) are readily convertible into cash, whereas inventories must first be sold, something that takes time and is by no means guaranteed. A ratio of 1 or more is generally considered desirable, but again, the chemical industry runs somewhat lower than this, and this ratio has also been declining.

Debt-to-Equity Ratio

Of the many financial ratios commonly used to help analyze company performance, one of the most important is the debt-to-equity ratio. It indicates the amount of debt the company has, and thus its basic strength, ability to raise more capital (debt), and the extent bankers will influence the company's decisions. It is difficult for a company to stay modern and competitive, and even more so to grow, without the considerable use of debt. Some debt is thus highly desirable for most companies, and when it is considered that on the average about 50% of the chemical industry's after-tax profit is used to pay dividends to the common stockholders, and a moderate amount more goes to preferred stockholders, not too much is left over. With the modest after-tax ROI realized by most companies the accumulation of capital for reinvestment is slow. This, coupled with borrowed money having its interest considered as an operating expense (and thus being deducted from taxes), and inflation reducing the real cost of debt repayment makes borrowing attractive if the interest rate is not too high. However, there is a limit: if the ratio of debt to equity becomes what the bankers (who hold much of the company debt) consider to be excessive, they will put pressure on the company and make its decisions so conservative that it is very difficult for it to stay healthy. In this book we shall define total debt as short-term debt, plus the current portion of long-term debt plus the long-term debt. The debt-to-equity ratio can be defined in many other ways, but long-term debt/equity and long-term debt/equity plus long-term debt are commonly used in addition to the preferred method. The 12 company Table 9-1 debt-to-equity ratios for 1982 and 1986 are

$$\text{debt-to-equity ratio} = \frac{\text{total corporate debt}}{\text{stockholder's equity}}$$

$$(1982) \ \frac{1.78 + 0.91 + 22.04}{45.05} = 0.54$$

$$(1986) \ \frac{3.27 + 0.85 + 20.87}{41.63} = 0.60$$

A debt-to-equity ratio of 0.5 has over the years been considered to be borderline; holding at this value is excellent, progressive management; above 0.5 needs close banker attention; >0.75 almost always brings considerable banking pressure, and >1.0 usually results in bankers' dominating management. Currently, with many large mergers and leveraged buy-outs these rules have tended to relax. Considerable debt is required to complete large, attractive mergers or to thwart unwanted takeover attempts. In this case the bankers will allow the excess debt, but only for a limited time. Companies like du Pont and the former Allied Chemical have been under such pressure because of their large mergers, and Union Oil and Phillips Petroleum because of antitakeover maneuvers. They become almost forced to sell off some operations in order to reduce the debt, no matter how worthy these divisions may be, and while this is going on funding internally generated investment opportunities is greatly reduced.

Debt-to-equity problems usually result from a company operating at a normal ratio of close to 0.5 and then suddenly having some problem or opportunity that requires considerable borrowing. This can be tolerated, but if the problem persists they not only loose their credit worthiness and can't borrow more money, but also have to soon let paying back debt become the major objective of the company. Few can prosper, and many not even survive on that basis. The 1982–1986 average debt-to-equity ratio for the 12 chemical companies of Table 9-1 was about 0.53 (calculated on the above basis), and had been rising.

Interest Payments.   The relationship between the yearly income and the amount of interest due on long-term debt and dividends for preferred stock says a lot about what will be left over for earnings per share of common stock. Analysts typically like to see a company generate enough annual income to cover its interest payments three to four times over. A company is considered highly leveraged that has a high proportion of bond interest and preferred stock dividends compared to the common stock dividends. In this situation, a slight increase in earnings can result in a dramatic increase in the ratio of earnings per share of stock. The opposite is also true, and a slight decrease in earnings could mean that earnings are eaten up by interest payments on debt, that is, the figure in the earnings per share spot on the income statement shows up as a loss (inside parentheses). In 1983 and 1984 the average of 60 chemical companies was 28 and 17%, respectively, of interest expense per operating income (before interest and taxes) (*Chemical Week* 1984).

Return on Investment or Sales

The most direct measure of a corporation's performance may be obtained by various ratios relating its income to sales or assets. One such simplified ratio often reported in the press is the *return on equity:*

$$\text{ROE} = \frac{\text{net profit after tax}}{\text{stockholder's equity}}; \quad \text{Table 9-1 (1982)} = \frac{2.750}{45.95} = 6.0\%$$

$$(1985)\ \frac{0.986}{40.82} = 2.4\%$$

$$(1986)\ \frac{3.414}{41.63} = 8.2\%$$

(See Figure 7-3 and Tables 7-3 and 7-4 for company and industry-wide values.) Stockholder's equity is a reasonable approximation of a company's net worth because it represents total assets − all liabilities, which means that it includes working capital and depreciated capital, less the company's debt (liabilities). This ROE will be somewhat higher than a rate of return ratio based upon undepreciated assets plus current assets (approximately, the total investment) since it does not include assets acquired from borrowed capital, but it does include the income from these assets.

Another common attempt to measure a company's rate of return is given in financial comparisons as the *return on sales*. It is widely used because sales and after tax profit are the principal financial items reported for any company, and a simple ratio gives some measure of profitability. It is intrinsically not too meaningful, and yet in actual practice it is a surprisingly useful and consistent measure. Values for various companies, and the industry average are shown in Tables 7-3 and 7-4 and Figure 7-4.

Pseudo-DCF

An approximation of discounted cash flow (DCF) may be obtained by defining the needed financial terms as follows:

a. Total investment = undepreciated assets + current assets.
b. Working capital = current assets.
c. Cash flow = net after tax income + depreciation.

If we assume a 20-year average project life with the factors shown on the balance and cash flow sheets for each year held constant over this period, the pseudo-DCF then uses the terms total investment at time 0, cash flow for 19 years, and the working capital returned on the twentieth year, plus the cash flow. Although not a very realistic DCF figure to compare with individual project DCF estimates, it at least allows a rough first approximation of the Table 9-1 12 corporations' average DCF. The 1982 through 1986 average pseudo-DCF values for these companies were 3.5, 4.1, 5.2, 2.8, and 5.7, respectively. Taking the actual five-year cash flows, the 1982 assets, and 1986 cash flow (and working capital return) for the remaining 15 years, the DCF was 4.9%. Similar pseudo-DCF values are plotted in Figures 7-12 and 7-13.

These ROE and DCF numbers are very low, but nevertheless fairly typical of most chemical companies. In many cases they have been reduced by high operating costs and company overhead, but more importantly, the reduction is caused by lower than anticipated sales prices and plant operating rates (because of limited markets). These low numbers provide all the greater reason why management desires new projects to have a higher DCF (i.e., hopefully over

15%) to cover the risk of any new project, and to attempt to raise the present low DCF values.

Cash Flow/Capital Spending; Book Value

Capital spending is an easy to understand line item in the cash flow spending table, as is the change in capital spending over the years. However, the ratio of cash flow/capital spending provides a more graphic figure to show the company's commitment to new projects, modernization, and environmental spending. This ratio is also plotted quarterly in the magazine *Chemical Week,* as shown in Figure 9-2. It is seen that the capital spending percentage has been erratic, but has generally held fairly steady during the 1980s at a value averaging about 1.0.

The *book value* of a company is the sum of the total assets, less intangibles (good will, etc.), less the current and long-term liabilities, including preferred stock (which in this case can be considered as if it were a bond). In other words, it is the value that would be received for the company if the debt were paid and the assets sold at their stated value. This number can be compared with the market value per share of stock times the total number of shares outstanding. If the result is lower than the company's book value (which as we have noted before is often undervalued), then the stock is selling at too low a price. That is, you could buy all of the company's stock for less than it would cost you to buy the company's assets. This is the basic situation that attracts many "raiders" who, by acquiring a company, will make money selling some or all of its divisions at close to (or more than) the real value of the assets in each group.

GENERAL

One can usually obtain an annual report by calling or writing the corporate secretary of the company you want to evaluate, or from stock brokerage companies. If you are a member of the Dow Jones News Retrieval Service you can call up the disclosure data base on your modem and download 10K and 10Q information (much more detailed financial information than annual reports)

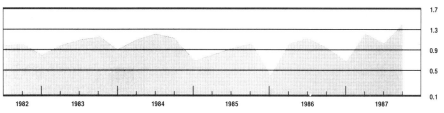

Source: *Chemical Week* 1988. Excerpted by special permission from *Chemical Week*, January 27, 1988. Copyright © 1988, by McGraw-Hill, Inc., New York.

Figure 9-2. Chemical companies' cash flow over capital spending.

directly into your computer. There are many good books on reading financial reports. Merrill Lynch, Pierce, Finner & Smith, Inc., has a brochure called "How to Read a Financial Report." John Wiley & Sons publishes a book by John A. Tracy with the same name. American Management Associations publishes a book by Donald E. Miller called *The Meaningful Interpretation of Financial Statements*. Dun & Bradstreet publishes an annually updated "Guide to Your Investments" by Nancy Dunnan that includes a section on interpreting financial statements. Barrons publishes a *Dictionary of Finance and Investment Terms,* a handy companion piece to any other reading material.

Reading an annual report is never easy, and in the end you seldom feel that you fully understand what is happening with the company. However, for a company that you are interested in it is almost always an engaging and worthwhile experience. The message from the president and other text help you understand what the company feels is important in the way of past and future actions. Some analysis and manipulation of the numbers (i.e., ratio analysis, etc.) lets you draw some of your own conclusions on what and how the company is doing, and to some extent indicates the quality of the management. This later point is the most difficult of all, since large corporations, just as all bureaucracies, tend to hide incompetence, marginal performance, and overconservatism. However, a close analysis of a series of annual reports can begin to show a pattern of static or declining earnings, few new projects, low research and development spending, competitive decline, and so on. Generally things will look fairly good, however, and reading the annual report may give you much better feelings of pride and knowledge of your own company, or a better feeling for where your investments should be.

REFERENCES

Chemical and Engineering News. 1987. Facts & figures. (June 8):46; 1986 (June 9):55; 1985 (June 10):41; 1984 (June 11):45.
Chemical Week 1984. Chemical financial review. (July 25).
Chemical Week 1988. (Jan. 27):38.
Compressed Air Magazine. 1987. Understanding financial reports. (June):10–15.
Fortune 500 1987. (April 27):355–412.

10

PROJECT MANAGEMENT

Many responsibilities of a chemical engineer have an important economic impact and an economic component, even though they are not direct economic functions in themselves. Such is the case with project management, where many job functions other than economics are involved, but where a large part of the work (e.g., cost control) has a direct financial reportability, and the successful completion of the entire project may have a significant economic impact upon the company or at least upon the division performing the work. For that reason project management will be briefly reviewed in this chapter, with its control function covered in somewhat greater detail.

MANAGEMENT PRINCIPLES

As a chemical engineer begins to be given more job responsibility and/or starts to advance through the supervisory ranks there are many new skills to develop in relation to management principals and abilities. In general with each step of promotion the amount of individual work performed by the engineer decreases, and supervisory time increases, somewhat as diagramatically shown in Figure 10-1. With promotion many engineers cling to their former level of personal work output, and do their new supervising function as an extra assignment. This can result in both poor supervision and job performance. Conversely, some engineers immediately want to be full time supervisors, and thus are often overbearing and harass the staff with detailed instructions. In this case the company has lost the services of a good engineer, and gained a very poor supervisor. A middle ground is obviously best with a strong suggestion to be sure to do an adequate amount of personal work, since this results in better supervision, greater output, and awareness of the group's problems—and the normal tendency is to over supervise.

Many engineers can naturally and intuitively switch into management and do an excellent job from the beginning. However, it is recommended that everyone can benefit from studying management requirements at every promotional step. The basics are very simple and primarily common sense, but a reminder about

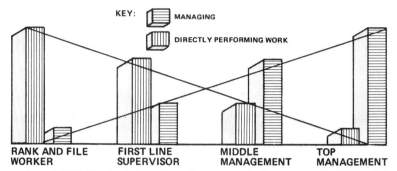

| RANK AND FILE | FIRST LINE | MIDDLE | TOP |
| WORKER | SUPERVISOR | MANAGEMENT | MANAGEMENT |

Source: Magad 1985. Reprinted from *Plant Engineering*, April 11, 1985. © 1985 by Cahners Publishing Company.

Figure 10-1. Reduction in personal work performed as managerial responsibilities increase.

successful techniques, standard methods and procedures, laws governing employee relations, and so on can be very important. There are frequent magazine articles on various components of management, and numerous books on the subject. Also, most companies are willing to send new managers to training courses or lectures on management, such as the excellent series put on by the American Management Association (Martin 1976). All of this self-study can result in better management skills, and even more importantly, increased confidence and satisfaction in doing a good job on a new assignment in a different but challenging field, management.

Project management is one of the managerial responsibilities that engineers are frequently assigned, and sometimes fairly early in their careers. Traditionally in the past management tasks on major plant projects have been principally performed by a general contractor. This practice is less common today, when most companies retain major management responsibility for their own projects. For the plant engineer this change in philosophy can mean an added challenge— he may be asked to assume new supervisory responsibilities. He will have to learn project management, and then accomplish the construction goals not only in strict accordance with the performance requirements but also within demanding budget and schedule limits. There are many excellent books (Archibald 1976; Guthrie 1977; etc.) and courses frequently appearing on this subject to assist in the learning process.

Management in general involves planning, organizing, staffing, influencing, and controlling. These elements are discussed in the following section based on the general outline shown in Figure 10-2.

Planning

Planning is the process of identifying objectives and proposing their methods of attainment. With project management it also means first *Defining the Scope of Work* (Dickson 1986). More than any other segment of project management,

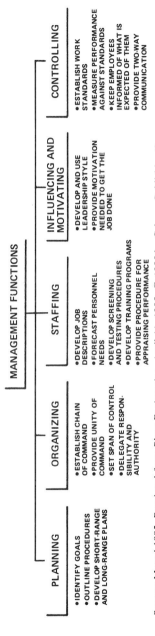

Figure 10-2. General management functions.

Source: Magad 1985. Reprinted from *Plant Engineering*, April 11, 1985. © 1985 by Cahners Publishing Company.

the scope of work is the key to a project's success. It must contain a precise description of the project, outlining the proposed work as completely as possible. The scope of work must give company management, the project team, and the contractors a clear picture of what is required of them, the services each will provide, and the type of support each can expect from the others.

The scope of work has a twofold function. It is an internal agreement with the company's management on the project's objectives, and also the basis for establishing what may be required from the company staff and potential contractors. For this reason, the scope of work should be developed with input, review, and approval by company management, engineering, research and development, maintenance, and the operations department. In addition, the scope of work may be the basis for bidding documents for prospective contractors that delineates construction details and the level of effort required to perform the work. For this and other projects, such as research and development or development studies (pilot plant operation, plant tests, etc.) or a major maintenance activity, the project detail must be completely specified for the company's own staff, including the coordination between the various groups involved.

Organizing

Organizing involves the assignment of duties and coordination of efforts to ensure maximum efficiency in the completion of the project. The following are some basic principles of organization:

- Proper organization requires a clear-cut chain of command from top management to the lowest levels. Organization charts help define the chain.
- Each worker should be accountable primarily to one person, even if on assignment from another group.
- A small number of people should report to one supervisor. Ideally six to eight is a good limit, but it can be more depending upon the types of activities and the people supervised. It is influenced by the predictability of the work, the quality of the workers, the complexity of the jobs, the supervisor's individual input, and the measurability of results.
- The manager must know how to delegate both the responsibility and authority for getting a job done. This delegation should be clear to everyone.

Staffing

To complete a project within budget and time limits, project managers need an integrated team that can work toward a common goal. A productive group does not happen spontaneously; it must be carefully put together. The goals of team building are to increase the ability of various staffs to work together effectively.

Each group in the organization should have the opportunity to derive satisfaction and pride in their work. With project management the team often consists of others from your organization who have duties or responsibilities for portions of the project, contractor personnel, and staff people from both your and the contractor's organization. The project manager attempts to organize for:

- Task accomplishment: completing the project work, solving problems, making decisions, establishing priorities, planning activities, allocating tasks, adjusting work and schedules as needed, and using resources effectively.
- Team relationships: building and maintaining effective communication procedures and an esprit de corps (between supervisors and subordinates and among colleagues and peers), improving work habits and efficiency, understanding and managing the team's needs and problems, clarifying misunderstanding and managing conflicts or problems quickly and efficiently.
- Team work functions: establishing the work assignments and execution procedures for each group (such as how purchasing coordinates with the designers, engineers, contractors, and production groups). It is important to ensure an efficient coordination among all of the staff involved, including management and the contractor.

The project manager who stresses the effectiveness of team building can usually obtain a higher standard of performance and quality from both his company people and the contractors doing the work.

Controlling

This is the process that measures work accomplishment, expense budgets, and schedules to ensure that the project is completed as close to the targeted values as possible. It involves the following considerations:

- Plans, policies, and procedures should be developed to provide a format for directing the work and measuring the results.
- Budgets, work standards, and schedules should be established for the staff, contractors, and all activities.
- Current costs, performance, and status should be constantly measured against the budgets.
- Problems and budget deviations should be frequently reviewed in order to promptly make changes and accelerate or slow down operations as needed.

This entire control function will next be examined in more detail to establish some of the detailed techniques, working methods, and procedures.

COST CONTROL

Code of Accounts

At the beginning of a project manager assignment one of the first actions that should be taken is to be sure that there is a functional "code of accounts" for the project. This may already exist, but if not, one should be established to ensure that there are complete code numbers to allow accurate and detailed accounting records to control and document the work. Each major grouping of expenditures that one can constructively visualize should be listed and categorized, and then given a numerical code. There is no hard and fast rule as to how many accounts there should be, or how the subdivisions are to be made, but most corporations have their own standard code of accounts which can be adopted to the needs of each project. A portion of a typical code of accounts is illustrated in Table 10-1. In this particular case, the 700 code series is reserved for instrumentation. The second and third digits are used for identifying the principle category subdivisions; thus 700 identifies general instrumentation, 722 control valves, and so forth. Further numerical coding may be used to identify sections of the plant (722-03, control valves in tank farm), or any other function, such as vendors, and so on.

Numerical coding may also be used to assist with time and spending control as a means of monitoring expenditures within any cost category or subdivision desired. Such a code can facilitate access to computer stored data for time estimates, shipping dates, purchase order and other document identification and status reporting. For example, the total 700 code account may be summarized or printed at any time, as well as subaccounts such as 721 (relief valves), the total account of a particular vendor, or the account ascribable to a particular material of construction (such as nickel).

Spending Curve

The complete project purchasing history should generally be monitored by a project spending graph. A typical curve takes an S-shape, as the example shown in Figure 10-3 where the coordinates can be dollars and time, or some other scale such as the fraction of the total budgeted cash spent to date versus the cumulative fraction of the total scheduled time that has elapsed (Haley 1987; Roth 1979). For each project the spending curve should be estimated and constructed after the best possible time and manpower schedule has been established. Construction costs may then be tracked on the graph to show how close the project is conforming to budgets and schedules. The shape of the curve in Figure 10-3 is quite typical of many projects, with very slow cash dispersement at first until the sudden spurt at the 40–60 time percent point when equipment delivery commences.

Effective cost control, of course, involves a great deal more than graphical

Table 10-1
Typical Code of Accounts.

700-00 Instrumentation

Acct. no.	Description
700-10	Design calculations
700-20	Specifications—general
700-30	Vendor correspondence
700-40	Inquiries
700-50	Operating Instructions and start-up data
700-60	Correspondence—general
701-00	Indicators—gauges
702-00	Controllers—recorders
703-00	Transmitters
704-00	Thermocouples—RTD
705-00	Analyzers
706-00	Transducers
707-00	Switches
708-00	Alarms—annunciators
709-00	Intercom
710-00	Panel—enclosures—mounting shelves
711-00	Graphics
712-00	Control computer
713-00	Management information computer
714-00	Input terminal
715-00	Printer
716-00	Video terminal
717-00	Modems
718-00	Air supply set
719-00	Orifice meter run
720-00	Valve manifolds
721-00	Relief valves
722-00	Control valves
723-00	Regulators
724-00	Piping
725-00	Wiring
726-00	Supports
727-00	Installation
728-00	Painting
729-00	Testing

Source: Lowe, 1980. Excerpted by special permission from
Chemical Engineering, July 14, 1980. Copyright © 1980, by
McGraw-Hill, Inc., New York.

tracking. For example, cost control records are maintained along with the project completion monitoring techniques discussed in the next section, so costs versus subprojects' completion can be closely watched. If deviations occur, as in Figure 10-3 where $5.7 million, or 57% of the budgeted funds have been expended,

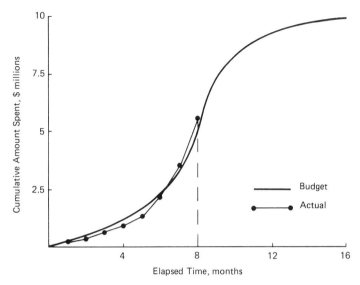

Figure 10-3. Typical project spending curve.

but only 50% of the funds were budgeted to have been spent, the project manager must determine the cause, and if an important deviation, attempt to get the costs back in line. Also, corporate management must be kept abreast of the progress of the project by means of periodic status, cost, and time over- or underrun reports and review conferences. This serves to help keep the project manager constantly monitoring the work for time and budget control, and initiating procedures for keeping the project on schedule.

Subprojects, Tasks

The method used to estimate the spending pattern for a project is basically the same as required for all aspects of project time and accomplishment control (including bar charts, CPM, etc.). That consists of dividing the project into as many substantial subprojects or tasks as can be envisioned, and then estimating the manpower, equipment, time requirements, dependency on preceding projects, sequence of other dependant projects, and all other information that may be useful for controlling the subprojects. Often the first subproject selection after this amount of detailing will appear to have the tasks too large and complex or interdependent, and you will wish to break some of them into additional subprojects or sub-subprojects. Sometimes the opposite will occur; the projects will be too small or trivial, and you will wish to combine some of them. In many cases of medium size or complex projects, or until you become quite fluent with scheduling, you will need to go through the subproject list with all of its labor, cost, and time components several times before you are satisfied that you have a reasonably optimum number of projects that are well balanced, and easy to track and control without excessive accounting involvement. From this project list you can then take off the labor and equipment requirements with time, and prepare the budgeted spending curve.

TIME CONTROL METHODS

An equally important aspect of project budgeting and monitoring is time control, to make sure that a project will be completed within the allocated time. Time overruns usually result in cost overruns, and for construction projects, plant start-up delays, which may cause costly production delays and failure to live up to product delivery schedules to customers. As was the case with cost control, a number of planning and monitoring tools are available for project progress monitoring, some of which are readily adaptable to computer usage. In fact, the complexity of most very large projects' control and monitoring normally demands the use of computers, and a thriving business exists in the appropriate computer software.

Bar Charts

The simplest time planning tool is the ordinary bar chart. A very simple example for an R & D project is shown in Figure 10-4, and a more detailed R & D project schedule is given in Figure 10-5. Very often bar charts have codes to show the amount or status of project completion, such as shading the bar along its length to indicate the percent of completion, or showing hollow diamonds along the bar for the fraction of project completion. Overruns can be shown by

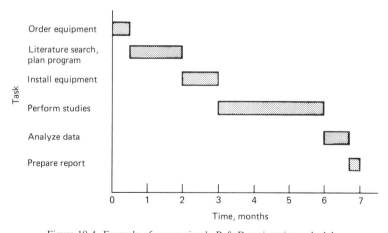

Figure 10-4. Example of a very simple R & D project time schedule.

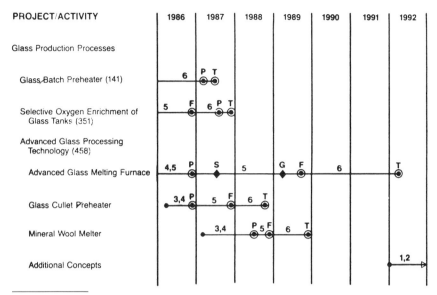

Figure 10-5. More complex bar chart for an R&D project.

Code
1. Assess R & D needs.
2. Evaluate existing information.
3. Conduct experimental data acquisition program.
4. Design, assemble, and conduct "laboratory-type" tests.
5. Design, construct, and conduct limited field experiments.
6. Conduct field tests to verify system performance.
S. Select most promising concept for further development.
G. Go/no-go decision.
F. Initiate field test.
P. Achieve proof of concept.
T. Transfer technology or information for application.
Source: Institute of Gas Technology 1987.

extending a dotted bar, or by hollow and shaded diamonds at the ends of the bar. Many similar legend devises can be employed for other information, if desired, such as showing engineering, procurement, and installation periods all as different shadings along the bar.

For small jobs the bar chart is an adequate resource allocation and time control tool. However, no matter what the job size or the complexity of the monitoring method, the resulting schedule is only as reliable as the projected timing of the individual component tasks. If, for instance, some of the equipment delivery in Figure 10-4 is stretched out, other equipment may still be available within the total projected time period, but jobs which depend upon the missing equipment will be correspondingly delayed, and so might the completion of the project. Thus, the effectiveness of scheduling depends upon the realism and accuracy of the initial time forecasts for the individual tasks. Bar charts, being simple, easy to construct, and read have become the basic method for time control of smaller projects. These charts also are quite useful in assisting with the preparation of the previously discussed spending schedules.

Network Diagrams; Critical Path Method (CPM)

Large construction projects involve many subcomponent jobs, as illustrated by the abbreviated list given in Table 10-2. Each component, such as the job described as "excavations and foundations" may in itself involve a complex network of operations that needs to be incorporated into the budget and timing schedules. Many hundreds of jobs may be involved, and a simple bar chart rapidly becomes unwieldly and difficult to construct.

In order to handle the project management of medium and larger jobs various graphical techniques have been developed, broadly called "network diagrams." The most well known and useful of these is the "critical path method, or CPM."

Table 10-2
Typical Abbreviated Job Sheet.

Job no./Job	Job Sequence	Job Duration, Days
1 Clear site	0:1	4
2 Excavations and foundations	1:2	9
3 Install support steelwork	2:3	9
4 Install stairways	3:4	3
5 Install platforms	3:5	8
6 Install handrails and platforms angles	3:6	7
7 Install equipment	4, 5, 6:7	4
8 Calibrate equipment	7:8	7
9 Install prefabricated ductwork	7:9	21
10 Install process pipework	4, 5, 6:10	10
11 Connect pipework to equipment	8, 10:11	4
12 Install service pipework	4, 5, 6:12	7
13 Connect service pipework to equipment	8, 12:13	3
14 Install instrumentation	4, 5, 6:14	14
15 Connect instrumentation to equipment	8, 14:15	4
16 Install electric apparatus	4, 5, 6:16	9
17 Connect electrics to equipment	8, 16:17	3
18 Insulate equipment	8:18	10
19 Insulate pipework	11, 13:19	14
20 Insulate ductwork	9:20	6
21 Testing equipment	1–17:21	4
22 Painting	1–13:22	7
23 Cleanup	1–20:23	1

Source: Lowe, 1980. Excerpted by special permission from *Chemical Engineering*, July 14, 1980. Copyright © 1980, by McGraw-Hill, Inc., New York.

It is a graphical procedure which may look somewhat like either a line (bar) chart or a process flow diagram, and where each line on the chart represents a subproject or task. The key is that the line and its "nodes" (starting and ending points) show its dependence upon the other subprojects (i.e., which subproject must be completed before another can start). This allows a knowledge at all times of what sequence of subprojects represent the longest time for the entire project to be completed. This is the "critical path." As the work progresses and some schedules slip or accelerate from the budget, the situation can be constantly reassessed and the then-current critical path determined. This allows the project manager to constantly adjust manpower and priorities in order to maximize the effort to keep the project on schedule.

Such network diagrams are not only for construction projects, they are also valuable tools for any project that combines multiple tasks. A project can be as simple as installing one piece of equipment or as complex as carrying out a multiyear research and development study, a major plant maintenance program (turnaround), or the construction of a new plant. Network diagrams help project managers monitor and control project costs, time, labor, equipment, and materials.

In its simplest form a CPM chart may be constructed as a modified bar chart, with an indication of the comparative time and sequence to complete each job, such as shown in Figure 10-6. All subprojects or tasks are plotted with their line length exactly scaled to the time involved, and a base time scale is maintained from the start to the finish. In addition to constructing and working with these charts manually, programs have been devised which permit the computer generation of a CPM plan, and the constant ability to detail which tasks are on the critical path, which have spare time, and so on, as the work progresses. Programs are also available to facilitate manpower and equipment scheduling based upon the CPM plan (for instance, workforce demand leveling), and to help monitor the progress of construction, pinpoint jobs that must be done in succession and simultaneously, and update the plan.

For larger and more complex projects the CPM charts are more frequently constructed as flow-sheet-type layouts. The jobs (or activities) are symbolized by lines with arrows which point in the general direction of the project time flow, but are not drawn on a time scale (see Figures 10-7 and 10-8). The project number is usually indicated above or below each line, and sometimes the job duration is listed on the other side of the line. Each arrow starts at a preceding node P and ends on a succeeding node S. The nodes are usually symbolized by small labeled circles and numbered in sequence; their exact position, and the arrow lengths are immaterial. Each activity ends at a particular node, and that activity must be completed before the next activity starts.

Note that in Figure 10-7 (and its job sheet in Table 10-3) there is no requirement or implication that jobs B, C, and D need to be started simultaneously;

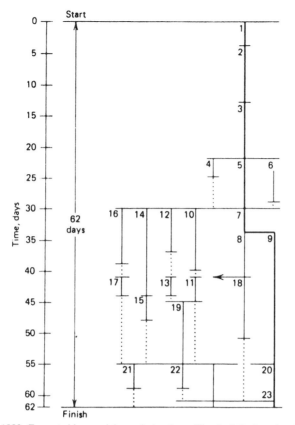

Source: Lowe 1980. Excerpted by special permission from *Chemical Engineering*, July 14, 1980. Copyright © 1980 by McGraw-Hill, Inc., New York.

Figure 10-6. Critical path bar chart diagram (of Table 10-2 job).

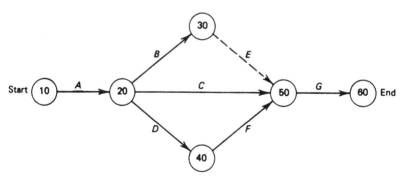

Source: Valle-Riestra 1983. Reprinted by permission from *Project Evaluation in the Chemical Process Industries*, 1983, McGraw-Hill, Inc., New York.

Figure 10-7. A simple flow sheet type of critical path diagram (for Table 10-3 job).

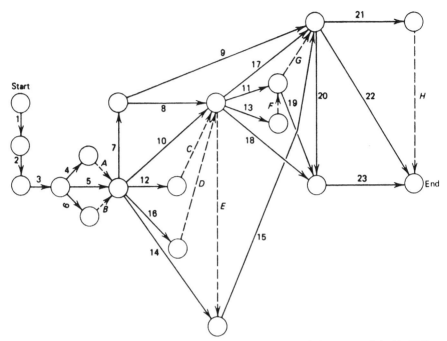

Source: Lowe 1980. Excerpted by special permission from *Chemical Engineering*, July 14, 1980. Copyright © 1980 by McGraw-Hill, Inc., New York.

Figure 10-8. Critical path flow sheet diagram (of Table 10-2 job).

only that they can not start before job *A* is completed. The length of the line can, if desired, as in the bar chart CPM, accurately indicate the duration of each task, and in some cases dotted lines are used to indicate waiting periods before the next job starts. Actually, these jobs could have been done during any portion of the dotted plus solid line, or even stretched out to occupy the entire period.

Table 10-3
Simple Job Sheet (Showing Nodes).

Sequence	Nodes	Job	Duration
A	(10, 20)	*A*	4
A:*B*	(20, 30)	*B*	4
A:*C*	(20, 40)	*C*	6
A:*D*	(20, 50)	*D*	12
E ("dummy")	(30, 50)	*E*	0
D:*F*	(40, 50)	*F*	4
B, *C*, *F*:*G*	(50, 60)	*G*	6

Source: Valle-Riestra 1983. Reprinted by permission from *Project Evaluation in the Chemical Process Industries*, 1983, McGraw-Hill, Inc., New York.

By plotting the jobs in this manner, the shortest period by which the entire project of Table 10-3 can be completed (22 days) is noted by the longest duration of any completely connected (from start to end) set of lines. This sequence of jobs is called the critical path, and any delay with any job on this path (A, D, F, or G) will directly delay the project. All other jobs have some leeway in their scheduling, at least based upon the original time estimates. Again referring to Figure 10-7, the activities A, B, . . . are labeled, for purposes of computer manipulation, on the basis of their numbered P and S nodes, as (P, S); thus activity F is labeled (40, 50). The node numbering is arbitrary, but for any activity the S number must be higher than the P number. Two activities may have common preceding and succeeding nodes, but not the same nodes. For example, jobs B and C cannot be started until A has been completed, and they both must be finished before G can be started. To avoid an anomaly, job B, for instance, is provided with its own S node, and dummy activity E (30, 50), having zero duration, is added to connect job B to the desired P node of job G. It makes no difference whether the dummy was added to B or C, but for the sake of internal consistency it is usually best to add the dummy to the shorter duration job. Thus, job B is scheduled to take less time than job C. Dummies also indicate required precedence.

The procedure for drawing a network diagram is to first have a complete breakdown of the work to be done, as previously noted, into a series of tasks with estimated time of duration, manpower, costs and sequence. From this you can start the network diagram from either the beginning or the end of the project, or both. There are frequently somewhat obvious large subnets you can quickly draw and then sort out the connections in the diagram where there are activities in progress simultaneously. You may even find some new work tasks that should be included or deleted. It is often helpful to do this at least initially with a time base (even for a large project) and with the presumption that each activity starts at the earliest possible time. With these time estimates on the network diagram it may become apparent that the entire project will take too long. At this point, you can identify particular activities that may be candidates for time compression (that is, tasks you believe can be done faster), or you may schedule more parallel activities. For medium sized projects one should construct a network diagram of perhaps 40 activities, or up to 60 if required. Such a diagram can normally be drawn reasonably quickly (i.e., perhaps 2 hours when you become somewhat experienced) and will fit a standard 17 × 22 in. sheet of paper. If any individual activity in this network is very large, the project manager may wish to make a separate network diagram for it. In this way, with a few hand-drawn networks of a reasonable number of activities each, even large projects can be handled without the use of a computer based network system, if desired.

A CPM diagram of the flow sheet type, but equivalent to the bar chart in Figure 10-6 is shown in Figure 10-8. The bar chart is easier to draw, and the project timing is more readily visualized since the bar chart includes a time scale and parallel jobs are more obvious. In the flow sheet CPM diagram, however the project dependency is more easily seen, and much more complex arrays can be covered. Each of the activities are labeled with their number assignments from Table 10-2. The dummy activities are also included to avoid duplication of activities between two specific nodes. Examples include activities 4, 6, dummy (A) and (B) in relation to activity 5 where the dummies are associated with the shorter duration jobs. Dummy E is not used to avoid activity duplication, but only to show the precedence of job 8 over job 15.

An example of a combined time scale and flow sheet CPM is shown in Figures 10-9 and 10-10. The present time is noted on the time scale, and in Figure 10-9 little triangles are located along each projects' line to show its current status, or the extent that the subproject is completed. Dotted lines are used to show slack time (or a dummy activity). Figure 10-10 is a redrawn version of Figure 10-9 in which the project manager has taken advantage of the slack time. All late (delayed) activities are drawn to show the work remaining to be done, and subsequent tasks are thus rescheduled in several cases.

Although activities A^*_2, B^*_2, $C_{1\text{ and }2}$, and D^* are in fact later than planned, the project has not yet suffered an irretrievable schedule slippage. The project now has two critical paths, whereas it previously had only one, and there is no longer any slack in the upper branch (A^*_2 and G^*). Because there is still one month of slack on the lower path (C), perhaps some of the resources allocated to it (or to task B^*_2) might be redeployed to one of the other critical path activities. It is vastly more difficult to complete a project on time with more than

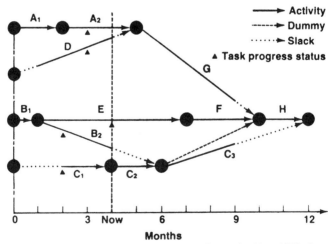

Source: Rosenau 1987. Reprinted with permission from *Chemtech.*, May, 1987. Copyright 1987, American Chemical Society.

Figure 10-9. Critical path flow sheet diagram with time scale.

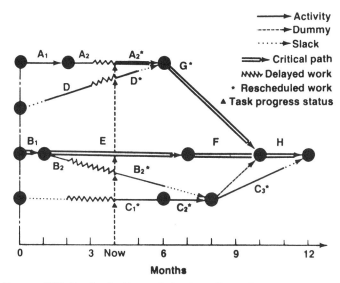

Source: Rosenau 1987. Reprinted with permission from *Chemtech*, May 1987. Copyright 1987, American Chemical Society.

Figure 10-10. CPM diagram showing completed and slack time.

one critical path, so it is unlikely that this project will be completed on schedule, although this is not yet definitely known.

Time Reduction.   The job durations in Table 10-2 were based upon a normal staff of workers and equipment, and the project time turned out to be 62 days. If this much time is unacceptable, consideration can be given to having some of the project's jobs "crashed," or extra workers, equipment, and overtime can be authorized to shorten the time. Some jobs cannot be crashed, and others do not require it.

It can be assumed that in most cases crashing is expensive and may result in difficult scheduling problems. Consequently, the cost benefits of a reduced time schedule must be balanced against the extra expenses. This usually results in a request for a rather limited amount of crashing, or that the crashed jobs be the ones that can be most economically performed. This might include first the jobs that are done alone, such as in Table 10-2 and Figure 10-6 jobs 1, 2, 3, and 23, or the ones with an obvious excess time requirement, such as 5, 9, and 20. Beyond that, the interrelationship of projects causes crashing to become more complex, and a number of trial-and-error interrelated estimates of the time savings and increased costs would be involved.

Manpower Leveling.   An additional benefit that can often come from CPM analyses is the possibility of leveling out the number of workers involved, or assistance in scheduling the more highly skilled or short-supply workers. The critical path diagram clearly shows the time dependency of the projects and gives an indication of where there is slack time and projects can be moved earlier or later, or spread out with a reduced staff. By so doing in the most judicious manner the total number of workers on the site at any time can be at least somewhat smoothed out, and the demands on the more critical or highly skilled workers also evened. In actual practice this factor is often so important that a number of trial-and-error constructions as with the total time estimates noted above, must be made, essentially working backward until a satisfactory manpower schedule can be devised with both the more skilled and total staffs. This may result in an increased project time schedule, but it will be more realistic, workable, and cost effective. Computer assisted CPM scheduling is especially useful in such manpower leveling and in running through the many economic comparisons needed for leveling and crashing.

PERT (Program Evaluation and Review Technique)

A variation on the critical path method employing statistical probability calculations is the PERT program. If reasonable estimates can be made of the possible range of completion times on each subproject, a statistically (beta distribution) probable "expected time, t_c" can be established:

$$t_c = \frac{t_l + 4\, t_m + t_n}{6}$$

where t_m is the most likely job completion time, and t_l and t_n are the low and

Table 10-4
Typical Data Provided in a Computer Printout for Computer-Based Network Reporting

Event				Start		Finish		
Start	Finish	Activity	Duration	E	L	E	L	Slack
1	2	a	4	0	18	14	22	18
1	3	e	10	0	0	10	10	0
1	4	h	6	0	11	6	17	11
2	5	b	3	4	22	7	25	18
3	5	c	2	10	23	12	25	13
3	5	f	10	10	10	20	20	0
4	6	j	3	6	17	9	20	11
4	7	k	7	6	23	13	30	17
5	7	d	5	12	25	17	30	13
6	7	g	10	20	20	30	30	0

Source: Rosenau 1987. Reprinted with permission from *Chemtech*, May 1987, pp. 288–292. Copyright 1987 by American Chemical Society.

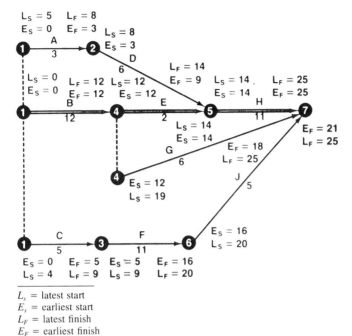

L_s = latest start
E_s = earliest start
L_F = latest finish
E_F = earliest finish

Source: Rosenau 1987. Reprinted with permission from *Chemtech*, May 1987. Copyright 1987, American Chemical Society.

Figure 10-11. CPM diagram showing complete, earliest, and latest information.

high time values, respectively, that you feel have less than a 5% chance of occurring. With these expected values for each project's completion time you can estimate a new most probable critical path time period as before, and establish the earliest expected completion time for each job, as well as the earliest and latest start and finish times. Similar probability calculations can be made to determine the odds of completing the project at any given total elapsed time.

It is seldom that completion time estimates of this accuracy are available, since most major project delays are from unforeseen events. However, some of the benefits of network analysis are from related factors to these PERT calculations. As an example, the earliest and latest completion time information from a CPM analysis are shown in a simple CPM chart (Figure 10-11). A computer printout for a different project is given in Table 10-4. Many computer software packages are available for PERT and CPM programs, including drawings and complete data analysis of all types (*Plant Engineering* 1988b; Katzel 1987). Most are labeled simply as PERT programs ·alone, since they easily have the capability to handle the probability analysis if data are available, and otherwise can treat the project as if it were a CPM network. There are many excellent reviews of computer aided project management (*Plant Engineering* 1988a).

REFERENCES

Archibald, Russell, D. 1976. *Managing High-Technology Programs and Projects.* John Wiley & Sons, New York.

Dickson, Thomas W. 1986. How to manage successful construction projects. *Plant Engineering* (March 27):56–57. Cahners Publishing Co.

Guthrie, Kenneth M. 1977. *Managing Capital Expenditures for Construction Projects.* Craftsman Book Co., Solana Beach, CA.

Haley, Charles L. 1987. Controlling project costs. *Plant Engineering* (Aug. 27):56–56. Cahners Publishing Co.

Institute of Gas Technology. 1987. R & D programs:208.

Katzel, Jeanine 1987. Microcomputer application software. *Plant Engineering* (May 10):68–108. Cahners Publishing Co.

Lowe, C. W. 1980. Get more control over any project with bar charts. *Chemical Engineering* (July 14):139. McGraw-Hill, Inc., New York, NY 10020.

Magad, Eugene L. 1985. Management principles for plant engineers. *Plant Engineering* (April 11):43–45. Cahners Publishing Co.

Martin, Charles C. 1976. *Project Management.* American Management Association, New York.

Plant Engineering. 1988a. Computer-aided project management. (Jan. 21):50–56. Cahners Publishing Co.

Plant Engineering. 1988b. Project management software. (Jan. 21):57–60. Cahners Publishing Co.

Rosenau, Milton D., Jr. 1987. Planning without pain. *Chemtech* (May):288–292. American Chemical Society.

Roth, Joanne E. 1979. Controlling construction costs. *Chemical Engineering* (Oct. 8).

Valle-Reistra, J. Frank 1983. *Project Evaluation in the Chemical Process Industries.* McGraw-Hill, New York:731.

11

PERSONAL INVESTING

WHY INVEST

Why should an engineer think of a personal savings and investment program? The company he works for will have a retirement plan and there is always social security. Furthermore, there is a never-ending demand for limited funds, so perhaps investing should wait?

Every situation is different, and perhaps that is the correct conclusion, but there are very compelling reasons to start a personal savings and investment program as early as possible in one's career. First, as discussed in an earlier chapter, tenure forever with even the best corporations is no longer assured, since the pattern has been established that there can be major layoffs, particularly with the very junior or senior staff, whenever greater "efficiency" is needed, or when mergers occur. Hopefully pension plans will remain, but many well-established engineers have lost their jobs, pensions, and security very unexpectedly. By the same token, with the massive governmental debt and the rising medicare costs, there is a reasonable possibility that engineers today may never see any significant social security benefits because of their high income level. The social security funds may be so depleted that they will only be available for lower income retirees, or the payments may be reduced for everyone to poverty levels. For these job tenure or retirement uncertainties alone, everyone, and particularly engineers with their good salaries, should start as soon as possible investing in their own emergency fund or retirement plan, preferably in an IRA program which we shall discuss later.

For a second but related reason, everyone needs some extra money for more normal emergency or special situations, or planned major expenses such as house or car purchases or the childrens' college education. A savings-investment program is the only way to ensure that these funds will be available when needed.

Finally, and perhaps not of interest to everyone, there is every expectation that with a careful investment program most engineers can be well-off to modestly wealthy by the time they retire. The power of compounding is so great that an early start with wise investments will generate very large returns. Besides each of the above reasons an extra benefit that usually accrues is that most people, once they get started, find that personal investing is very interesting, challenging, and rewarding. It even very often helps in one's career advancement.

HOW TO INITIATE AN INVESTMENT PLAN

Budgeting

The first step in any investment program is to prepare and execute a budget. Personal budgets were discussed in Chapter 8, and that material should be reviewed in this context. The major function of the budget in this regard is to initiate a definite and substantial savings routine. Sometimes this can be done most easily by a withholding program of some sort at work, such as to a credit union, possibly a company stock purchase plan, or even by allowing an excess income tax deduction. It is best to do the saving on your own, however, if you can so that you more completely control the funds, but any method will work just fine. The amount saved does not need to be too large at first, but it should be as much as possible on a regular basis, and it can be increased later when one's income increases and spending demands become somewhat less urgent.

A corollary to a savings plan is the reminder to not incur debt beyond a fixed and budgeted amount, such as for a car, home, furniture, and so on. With your good engineering job, credit will be given all to readily, and the repayment and interest can easily get out of hand. Some credit is an excellent part of your investment program since inflation reduces the present-day value of what you repay, and some (all, if carefully done) interest reduces income taxes. With other borrowing (such as on a house) the loan acts as a means of obtaining leverage on the investment, and thus can greatly increase its value. However, all borrowing must be carefully planned and controlled or else it can ruin the entire budget. Particular care must be taken with credit cards and revolving charge accounts since they are the most expensive form of credit on the market, the most convenient to use, and simplest to get. The problem is that too many consumers get in over their heads, but if used wisely credit cards are a good way to track expenses. They allow you to buy quality merchandise where you might otherwise shop for cheaper items; they give you monthly itemization of your credit and expenses, and allow some activities not possible or convenient with cash (car rental, motel reservations, etc.). And once your credit is established for the card, it's enhanced for other purposes as well.

Initial Investment Program

Once a budget is established an investment plan should be formulated, first with the initial or high-priority items and later as an outline for one's entire working career. Obviously each plan has to be very individualistically based upon a person's income, amount saved per year, age, family needs, and personal desires on balancing security, risk, and the time required to manage the investment accounts. There are an infinity of investment possibilities, but for many people three basic components should be the first items considered:

Cash. Everyone should have as their first financial objective to always have a reasonable amount of readily available funds for emergencies, special opportunities, and general peace of mind. It can be as little as $2,000 early in one's career, and perhaps stabilized later at $10,000 or more, depending upon your net worth. It should be readily available (at least during any weekday, or constantly by writing a check), totally safe and secure, and earning some interest. This could be as a money market account in a mutual fund, bank, or savings and loan (the later two should be guaranteed by the government's SDIC up to $100,000). Most money market funds have limited free check-writing privileges. Your "cash" could also earn slightly higher interest as a "CD" or term account (i.e., such as a short 15- or 30-day period that can be "rolled over") in a bank or S & L, or as "commercial paper" or the like. The money isn't quite as available, but in an emergency before the term expires it can be easily "borrowed against" with some loss of interest.

IRA Account. Although not always the second financial item of importance, the individual retirement account (IRA) should be a must on almost everyone's list (Sing 1987*b*). The government allows up to $2,000 per year to be invested by everyone in an IRA account where the earnings are not taxed until you retire or cash-out (withdraw) the account at an earlier date with a slight penalty. Also, if your adjusted gross income is less than $25,000 ($40,000 if married and a joint return) your entire contribution may be deducted from your taxable earnings (Perlmutter 1987). Part of the contribution may be deducted with gross adjusted income between $25,000 and $35,000. These funds can be invested in essentially any manner that you wish, although the vast majority of people use mutual funds. The pure mathematical advantage of compounding where 20–35% of your earnings (i.e., your income tax rate) are not deducted, and the extremely low present-day value of taxes paid many years hence make this an unusually attractive investment. It alone can make you a near-millionaire upon retirement if it is started early and carefully managed.

Purchasing a House. Again, the exact priority of purchasing a house depends upon one's situation and finances, but it should be very high on the priority list. It is an individual's easiest example of a large leveraged investment, it has excellent tax advantages, and it allows one to accumulate equity with what would otherwise be a pure (rent) expense. There are two main factors to consider: (1) You must find a house with a down payment you can afford and where the monthly payments (repayment of principal, interest, property taxes, insurance and maintenance) are also within your budget. This usually takes quite a bit of searching—and compromising. But remember, in most cases your appreciation (profit) on the house allows you to "move up" to more desirable houses reasonably rapidly if you wish. (2) You must be as certain as possible that your house will increase in value. This usually happens automatically, but you must be sure that the neighborhood has little likelihood of deteriorating for any of many causes. You can also usually spot the possibility of rapid appreciation, such as with a new home in a nice tract, a lower valued or run-down house (that you intend to fix up) in a nice neighborhood, and so on. The house must also be statistically easily saleable, such as with at least three bedrooms and two baths and no elaborate swimming pool or barbecue. With a detailed search program to thoroughly understand the town's layout, housing values, trends and school locations you can usually be quite comfortable with this decision.

Obtaining the loan on the new house is usually fairly simple because of your good salary and position, and there are many potential lenders. However, merely looking for the lowest interest rate or the best monthly payment schedule, while still the correct thing to do, is no longer as simple as it once was. Home mortgages have undergone radical changes with the introduction of adjustable-rate loans. The choices are many, and it pays to shop around for the best rates, terms, and closing costs. There are a multitude of loans with many variations, and a lender can often tailor a program to the buyer's needs. Just as confusing because of the choices and needs is insurance. The home buyer is required to have certain kinds; others are provided at minimum levels or are a personal choice.

The purchasing of a home requires considerable work and is a major decision and responsibility, but it is an excellent hedge against inflation, and almost always a profitable investment even for short periods. The main caution is to think hard about possible declining values through a general economic decline of the community as a whole, or in the specific neighborhood involved. If values do appear uncertain, the prices are too high, or your duration at that location may be very short, renting would be the most prudent approach, but when ever possible purchasing a home not only provides better living accommodations but also becomes one of your best investments.

INVESTMENT STRATEGY

Diversification (Balanced Portfolio)

The most important guideline in any investment program is to have a variety and balanced range of investments (King 1987; Greene 1988). All investments have some risk of major loss, and each offers a good or poor rate of return at different times. More specifically, all investments offer various levels of:

Risk—you might loose some or all of your money.
Return—annual yield or dividends plus appreciation vary widely.
Liquidity—how quickly can you convert your investment into cash.
Level of personal management—how much oversight by you is necessary and what are the possibilities of quickly needing to buy or sell.
Protection against inflation.
Tax effect—how much of your investment will go to Uncle Sam.

Ownership of a number of different investment types provides an improved probability that no matter what happens to the economy or your lack of attention to "portfolio" management some of your investments should do well, and the average will be a more conservative but steadier gain. In general this means that when your net worth is large enough some of your funds should be spread among cash or money market accounts, real estate, insurance, bonds, stock, precious metals, and finally in more speculative investments. As your investment portfolio grows you should try to add diversification somewhat in the sequence listed above with variations and the relative percent of each depending upon your own personal interests and the economy. For example, for those who like a more vigorous and personally demanding program, diversification may stop to some degree and concentrate on real estate. For those with less time available (or interest) stocks or bonds (generally as funds) can be the major vehicle. Depending upon the economy, precious metals should perhaps be the major emphasis, at least for a certain period. Whatever the situation, a reasonable amount of diversification is a very wise and necessary investment strategy.

Economic Cycles

To a surprising extent the economy of nations runs in a cyclic manner. The cycles may vary in length and intensity, but the sequence and pattern often remains very much the same. Figure 11-1 presents an idealized representation of these cycles, with brief notes on the events occurring in the economy and the most appropriate investments for each period. Unfortunately no one can say with certainty where the cycle starts for some of the periods, or predict when it will change and the extent of the change. However, as a general guide a feeling for the usual cyclic sequence can considerably help in making investment decisions. It definitely shows that for active portfolio management and the greatest profits, fairly frequent shifts in investments are required based upon the changing economy alone. If invested funds are not to be needed for a long time, then a more passive, nonchanging position may be warranted, even if probably not quite as profitable.

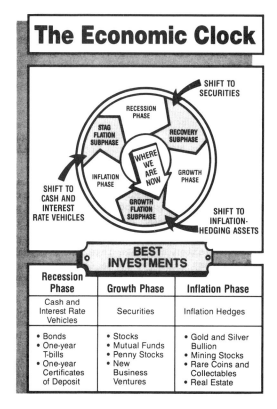

The Economic Clock

The Inflation Phase:
To sustain economic growth, large amounts of money are allowed into the economy. But soon the supply of money outpaces the supply of goods and services, and inflation accelerates. When it gets bad enough, inflation replaces all other economic concerns in the minds of the people, the press, and the politicians. At this point, inflation-hedging investments begin an explosive upward trend.

The Recession Phase:
The Recession Phase is brought about after the government has put extremely tight controls on monetary growth in order to cool down a previous inflation. Interest rates climb, loan capital becomes extremely tight, businesses fail, unemployment skyrockets, and virtually all the investment markets become extremely sluggish.

The Growth Phase:
To counteract a recession, the government allows the money supply to grow. Interest rates fall, consumer demand picks up, and business starts to expand. At its peak, the Growth Phase achieves low unemployment and a hot Dow market. In other words, happy days are here again and fears of inflation are placed on a back burner in the public's mind.

BEST INVESTMENTS

Recession Phase	Growth Phase	Inflation Phase
Cash and Interest Rate Vehicles	Securities	Inflation Hedges
• Bonds • One-year T-bills • One-year Certificates of Deposit	• Stocks • Mutual Funds • Penny Stocks • New Business Ventures	• Gold and Silver Bullion • Mining Stocks • Rare Coins and Collectables • Real Estate

Figure 11-1. The economic cycle.

As an example of such decision making, in early 1987 the country had been experiencing many years of sustained economic growth, the stock market was very high, and inflation (Figure 11-2) and interest rates (Figures 11-3 and 11-4) were relatively low. However, the national debt had grown very large and the government appeared to be unable to seriously reduce their immense yearly deficit spending. The value of the dollar had rapidly declined on world markets (Figure 11-5) and the U. S. balance of payments (imports minus exports) was extremely large. Then, from October 5 to 19, 1987 the stock market dropped about 900 points (34%), including about 500 points on October 19 (Figure 11-6). There was little time to react, and automatic sell orders were not possible on October 19, so the majority of stock investors and mutual funds lost heavily. Considerable turbulence in the market followed, and it became unclear as to where the economy was on the cyclic chart. Those who invested in precious metals (or metal funds) early in 1987 did very well for the year, as did real estate and some commodities, such as chemicals (Savage 1987), thus pointing

Percent

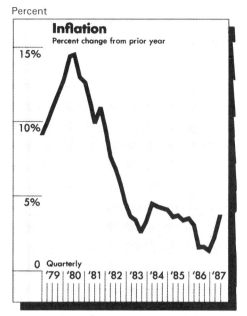

Figure 11-2. Rate of inflation. Percent change from prior year.

Percent

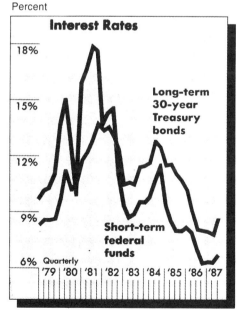

Figure 11-3. Interest Rates.

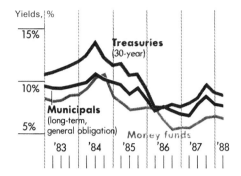

Figure 11-4. Bonds. yields.

out the wisdom of some diversification and hedging on the rising but worrisome stock market. The very large 1987 stock loss, preceded by a substantial 1987 bond loss because of rising interest rates (Labich 1987), provided a graphic demonstration of the risks with investments, and the amount of care and attention required. Even a major decline such as this should not deter one from

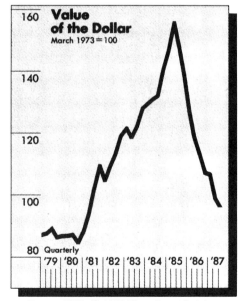

Figure 11-5. Value of the Dollar. March 1973 = 100.

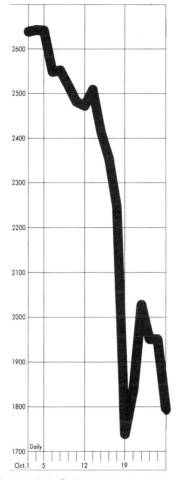

Source: *Fortune* 1987. © 1987 Time, Inc. All rights reserved.

Figure 11-6. 1987 stock market crash (Dow Jones industrial average).

making investments, however, for over a reasonable period the prudent investor will come out well ahead, but it does reinforce the need for study, care, and caution.

Inflation

In any investment program it is necessary to consider inflation. For instance, if you had been earning $2,500 a month, how much would you need to live the same way 20 years later if the annual inflation rate was 5.5%? Answer: $7,295 a month. Inflation was quite modest in the mid-1980s, but it had been in double digits during 1979–1981 (Figure 11-2). Whatever the rate it is important to protect your capital by keeping the value of your assets growing faster than inflation. The net return on your investments is the actual rate minus that year's inflation, and at worst you need this to at least be a positive number.

INVESTMENT TYPES

There are an extremely large number of methods of investing your money and attempting to make it grow. As your net worth, investment knowledge, and interest grow you may wish to try many of them. However, only a comparatively few are in wide and common use, so we shall limit our discussion to them. Some are specific investment opportunities of their own particular type, and others such as the next section are merely methods of making your investments.

Mutual Funds

A mutual fund (*Mutual Fund Forecaster* 1987) is a corporation (or business trust) that sells shares of its stock to the public, and invests the proceeds in a wide variety of securities, such as stocks and bonds of other companies. Mutual fund shareholders own part of the combined income, profits, and losses of the fund's investments. At the end of each business day a fund totals the value of all of its investments and other assets, deducts its liabilities, and determines the precise value of each outstanding share. That value is called the "net asset or share value." It includes the dividends, interest, or changed value of the securities it holds (stocks and bonds are traded on "exchanges"—i.e., New York, American, Pacific, etc.—daily and their price constantly changes according to the demand and expectations for them).

Funds are broadly classified into two categories, open-end and closed-end. An open-end fund is nearly always "open" to investors, who can, at any time buy new shares from the fund, or sell (redeem) shares they already own back to the fund at the net asset value. Closed-end funds make a single public sale of shares to the public, and then "close" their doors to new investors, and will not repurchase shares from old investors. However, closed-end funds trade on stock exchanges or in the over-the-counter market just like other stocks or bonds. Investors buy and sell them through a stockbroker (who will charge his usual commission) or a fund manager at whatever price the market determines. This is a disadvantage but not necessarily a major one for a fund that is performing very well.

The following paragraphs itemize some of the advantages and features of mutual funds. Of course, the positive attributes are unimportant if a fund isn't well managed and doesn't invest its money profitably. The task of selecting the best individual funds still requires considerable research and judgment.

Broad Coverage.   Without question, the single greatest advantage of a mutual fund is that it is a portfolio of many securities. Safety improves with numbers, and portfolios consisting of many securities are safer than ones with limited numbers. Most investors, especially those with limited assets, are unable to purchase a large number of securities, but mutual funds provide partial ownership in dozens, or even hundreds, of different stocks, bonds, and other securities.

Liquidity, Management.   The ability to sell an investment at a moment's notice is an important attribute. Investors can sell some or all of their shares back to an open-ended mutual fund often on a single day's notice. The funds provide professional portfolio supervision for investors who are unable or do not desire to manage their own portfolios. Nevertheless, professional management is probably the most overstated advantage of mutual funds. In truth, management is "professional" only if it generates a reasonable return on an investment portfolio. Some fund managers do, indeed, consistently beat the market through superior stock selection, market timing, or portfolio strategy. Others, however, have inferior track records, so a fund's prior history should be carefully examined.

Ease of Investment.   Buying a fund is as easy as opening a bank account. All funds have simple mail organized forms and brochures. Most funds offer several types of accounts, including Individual Retirement Accounts (IRAs) or company plans such as SEP IRAs or Keoghs. Also the mutual fund industry has expanded and become extremely diverse. Funds are now available that invest in almost every type of security imaginable. A representative, but by no means exhaustive list includes: blue chips or growth stocks, small company stocks, over-the-counter stocks, preferred stocks, corporate bonds, government agency bonds, Ginnie Mae bonds, U. S. Treasury bonds, gold stocks, health-care stocks, computer stocks, utility stocks, stocks of many other specialized industries, foreign stocks from all areas, treasury bills, bank CDs, commercial paper, real estate trusts, and so on. Regardless of what type of investment you seek, there is likely to be a mutual fund to serve you.

Many mutual fund organizations allow investors to switch from one fund to another, or even to allocate their investment among multiple funds. This service, usually available at no charge, enables investors to switch objectives with a minimum of bother. Finally, fund managers do most of the paperwork. Capital gains and losses on the purchase and sale of securities, as well as dividend and interest earnings are neatly totaled for each stockholder at the end of the year for his tax purposes. Funds also do all of the day-to-day paperwork chores, including handling of stock certificates, brokerage statements, transfer agents, stock offerings, and so on. All investors have to do is give their money to a fund and it does the rest.

Costs. Small investors may have to pay commission rates equal to several percent of the amount of their investment when they buy and sell securities directly. Mutual funds buy and sell in large blocks and obtain lower trading costs, the benefit of which passes through to shareholders. However, some mutual funds impose a "sales charge" or "load" on investors who buy new shares. Sales charges range from zero to about 10%, although several hundred mutual funds, including virtually all money market mutual funds, are "no-load" and do not charge any initial sales fee at all. When you sell shares back to a fund there may be another charge, a redemption fee of from 1 to 5.3% of the amount of sale, although most funds have no such charge. Also, most funds permit investors to automatically reinvest capital gain and income distributions in new shares with no sales charge. However, a few funds levy fees up to 7.25% of the total amount of investment. There are also often "hidden" load charges used only to advertise and promote the fund. Most do not yet levy these charges, but more and more are adding them, and they can range up to 1.25% per year. Finally, mutual funds incur various day-to-day operating costs which are paid from fund assets, and therefore reduce stockholder's returns, typically ranging

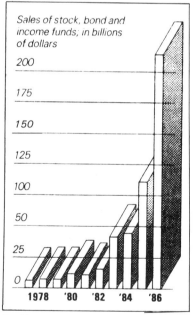

Source: Sing 1987. Copyright 1987, *Los Angeles Times*. Reprinted by permission. Patricia Mitchell illustrator.

Figure 11-7. Mutual fund sales.

from 0.25 to 1% per year. In addition, some mutual funds contract with a portfolio management company to make all investment decisions, which also usually costs from 0.25 to 1% of fund assets per annum. Without any question, the no-load funds as a group should be preferred, and have out-performed those with various loads.

Fund Growth.   Mutual funds have recently become the most popular investment for the American public. Fund sales have increased as seen in Figure 11-7, indicating a very rapid surge in popularity. The type of funds most widely purchased are shown in Figures 11-8 and 11-9, which provides some detail on the popularity of various types of funds. It is seen that the bulk of the purchases are for bond funds, which offer a much lower potential gain, but perhaps are a more steady income source than stocks. It appears that the average mutual fund investor is generally conservative.

As an example of the effect of interest rates on bonds, if a bond was originally issued for $1,000 at 10% interest, the $100 per year interest is fixed for the life of the bond (i.e., 20 years or so). Then, if two years later the interest rate of everything else rises to 12% and one needs to sell the bond, it would usually

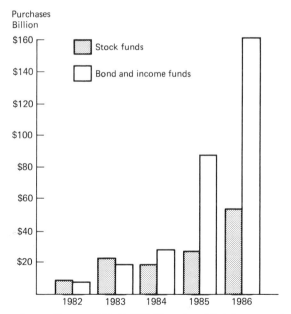

Source: *Fortune* 1987. © 1987 Time, Inc. All rights reserved.

Figure 11-8. Mutual fund distribution.

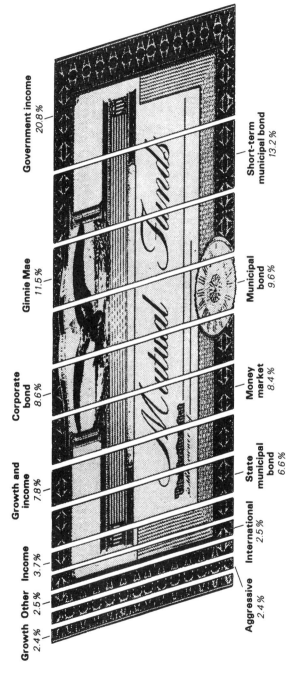

Source: Sing 1987a; Illustration, Patricia Mitchell. Copyright 1987 *Los Angeles Times*. Reprinted by permission.

Figure 11-9. The mutual fund market breakdown. *Net sales by category based on 1986 total of $189.98 billion.*

only be worth about $^{10}\!/_{12} \times 1,000 = \$833$. The actual two-year gain would be $2 \times \$100$ interest minus $\$167$ capital loss $= \$33$. This would mean that the bond actually only averaged 1.65% interest on the principal, and not the 10% anticipated.

Considering that commercial interest rates (Figure 11-3) and bonds (Figure 11-4) declined from about 18 to 5% over the period 1981–1987, bond values appreciated significantly. However, this changed during 1987 when interest rates started to rise, and mutual bond fund holders received less appreciation on their investment than they hoped for, but still slightly more than with stock funds. As interest rates started to increase they would have been better off to utilize their fund's switching privilege to money market funds (where interest is always near, but usually 10–30% below the current corporate rate), or inflation hedges such as gold funds. It is always difficult to make such decisions, and most investors did not switch, resulting in considerable dissatisfaction with the bond funds for that period.

Stock funds also have some probability of losing money during any cycle of the economy. The most recent ''bull market'' or growth phase of the economy started in 1982 and lasted until 1987, during which time nearly all stock funds made high yields with the steadily increasing stock market (Figures 11-10 and 11-11) and thus increasing stock prices. Most stocks also pay some dividends (similar to interest, but not fixed and subject to change by the company's board of directors), but the amount is small, as indicated by the chart in Figure

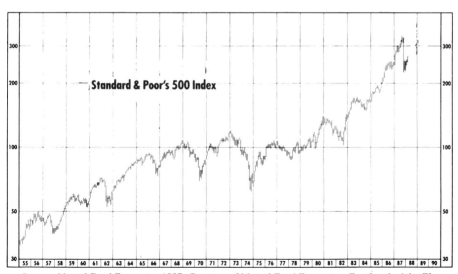

Source: *Mutual Fund Forecaster* 1987. Courtesy of Mutual Fund Forecaster, Fort Lauderdale, FL.

Figure 11-10. Standard & Poor's 500 Index Stock market averages.

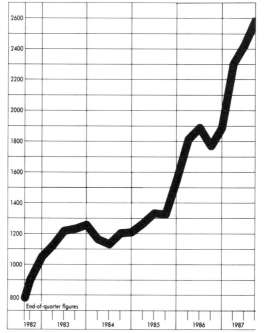

Figure 11-11. Dow Jones industrial average.

11-11. Considering that this figure includes many utility companies which pay high dividends, roughly equal to the then-current corporate interest rates, the average nonutility company dividend rate is less than 2%. Consequently, the earnings from most stock investments comes primarily from the increase in stock price, which benefits during a growth period not only from the companies' increased earnings, but also from the amount people are willing to pay for those earnings, or the (stock) price-to-earnings ratio, as seen in Figure 11-13. The Dow-Jones industrial average, or the Standard & Poors 500 compilation of stock prices of a number of companies, also shows the same growth, and indicates that an average stock purchased in 1982 would be worth about 2½ times that amount in 1987. Not all stocks or funds did that well, but some did considerably better. An example of the various funds available, and their performance to 1987 is listed in Table 11-1. The very wide spread in earnings, and the generally poor average performance indicates why considerable research is necessary before purchasing mutual funds. A further comparison between various stock and bond funds is given for a slightly earlier period in Table 11-2. As previously

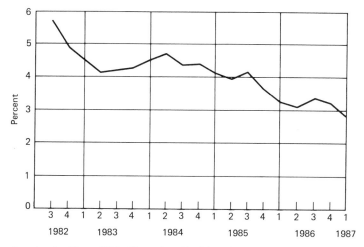

Source: *Los Angeles Times* 1987. Illustration, Patricia Mitchell. Copyright 1987. Los Angeles Times. Reprinted by permission.

Figure 11-12. Dividend yields of stocks, *by quarter.*

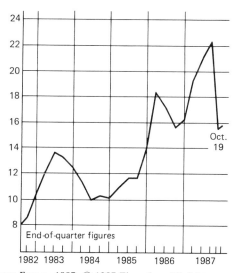

Source: *Fortune* 1987. © 1987 Time, Inc. All rights reserved.

Figure 11-13. Price earnings multiple of Standard & Poor's 500-stock index.

Table 11-1
Mutual Fund Performance
(Appreciation plus reinvested income and capital gains)
Quarter Ended March 31, 1987

Top Performers		*Worst Performers*	
Van Eck Gold/Resources	+64.76%	Equity Strategies	−2.12%
New England Zenith	+64.74	Industrial Govt.	−1.53
Alliance Technology	+61.26	First Inv. U.S. Govt. Plus-I	−1.40
Strategic Investments	+60.78	Benham Target 2015	−1.26
U.S. Gold Shares	+60.43	MFS Lifetime Govt.	−1.11
Financial Port-Gold	+58.65	Maxim Bond	−0.86
Franklin Gold Fund	+58.10	Benham Target 2005	−0.76
U.S. New Prospector	+58.02	Seligman U.S. Govt.	−0.31
U.S. Prospector Fund	+55.56	Benham Target 2010	−0.28
Midas Gold Shr. & Bullion	+54.50	Scudder Target U.S. Govt. 1990	−0.10
S&P 500 (dividends reinvested)	+21.35		
All funds average	+14.66		

Year Ended March 31, 1987

Top Performers		*Worst Performers*	
New England Zenith Cptl Growth	+142.27%	88 Fund	−26.70%
USAA Gold	+119.70	American Heritage	−16.07
Van Eck Gold/Resources	+113.53	Div/Grow-Laser Adv Tech	−12.21
U.S. New Prospector	+106.04	Bowser Growth	−12.06
Financial Port Gold	+97.05	Fidelity Sel Life Ins.	−7.47
U.S. Prospector Fund	+96.49	Steadman Amer Industry	−6.84
IDS Precious Metals	+93.56	Fidelity Sel Electronic	−6.60
Colonial Adv Str Gold	+92.71	Rochester Growth Fund	−6.47
Vanguard Special Gold	+90.23	Industry Fund of America	−5.60
Hutton Inv. Sr Prec Met	+90.10	ISI Income Fund	−5.18
S&P 500 (dividends reinvested)	+26.21		
All funds average	+18.18		

Five Years Ended March 31, 1987

Top Performers		*Worst Performers*	
Merrill Lynch Pacific	+445.35%	44 Wall Street	−48.44%
Fidelity Magellan	+380.82	American Heritage	−14.36
BBK International	+379.40	Industry Fund of America	+0.50
Looms-Sayles Cap	+374.41	Interstate Cap Growth	+2.54
Vanguard World-Inter Growth	+354.83	Steadman Amer Industry	+3.26
Alliance Technology	+348.30	Sherman, Dean Fund	+5.98
Fidelity Sel Health	+327.55	First Inv Natural Resources	+6.55
Putnam International Equities	+323.67	Steadman Oceanographic	+14.94
New England Growth Fund	+284.76	American Inv Growth	+20.00

Table 11-1   (Continued)
Five Years Ended March 31, 1987

Top Performers		Worst Performers	
Vanguard Qual Dvd.	+284.41	Afuture Fund	+47.83
S&P 500 (dividends reinvested)	+224.25		
All funds average	+165.65		

Group Performance (periods ended March 31, 1987)

	Quarter	Year	Five Years
Gold-oriented funds	+49.40	+84.89	+149.88
Health funds	+26.56	+28.41	+245.88
Science & technology funds	+26.39	+24.30	+165.50
Natural resources funds	+25.58	+36.82	+101.00
Small company growth funds	+22.73	+12.87	+159.60
Capital appreciation funds	+22.52	+20.32	+172.15
Growth funds	+20.20	+18.22	+171.49
Specialty funds	+20.00	+14.29	+207.37
Growth & income funds	+16.00	+18.61	+184.77
International funds	+15.20	+42.58	+269.03
Global funds	+15.09	+28.25	+220.60
Option growth funds	+13.89	+13.07	+126.25
Balanced funds	+11.77	+16.97	+185.35
Option income funds	+11.06	+13.72	+106.02
Equity income funds	+10.45	+14.26	+183.08
Income funds	+5.86	+11.14	+140.08
Utility funds	+3.50	+11.66	+173.60
Fixed income funds	+3.01	+9.08	+124.71
All funds average	+14.66	+18.18	+165.66
S&P 500 (dividends reinvested)	+21.35	+26.21	+224.25

Source: *L.A. Times* 1987. Copyright 1987, *Los Angeles Times.* Reprinted by permission.

noted, the bond funds did well for this period, but on average only produced about one-third the yield of stock funds.

Direct Stock and Bond Purchases

Mutual funds have the various advantages previously discussed, but there are many situations where a direct purchase of stocks or bonds is the best investment strategy. Many times you know a company well (it might be the one you work for) and you have high hopes for its growth and stock appreciation. By a direct stock purchase you will automatically receive the company's annual report, which may be of considerable interest. Most companies also have a dividend reinvestment program so that your income is automatically compounded quarterly. This is particularly beneficial for utility stocks that pay

Table 11-2
Best-Yielding Bond Funds — 1986

	% Current Yield			% Total Return (Or Loss) Through May		% Sales Charge
High-Yield Corporates	May	April	May 1985	One Year	Five Years	
1. Putnam Yield Trust 1	12.9	12.8	13.5	12.6	118.7	6.75
2. Oppenheimer High Yield	12.6	12.8	13.0	9.1	91.3	6.75
3. Bull & Bear High Yield	12.6	12.6	13.5	18.6	—	None
4. Paine Webber High Yield	12.5	12.5	—	14.8	—	4.3
5. National Securities Bond	12.4	12.2	13.0	9.0	88.6	7.25
6. Vanguard Fixed Inc.—High Yield	12.3	12.3	13.6	20.7	125.8	None
High-Grade Corporates						
1. Vanguard Fixed Inc.—Inv. Grade	10.8	10.7	11.9	16.8	114.2	None
2. Security Income—Corp. Bond	10.0	9.9	11.7	9.1	106.8	4.75
3. Babson Bond Trust	9.9	9.6	11.0	17.8	115.4	None
4. Pioneer Bond	9.9	9.7	9.1	4.9	93.0	4.5
5. Dreyfus A Bonds	9.5	9.3	10.5	18.5	114.4	None
6. Merrill Lynch High Quality	9.5	9.3	10.6	12.4	114.6	4.0
U.S. Government Bonds						
1. Fund for U.S. Gov. Securities	11.4	11.4	11.6	11.8	116.0	None
2. Dean Witter U.S. Government	11.4	11.5	—	9.5	—	None
3. Alliance Mortgage Securities	11.3	11.2	11.9	7.1	—	5.5
4. Putnam U.S. Government	11.2	11.1	—	5.7	—	4.75
5. Smith Barney U.S. Government	11.1	10.9	—	5.9	—	4.0
6. AMEV U.S. Gov. Securities	10.9	10.7	11.3	12.7	97.7	4.5
High-Yield Tax-Exempts						
1. Value Line Tax-Exempt High Yield	8.9	8.9	—	13.8	—	None
2. Colonial Tax-Exempt High Yield	8.4	8.4	—	11.3	—	4.75
3. Vanguard Muni—High Yield	8.2	8.0	9.1	18.3	94.3	None
4. Stein Roe High Yield	8.2	8.1	8.9	18.8	—	None
5. Merrill Lynch Muni—High Yield	8.2	8.0	9.1	12.7	96.2	4.0
6. Fidelity High Yield Municipals	8.0	7.8	8.8	17.5	104.5	None
High-Grade Tax-Exempts						
1. USAA Tax Exempt High Yield	8.6	8.4	9.1	15.2	—	None
2. NEL Tax-Exempt	8.3	8.2	6.8	12.7	107.4	4.5
3. Benham Nat. Tax-Free Long Term	8.3	8.2	—	17.5	—	None
4. Keystone Tax-Exempt Trust	8.1	7.9	8.7	15.1	86.2	None

Table 11-2 (Continued)

High-Yield Corporates	% Current Yield			% Total Return (Or Loss) Through May		% Sales Charge
	May	April	May 1985	One Year	Five Years	
5. Vanguard Muni—Long Term	8.0	7.8	8.8	17.0	91.5	None
6. Safeco Municipal	7.9	7.8	8.7	17.1	—	None
Intermediate-Term Tax-Exempts						
1. USAA Tax Exempt Intermediate	8.2	8.1	8.6	12.0	—	None
2. Vanguard Muni—Intermediate	7.7	7.6	8.3	14.0	76.2	None
3. Dreyfus Intermediate Tax-Exempt	7.6	7.5	7.2	12.4	—	None
4. Fidelity Limited Term Muni	7.0	6.8	7.4	13.7	80.5	None
5. Calvert Tax-Free Limited Term	6.5	6.6	6.5	7.0	53.1	None
6. Vanguard Muni Bond Short Term	6.2	6.1	6.6	6.8	43.6	None

All figures are after sales charges, if any; gains or losses are with all dividends reinvested.
Source: *Money* 1986.

Best-Performing Income Funds
Funds that invest in high-yield stocks and bonds-1986

One Year	% Total Return (or Loss) Through July		% Current Yield	Money Fund Switching	% Sales Charge	Minimum Initial Investment
	One Year	Five Years				
1 Loomis-Sayles Mutual	47.9	185.9	3.6	Yes	None	250
2 Stratton Monthly Dividend	34.6	165.2	7.2	No	None	1,000
3 Dreyfus Convertible	32.7	121.4	6.0	Yes	None	2,500
4 Financial Industrial Income	30.0	169.4	4.6	Yes	None	1,000
5 Natl. Securities Total Income	27.5	141.0	5.7	Yes	7.25	250
6 Value Line Income	27.2	103.4	6.1	Yes	None	250
7 Steadman Associated Fund	26.4	65.1	3.9	Yes	None	100
8 Vanguard Wellesley Income	25.8	155.5	8.3	Yes	None	1,500
9 Bullock Balanced Shares	25.7	144.9	4.1	Yes	8.5	500
10 Vanguard Wellington	25.5	149.7	6.1	Yes	None	1,500

Source: *Money* 1986.
Types: Gro-Growth; G&I-Growth and income; Inc-Income; Max-Maximum capital gains.

very high dividends, and the dividend reinvestment means that you essentially obtain the then-current interest rate. By judicious shopping you can find excellent utility companies with high rates, a steady history of dividend increases, and with discounts on the dividend stock reinvestment price. For a long-term investment, or if caught at the correct point on the cycle for shorter periods, this can make an unusually good investment. For such stock purchases it is also recommended that a "discount" stock broker be used to purchase and sell the stock, for this will result in considerable savings. It is rare that the "services" offered by the regular brokerage houses (to justify their higher transaction costs) have significant advantages for the diligent investor.

With direct purchasing of most stocks care must be taken to both buy and sell at the appropriate time. You generally look for bargains or sound investments, but you should be prepared to be wrong by deciding in advance how much of a loss you are willing to take before you resell. Often investors will put in an automatic sell order at 10% (or some such fixed number) below what they purchased the stock for, and then adjust it periodically as the stock rises. For some stocks you may also wish to place an automatic sell order once you have realized the gain that you desire. With other stocks you do better holding them through a cycle or for a long time, but to make your investment work for you most effectively with most stocks you generally must buy and sell as the situation demands and move from one growth opportunity to another.

Several cautionary notes should be given regarding stock purchases. First, there is a great deal of literature (Securities and Exchange Commission 1987), and there are many news letters giving investment advice and recommendations on stock transactions. Every serious investor should read a reasonable amount concerning current market conditions, so the recommendations should be considered. However, most of the advice has not proven to be too successful and often is biased, so your own independent study and judgment should be preferred. Also, following the "herd" is often reacting too late and can result in poor performance. The second caution is to be wary of major losses, so that if you follow the market or a particular stock into what might be the final phases of its rise, you should be ready to quickly sell when the decline starts, and before you have lost too much. This is tricky, and generally means that you should do some advance "hedging," and move part of your investments to safer areas. Without special care it is even easier to loose money in the stock market than it is to gain. The stock market goes up and down with the business cycle, and although broadly rising, has not nearly kept up with inflation over the years. (See Figure 11-9. Note that the cycles have recently been 3, 3, 2, 4, 4, 4, 4, 4, and 5 years in length.) Stocks obviously represent excellent investment opportunities, but must be managed very carefully.

Bonds

Bonds are generally a more stable investment, and the interest income from them is fixed and definite unless the company or issuing group goes bankrupt or defaults. The bond value itself rises and falls with prevailing interest rates as previously discussed, so one easily can loose heavily in case of a required sale at an inopportune time, but for a long-term investment or during falling interest rates they are very good, and some probably should be in every investor's diversified portfolio.

There are many types of bonds, ranging from "junk bonds," convertible debentures, subordinated debentures, and regular bonds. Some federally issued or guaranteed bonds pay somewhat higher interest rates than corporate bonds, such as the Ginny Mae and Fanny Mae Government housing loan bonds. Considerably lower rates are given to the most secure of all direct government borrowing, Treasury bills and Municipals, as shown in Figures 11-3 and 11-4. The interest from Treasury bills or notes are also tax-exempt from federal taxes, while some municipal bonds are exempt from both state and federal taxes, which further enhances their value. It is also seen from Figure 11-4 that money market funds command the lowest interest rate of all. There is a moderate spread in interest rates between corporate bonds and short-term Treasury bills.

Precious Metals, Collectables

A time-honored form of investment since the earliest beginnings of metallurgy has been with precious metals or collectables, and they still deserve a place in almost everyones investment portfolio. Gold, silver, and platinum all have industrial, coin (except platinum), and jewelry uses, as well as being monetary symbols. They thus serve primarily as an inflation hedge since people feel that because of their scarcity, preciousness, and inertness they will always retain their value, and at least increase in price with inflation. They also have a price rise and fall history that accompanies the economic cycle, as seen in Figures 11-14, 11-15, 11-16, becoming very popular in inflationary periods, and falling in value during growth periods.

The metals may be purchased directly as coins in modest quantity, or as bullion in larger amounts. Some banks and sales organizations will hold the metals for you, or deliver them as you wish. If you do hold the metal, it will have to be reassayed when you sell (not so for coins), and it will take more time. There is a small spread between the selling and purchasing prices, which are quoted daily in the various "metal exchanges" (i.e., New York, London, Zurcih, etc.).

A second means of purchasing the precious metals is by buying stock in mining companies producing the metals, or in mutual funds specializing in such companies. It is to be noted from Table 11-1 that such funds were the leaders in appreciated value during 1987 as the metal prices began a moderate rise. The metal funds prices rose even more rapidly than the metals themselves during this period, attesting to their popularity.

Collectables can be any thing that you feel may increase in value with time, with perhaps the most common of the widely held materials being "rare" coins or stamps. These have fairly well-established values and can be bought and sold relatively easily, and even held by dealers (especially gold and silver coins) for you if you wish. Other collectables such as all types and forms of art, diamonds

$/ounce

Figure 11-14. Gold prices.

$/ounce

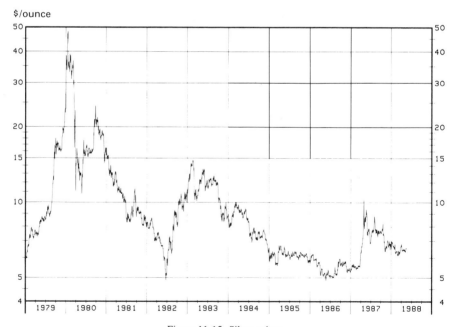

Figure 11-15. Silver prices.

$/ounce

Source of Figures 11-14, 11-15, 11-16: Charts courtesy of Harry Browne, Special Reports, Inc., Austin, Texas.

Figure 11-16. Platinum prices.

or gem stones, and so on have good possibilities of appreciation, but are perhaps less certain of their value for a reasonably quick sale. These items, and many other forms of collectables can have the dual advantage of being good inflation hedges, perhaps gaining substantially in value, and for some people being much more enjoyable to collect.

Real Estate

The purchase of a house has previously been discussed, but once one's net worth begins to rise, other investments in real estate becomes a serious possibility. Houses, apartments, land, housing, commercial, or agricultural developments, and so on all have the same high-leverage, easy credit advantages as your personal residence. Real estate is a good inflation hedge, and for the more physically active investor very attractive profits can be made. The rules are generally the same as with your house: look for property that you feel will appreciate in value, and where you can handle the initial payment, monthly payments, and other "carrying" costs. For rental properties you hope to find property with a positive cash flow, where the rental income will pay all of the costs and make some profit. This is often difficult, so in some cases where you expect good appreciation and wish to hold the apartments or property for awhile, you must be prepared to pay the negative cash flow costs.

If you are purchasing property to fix up, improve, divide (land), or otherwise sell fairly soon, you can budget the various expenses involved, but you must be careful in the terms of the sale. It is usually easy to sell with a lower down payment than you gave, and take your profit in first or second trust deeds (loans with the property as collateral). This can provide excellent profits and income, but it depletes your available capital. Ideally you attempt to sell the property with a higher down payment, or even all cash, so that you realize your profit right then, and have more money to make your next transaction.

These examples are merely two in a myriad of possibilities of investing and profiting in real estate. Many people find real estate to be very challenging, to involve more activity and personal participation, and to be fairly easy, and consequently have from 30 to 60% of their investments in real estate. It is not for everyone, but is at least worth careful consideration periodically through your investment career.

High-Risk Investments

There are many investments one is occasionally exposed to which must be considered very high risk, but which may offer commensurately higher rates of return, and might be very interesting. An example of this is oil drilling funds or groups. Obviously the risk is high, since true "wild cat" wells, drilling in previously untested areas, statistically only find oil or gas about 1 time in 14, and even if they do, the amount may not be sufficient to pay very good dividends. On the other hand, the occasional well is a "gusher," or produces very well. Considerably better odds may be obtained with funds that are drilling in more developed areas, are part of a multiwell program, or are buying (old) existing production, but there is still always uncertainty, and the potential payoff will not be nearly as high. Such drilling investments, however, if carefully considered as part of a balanced investment portfolio can have excellent tax advantages, obtain a reasonable-to-good (statistically) return, and be quite an exciting investment, particularly if you visit the site and somewhat participate in the activities.

A more frequently encountered high-risk investment is the occasional friend, family member, or just someone you have heard of who wants to start a new business, and would like you to invest some money in the venture. Almost no matter what the project you have to consider that the odds are nearly 100% that you will loose all of your invested money, or that the payoff will be negligible. The average chances of success for the "start-up" investor are very low. However, you may consider it somewhat of a donation to a worthy cause and be willing to take the risk, or you may be able to personally contribute some assistance or services that will make the project more interesting to you and perhaps reduce the risk. Many such new ventures are of broad general benefit to the country, especially the technical, production, or manufacturing ones, since this is where most of the new jobs are generated for our economy, and where new ideas and technology get started. Thus, such ventures are very much worth supporting, but on a cold, statistical basis you must consider it as almost a sure financial loss. As with the oil drilling, however, there is some chance of success, and the more you participate the more interesting the venture will be to you. For a balanced portfolio, when you have adequate net worth, modest investments of this type may be rewarding.

Obviously there are many, many types of investments not considered here, and once your investment program has started you will be exposed to, and wish to research many others. The problem always is that there are more investment opportunities than there are funds available, so it is a constant necessity to study, carefully consider, balance, invest, watch closely, and then hope for the best. No one ever makes all of the best decisions, and there are always the occasional major losses, but by being diligent it is amazing how well the program works out and how high the average rate of return becomes.

Life Insurance

Insurance has not been considered up to this point in that it basically is not an investment but a pure expense. The cheapest form of life insurance is "term" insurance, where the policy has no monetary value, and is merely a yearly expense. In times past this was the only type that should be considered by a prudent investor, since all other types of policies paid the equivalent of very low interest rates, generally far below what could be obtained in ordinary money market funds. Since the advent of the 1987 tax law, however, the insurance companies have developed a new form of policy that acts as life insurance, money market fund, and earns money tax free until paid out. It can be borrowed against at reasonable rates and has become one of the best "tax shelters" available. In an investment portfolio it would be somewhat equivalent to a combination of cash and a conservative IRA investment, along with term insurance. Particularly when an investor has greater income and tax costs this form of insurance becomes an excellent investment. It can be obtained with a lump sum payment in advance for the policy, which can be increased later as desired, or by conventional monthly or yearly charges.

METHODS OF INVESTMENT

Company Stock Purchase Plans

The investments that have just been discussed can be made by anyone or any group. Most people prefer to make their investments entirely by themselves, but there are other methods to consider. The easiest, but most restricted is through a company stock purchase plan. Some companies attempt to encourage employees to become stock holders by means of a special purchase plan where they match your funds when you buy stock, or sell it to you at a discount. Such

Table 11-3. Investment Groups.

How a little money can turn into a lot

GROSSE POINTE, Mich.—Back in 1940, when the Great Depression was a recent memory and the post-war boom a dream, a group of amateur investors began contributing $10 to $20 a month each and investing it in stocks.

They kept at it.

A few nights ago, the Mutual Investment Club of Detroit, now 19 members strong, held a dual celebration: Its 43rd anniversary, and its new stature as a millionaire, its assets having reached $1,128,000.

Following dinner at the Country Club of Detroit, retreat of the Fords and other great industrial families, it was also revealed that deposits totaled $145,716 over the years, and withdrawals were $319,229.

By 1969, after having deposited $75,327 over the years, the portfolio was worth $535,604. A catalytic mass had been established; the dividends—always reinvested—now totaled more each year than members' contributions.

Then bad times hit and tested the club's philosophy. Brokers went broke, investors lost fortunes, many investment clubs disbanded, and millions of people who had dreamed of fortunes now sought only to survive the present.

The credo of the Mutual Investment Club—now the underlying philosophy of the entire investment club movement—does not allow for faint-heartedness. Down markets, it states, are buying opportunities.

The record seems to show the club may have wavered, but just a bit. In 1972, the accumulated deposits of members fell for only the third time,

All this was reported by George Nicholson Jr., the club's broker from the beginning and a member soon after, a man of rare perception and intellectual versatility, and—with some justification, it turns out—vast optimism.

Two of the six founding members—Norman Hill and Lee Jacobson—remain active in the club, and nine of the other members have been with the club almost from the beginning.

In all probability, Mutual Investment will outlive its founders, because young investors have caught the spirit. O'Hara's two sons are members. Kenneth Janke, president of the national association, is a member, and so is Kenneth Janke Jr., 23.

Members have been successful in business. One owned five Howard Johnson franchises. Another built a company so successful the club

The latter sum, explained Thomas O'Hara, a member from the first year, was used to found private businesses and careers, finance educations, buy houses, finance vacations and assure financial security in retirement.

"The financial record is an amazing indication of what amateur investors can accumulate at the rate of $10 or $20 a month," said O'Hara, who now serves as chairman of the National Association of Investment Clubs (NAIC).

At the end of the first year, the young club showed a liquidating value of just $812, an amount whose size was offset by monumental—and some felt, unwarranted—hopes.

The hopes were justified. By the end of 1946, club members had deposited $5,080, withdrawn $2,519, and still had a liquidating value of $15,347. After a post-war slump, the value rose to $27,471 in 1951, and to $33,222 by 1952.

Despite repeated withdrawals, the club's value doubled again by 1955, once more by 1959, and again by 1965, when the portfolio's liquidating value was put at $246,738.

since 1941, to $90,819 in 1972 from $93,492. Nearly $44,000 was withdrawn.

It didn't happen again. Though the stock market lost much of its appeal during the inflation of the 1970s, members continued to increase their annual contributions, generally by about $7,000 or $8,000 a year.

But their investments suffered. From the 1969 peak the portfolio value fell to $394,434 in 1970, recovered to move than $430,000 in the next three years, and declined again to $387,959 in 1974.

But the investment philosophy paid off: The club had established positions when prices began rising again. Through the next few years members were able to "grow" their shares while also withdrawing $12,000 to $38,000 a year.

By 1979 the portfolio was valued at $759,038, and by the end of 1982, when the market rediscovered fundamentally sound stocks, the portfolio was above $1 million—$1,124,038 on Feb. 28, 1983, to be exact—on deposits of $145,716.

invested in it. They have been successful in professions, too. O'Hara, whose full-time work was running and building the national investment club association, was a director of the New York Stock Exchange.

The club has been the model out of which the investment club movement has grown and, says O'Hara, "reached better than 3 million people across the world."

It did it of course with considerable study and insight. All members chart and analyze stocks using NAIC guidelines, and then participate in discussions. But those conservative principles helped.

Remember them, advise the people at Mutual Investing Club of Detroit:

•Invest a set sum once a month regardless of market conditions.

•Reinvest dividends and capital gains immediately.

•Buy growth stocks—companies whose sales are increasing at a rate faster than the industry in general.

Source: Cunniff 1983. Courtesy of McClatchy News Service (McClatchy Newspapers, Inc.).

programs can be a real bargain with a growing company for long-term investment, or if you are allowed to sell some or all of the shares whenever you think that the market is at a high. The programs can also not be too good an investment if the discount is not too large, and you are too limited in your ability to sell. Remember that stocks pay a very low dividend, and the stock market generally does not keep up with inflation over a long period. However, it is good to own some stock in the company that you work for, since it tends to make you a more interested and knowledgeable employee (i.e., you get annual reports and know more of its financial position).

Investment Clubs

A method that many people find more enjoyable and comfortable with their investing is to join investment clubs or groups. Generally there are such clubs where you work, with meetings once per week or month at lunch, after work, or in the evening. Various members will report on situations that they have studied and the group will decide on what investments to make with funds contributed from each member. Buy, sell, and withdrawal decisions will likewise be mutually made. It is generally enjoyable, and such groups have a pretty good success record, as indicated by the news article shown in Table 11-3. In a similar manner many people like to work in groups for their real estate investing.

Investment Counselors

An investment method that is touted by the investment brokerage houses, and generally used with good success by sports and public figures is the use of the financial advisor. It is certainly logical that people who specialize in investments should be more knowledgeable and skilled with them, but whether the average small investor can make their money (and the agents their commissions) from such a service is very uncertain. Remember that the brokerage houses benefit from buying and selling stocks and bonds, and whether you make a profit or loss on each transaction is really not that important to them. They may want to keep you as a customer, but their advice has to be highly biased. Likewise, other professional counselors generally have to be very conservative, since to have you suffer any major loss would be devastating to them. The net result in almost all non-multimillionaire situations is that investment counselors are of little use, or even detrimental, to an engineer attempting to build up his estate. Advisors certainly have their place, and their information should be listened to carefully, but in general you are on your own, and can do a better job by yourself.

REFERENCES

Browne, Harry, Special Reports, 1988 (May 18):115 P.O. Box 5586, Austin, Texas 78763; 1-800-5315142.

Capital Gains. 1987 (February). 600 Shelard Plaza North, Minneapolis, MN 55426.

Chemical Week. 1987. Precious metals. (June 24):47–53.

Cunniff, John. 1983. Investment clubs. *Fresno Bee* (April 10):F-7.

Fortune. 1987 (March 16; Oct. 26; Nov. 23); 1988 (Jan. 4) Time, Inc.

Greene, Mark R. 1988. *Financial Planning*. American Chemical Society, New York:178.

King, William W. 1987. Investing under the new tax law. *Chemical Engineering* (Feb. 16):173–176.

L. A. Times. 1987. Stock market (March 8). Copyright 1987, Patricia Mitchell.

Labich, Kenneth. 1987. A $100 billion bath in bonds. *Fortune* (May 25):34–35.

Money. 1986. Fund watch. (July):148; (Sept.):56.

Mutual Fund Forecaster. 1987. Beginner's guide to mutual funds. 3471 N. Federal Highway, Fort Lauderdale, FL 33306:25–32.; (305) 5639000.

Perlmutter, Rod. 1987. IRA deductions. *The Fresno Bee* (March 6):C-6.

Savage, Peter. 1987. The chemical industry rides out the storm in Wall Street. *Chemical Week* (Oct. 26):15–16.

Securities and Exchange Commission 1982. What every investor should know.

Sing, Bill 1987*b*. IRA's. *L. A. Times* (April 5):IV 1–8.

— 1987*a*. Mutual funds. *L. A. Times* (Feb. 22). Copyright 1987, Patricia Mitchell.

12

EMPLOYMENT CONSIDERATIONS

SEEKING EMPLOYMENT

The most important economic problem facing all engineers is to find and maintain a job that will be financially rewarding, challenging, and enjoyable. There are many ways to seek employment, and essentially all of them should be considered, since jobs are seldom easy to find, and the greater the choice one has, the better. There is no fixed sequence of best sources but it might be suggested that these three top many lists: (1) Contact your friends in the industry to spread the word that you are looking, and see if they have any ideas. (2) Write letters, and send your resume, to any and all companies where you might want to work, where you think there might be openings in your field, or which you know of as a generally good employer of chemical engineers. This list would include many nonchemical or CPI companies. (3) Your graduating college placement office and former professors. They are usually set up to handle both new and old graduates, and are well staffed, organized, and equipped to provide excellent assistance. They also usually have literature and seminars on how to prepare for and conduct your recruiting and interviewing activities. These aids are almost always based upon years of successful experience and are excellent in their content. They should be carefully studied and followed, along with, of course, the specific rules and procedures for their recruiting assistance.

As a next step, there are a number of excellent references on job hunting, including journal articles (Curry 1987; Dennis 1984; Hileman 1987; Jackson 1987; Graham 1983; Schultz 1984) and books or pamphlets (Beakley 1984; CPC Annual 1987; Career Opportunities 1987; Easton 1984; Peterson's Guide 1987). There are employment advertisements for open positions, and you can advertise your availability in magazines such as *Chemical Engineering News* (American Chemical Society, ACS, publication), *Chemical Engineering Progress* (American Institute of Chemical Engineers, AIChE, publication), and *Chemical Engineering* (McGraw-Hill), as well as the large city newspapers in your broad geographical areas of interest. Both societies also have "employment clearing houses'' at their meetings where recruiters looking for employees and engineers seeking employment can get together. The AIChE also has a job referral service (AIChE 1988). The references, advertisements, and recruiting services are in no way positive, but in many circumstances they can help and be well worth the extra effort involved. As a final suggestion, employment agencies are usually not too productive, but sometimes they can turn up excellent job opportunities, often in related but somewhat different fields. Their fee is usually paid by the employer, so there is little to loose by applying, just so that you don't count on them to heavily.

Once an interest has been shown by a company, an interview will be arranged. In general you should interview as many recruiters as you have a reasonable interest in their companies, and is practical. If there are limitations, naturally try to select those companies where you feel you have the greatest chance and/ or you have the most interest. All recruiters will require a written summary of an applicant's qualifications, which entails preparing a resume. Some of the general factors to be considered, and a typical example of a resume are shown in Table 12-1. As a second step, an interview will be required, and Table 12-2 lists a brief checklist of things to think about before and during the interview. When preparing a resume, or interviewing, it would also be highly advisable to first reconsider your general job interests, and the basic types of jobs that may be available, such as listed in Table 12-3.

Resumes

Many different resume formats are equally good for introducing yourself to a prospective employer. The only basic requirements are that they should be brief, neat, easy to read, and present all of the desired information. To be most effective they should also clearly present the more important accomplishments and skills of your career to date. Some of the points to consider are: (1) most job openings are highly competitive; (2) assuming that you will present yourself favorably, the best way to make sure that the interviewer remembers you is to attach a picture on your resume; (3) any work record is impressive, and of course one should stress the most successful technical, mechanical, or managerial aspects of prior work; (4) community or school extra curricular activity records can be impressive; and (5) note all scholastic achievements, including grade point average if above a B level. This is a critical factor for various job categories.

Job Interviews

The interview is an essential part of obtaining a technical job, since before being hired you will often be interviewed twice; first with their traveling interviewer, and then if you pass that screening, at the actual place of employment. The

Table 12-1
Typical Resume Format.

NAME
ADDRESS (home and permanent)
TELEPHONE NUMBER (Where you can be reached
now and later if you move)

Position Desired: Chemical Engineer (with a stated goal in R & D, Sales, Production, etc.)

Sex: Male or Female   Citizenship: (U.S.A., or if other, status)

Birthdate: Date Available:

Educational Training:
College or University   *Location*

Degree   *Date*   *Grade Point Average*
                                             (if above 3.0)
                                             Ch. Eng. –
                                             Overall –
Other courses taken

Previous Employment:

Employer   *Location*   *Dates*   *Duties*
                                      From   To

List all technical and mechanical work experience, and other that shows a strong work
background

Technical Organizations:
     Am. Inst. Chemical Engineers; Am. Chem. Soc., etc. if applicable

Honors, Activities, Publications:
     Limited listing of school and community activities or honors, including items to help an
interviewer remember you, and to make you stand out when the resume is being read.

Summary of Experience:
     Few sentences on some of your areas of specialization; general technical experience and
successful areas of interest, including course work and special projects.

References: Provide technical or mechanical references only, such as former supervisors or
professors

A picture is optional, but highly desirable if interviewing with many other applications.

Table 12-2
Brief Job Interview Tips.

Interviewing is highly competitive, so you must work at it. This requires a good resume, and it means studying the visiting companies to the extent possible[a] so that you will understand what the company does, some of its statistics, what it is looking for, and the types of chemical engineering jobs available. You can then match this with your interests and abilities. Discuss the openings with the interviewer and be interested, knowledgeable, and enthusiastic. Talk to him enough so that hopefully you will make a good impression and that he will remember you, but do not be a show-off, or overbearing. Be prepared to concisely describe some of your accomplishments, technical specialities, experience, skills, and your interests. This should include a discussion of areas which you have done well or specialized in, or which your former company or university is particularly strong in, such as computer process control. Also be prepared to ask pertinent job related questions to better understand what may be offered and to demonstrate your interest. Dress well, but conservatively and not overdressed, and be sure to be punctual and patient. Prepare for the interview as you would an examination; have a plan of how you want to present yourself and what you want to find out. Interviewing can be a very educational and interesting experience, as well as a means of obtaining employment. It is often the only time early in your career that many of the detailed goals, operations, and organizational structure of companies will be discussed with you.

[a]Check the company's listing in any of: *Million Dollar Directory; Standard and Poor's Register of Corporations, Directors and Executives; Thomas Register of American Manufacturers; Moody's Industrial Manual; California Directory of Manufacturers; Directory of Chemical Producers*, etc.

interviewers at both locations will have planned to inform you about the company, and the general type of work that you would do if employed. They will also attempt to evaluate your qualifications to do their type of work. At the plant there will often be many ''interviewers''; one professional, and others who may be your potential supervisors or coworkers, perhaps more interested in evaluation (and sometimes not interested at all) than in informing you of working conditions. All will have a voice in whether you will be offered employment, so you must patiently repeat your qualifications and attempt to ''sell'' yourself to each person. With luck each interview should be an interesting experience, and most of the interviewers will be friendly and encouraging. Careful preparation will allow you to obtain the most from each one. The more interviews you have, the better you should be able to handle them, the greater your experience and learning from them, and the wider your choice and better the chance of finding rewarding employment.

CHEMICAL ENGINEERING JOBS

Chemical engineers have a relatively versatile technical education, with a good-to-fair knowledge of chemical engineering, chemistry, mathematics, science, physics, other types of engineering and economics. Because of this the employment opportunities are quite varied, and graduates take jobs doing a wide range of work and in many industries besides the chemical or process industries. During the past several years the majority of new graduates found work in

Table 12-3
Typical Chemical Engineering Job Types in the Chemical Industry.

Job Type	Duties	Job Requirements
Plant production (operations)	Direct production responsibilities. Supervise or assist with operation of the plant, process, or equipment.	Interact closely with operators and supervisors. Interest in supervision, economics, physical operations, and production. Moderate technical requirements.
Plant technical service	Assist the production department with trouble-shooting, modernization, etc.	Interact with technical and production departments. Practical problem solving. Medium technical requirements.
Engineering (engineering co., or company engineering department)	Any one or more of: making detailed plant designs, cost estimates, economic evaluations, construction supervision, etc.	Interest in design, costing, evaluation, or construction detail. Moderate technical requirements; often extreme specialization.
Technical sales	Sell the company's products.	Gregarious personality, and like to talk, travel, and meet people. Low technical requirements.
Research and development	Develop new processes and products or improve old ones.	Creativity and interest in problem solving. High technical requirements.
Management	Supervise staffs in any of the above or other areas.	Reasonable ability to supervise and work well with subordinates and superiors. Ability to communicate, schedule, plan, organize, make decisions, manage, and control costs. Variable (but reduced) technical requirements.

nonchemical industries, but with jobs where the chemical engineering education was applicable, as indicated in Table 12-4. Typical salaries and for new graduates are given in Table 12-5.

All through the 1980s finding employment has been comparatively difficult for new chemical engineering graduates, as indicated by Figure 12-1 showing the high unemployment rate about three months after graduation. Employment at graduation was even lower, such as in 1986 when only 38% of the chemical engineers were employed at that time. This unemployment number was probably

Table 12-4
Employment Statistics.
Industries Employing Graduating BS Chemical Engineers
(% of Total New Jobs).

Industry/Year	1977–1981	1982	1983	1984	1985
Chemical and petroleum	75–80	70	55	40	45
Electronics	0	15	30	17	10
Other	20–25	15	15	43	45

(Greater detail)	1984	1985
Chemical	28	31
Oil and gas	12	14
Electronics	17	10
Consumer goods	12	11
Food processing	4	3
Utility	2	2
Others (government, environmental, defense, etc.)	25	29

Source: Mathis 1986. Chemical Engineering Progress, July, 1986. Reproduced by permission of the American Institute of Chemical Engineers. Farrell 1985. Excerpted by special permission from *Chemical Engineering*, April 15, 1985. Copyright © 1985, by McGraw-Hill, Inc., New York.

cut in half after six months, and very likely was down to only 10–15% unemployment after one year. Job offers to new graduates (while still at school) decreased 20% in 1985, 35% in 1986, 40% in 1987, but were up 10% in 1988 (Jones 1987; *Chemical Week* 1984).

The basic problems causing the job shortage have been twofold; (1) an excess graduation of new engineers (see Table 12-6), and (2) a reduction in employment by the chemical industry. Figures 12-2 and 12-3 show the recent total industry employment, while Table 12-7 shows the employment of scientists and

Table 12-5
Average Salaries for New
Chemical Engineering
Graduates.

	1986–1987	1987–1988
BS	$28,844	29,820
MS	33,200	34,776
PhD	42,900	

Source: *Chemical and Engineering News* 1987a. Reprinted with permission from *Chemical and Engineering News*, May 18, 1987. © 1987, American Chemical Society.

Percentage of new bachelor's degree recipients who were
unemployed and seeking full-time employment

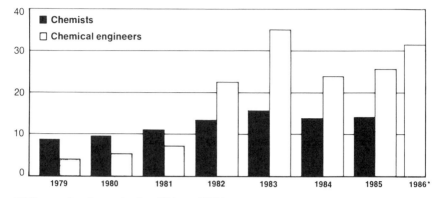

*62% unemployed at graduation. (Shinnar 1987.)
Source: Hileman 1987. Reprinted with permission from *Chemical and Engineering News*, October
28, 1987. © 1987, American Chemical Society.

Alternate Statistics Year	Chemical Engineers' Unemployment three months after Graduation
1982–1983	57%
1983–1984	42%
1986–1987	40%

Source: Farrell 1985. Excerpted by special permission from *Chemical Engineering*, April 15, 1985. Copyright
© 1985, by McGraw-Hill, Inc., New York.

Figure 12-1. Chemical engineering graduate unemployment (approximately three months after
graduation; perhaps twice these numbers at graduation).

engineers in the chemical industry over the same period. It appeared through
the 1970s that the need for massive alternate energy projects, space and high-
technology fields, and environmental problems would require large numbers of
new chemical engineers, so salaries rose rapidly, jobs were plentiful and
basically, large numbers of new engineers were "recruited." Then, by the time
chemical engineering enrollments in colleges began to greatly increase, oil prices
fell dramatically, the energy crisis was over (for that period), and the new
projects and jobs did not materialize. At the same time economic slow-down
and increased foreign competition caused the U. S. chemical industry to dramatically reduce its number of employees and its new hiring. It has been estimated
(Donaldson 1986) that perhaps 20,000 chemical engineers in excess of the
industry demand were graduated in the first seven years of the 1980s.

Table 12-6
Chemical Engineering Graduates.

Academic Year	Degrees Awarded in Chemical Engineering		
	Bachelors	Masters	PhDs
1965–66	2,848	994	354
1966–67	2,869	949	305
1967–68	3,211	1,156	367
1968–69	3,557	1,136	409
1969–70	3,720	1,045	438
1970–71	3,615	1,100	406
1971–72	3,663	1,154	394
1972–73	3,636	1,051	397
1974–75	3,142	990	346
1975–76	3,203	1,031	308
1976–77	3,581	1,086	291
1977–78	4,615	1,237	259
1978–79	5,655	1,149	304
1979–80	6,383	1,271	284
1980–81	6,527	1,267	300
1981–82	6,740	1,285	311
1982–83	7,145	1,304	319
1983–84	7,475	1,514	330
1984–85	7,146	1,544	418
1985–86	5,877	1,361	446

Source: Hileman 1987a. Reprinted with permission from *Chemical and Engineering News*, May 18, 1987. © 1987, American Chemical Society.

The result of this has been a major shift in employment to nonchemical or petroleum industries as shown in Table 12-4. Chemical engineers are fortunate with their multisubject education making them well qualified for jobs in many other fields, and this broadening is increasing (Basta 1987; Farrell 1985). Even in the mid-1970s (when chemical jobs were plentiful) it has been estimated that perhaps 30% of the new graduates went into other fields, and of those remaining after 10 years in the profession, perhaps 20% switched into other areas (Shinar 1987). All of this means that a chemical engineer can and almost always will find rewarding, well-paying employment, but that most will have to search much harder for this employment, and consider other related fields. Basically, seeking employment will have to be treated as a job in itself, and many employment letters will have to be sent out to prospective companies, employment want-ads answered, and close contact kept with the school employment office, friends in industry, professors, and so on, which might lead to employment prospects. It will be a chore, and often discouraging, but the odds are still very good for ultimate success.

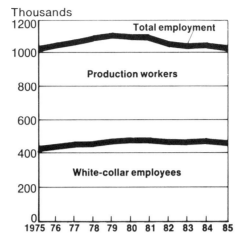

Source: *Chemical and Engineering News* 1986. Reprinted with permission from *Chemical and Engineering News*, February 3, 1986. © 1986, American Chemical Society.

Figure 12-2 Chemical Industry Employment Breakdown. (Chemicals and allied products)

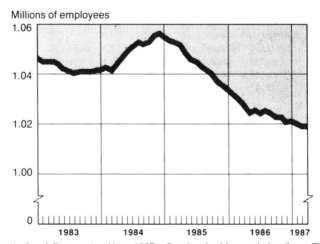

Source: *Chemical and Engineering News* 1987a. Reprinted with permission from *Chemical and Engineering News*, May 18, 1987. © 1987, American Chemical Society.

Figure 12-3 Total Chemical Industry Employment.

Table 12-7
Employment of Scientists, Engineers.

Thousands[a]	1986[b]	1985[b]	1984	1983	1982	1981	1980	1979	1978	1977	1976
Total, all industry	580.2	563.9	544.5	522.1	509.8	487.8	450.6	423.9	404.4	382.8	364.4
Chemicals and allied products	71.3	67.0	67.1	66.0	61.6	54.7	51.4	50.0	48.3	46.4	44.4
Industrial chemicals	26.8	25.0	26.7	27.2	25.9	21.6	20.9	21.4	21.3	20.6	20.1
Drugs and medicines	33.3	30.7[c]	30.1[c]	28.2	25.6	23.3	21.6	20.8	19.5	17.8	16.6
Other chemicals	11.2	11.3	10.3	10.6	10.1	9.8	8.9	7.8	7.5	8.0	7.8
Petroleum refining	10.3	13.4	13.2	14.7	15.6	13.0	10.8	10.1	9.9	8.9	8.6
Rubber products	13.2	13.4	na	na	8.1	10.3	na	8.1	7.9	9.1	8.6

Source: *Chemical and Engineering News* 1987b. Reprinted with permission from *Chemical and Engineering News*, June 8, 1987. © 1987, American Chemical Society.
a. Full-time equivalent number of R&D scientists and engineers in industry, as of January of each year.
b. Preliminary.
c. Partly estimated. na = not available separately, but included in total.

In the chemical process industry per se, and somewhat similarly in other industries, the more important basic job categories are listed in Table 12-3. A simplified, generalized path of promotional opportunities for these jobs is briefly outlined in Table 12-8. Some companies tend to promote promising employees into different types of work (i.e., from research to production to sales, etc.) while others maintain advancement within the same job category. Some engineers stay with one company their entire career; perhaps the majority seek more rapid advancement by finding promotional opportunities in moving from company to company, and even to differing job types. You may find excellent job satisfaction in various of these work categories, but often one's skills will excel in one more so than the others, and generally this field will bring greater

Table 12-8
Typical Abbreviated Managerial Path of Advancement Chart for Chemical Engineers in the Chemical Industry.

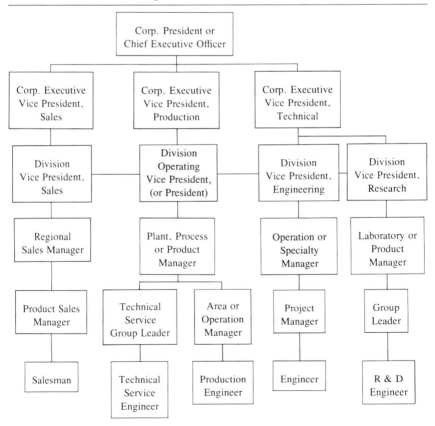

advancement and pleasure. Consequently, an engineer should carefully consider his skills and interests before accepting employment, but if there is no preference, or a strong interest is felt in several categories, one should plan on not spending too long in trying different job types (unless there is considerable success with each), so as to not loose promotional opportunities in one's strongest area.

ECONOMIC PRESENTATIONS

Informal Reports

All chemical engineers have a constant need to informally or formally, but effectively communicate with others, such as through conversations, memos, reports, or presentations. Perhaps the most frequent and common means of written communication is the brief, informal interoffice memorandums, or "memos" meant primarily for your supervisor and the file. Most of the economic estimates will end up as memos, and that is why memos are being considered here. A suggested outline of the content and a check list for memos is given in Table 12-9. These informal reports can play an important roll in determining one's reputation and advancement in a company, as well as the success obtained with recommendations being commercially implemented. Every engineer should work on them carefully until he or she feels fluent with their preparation and is confident that they are what the supervisor wants. It may be a little embarrassing, but having one's spouse or a friend help with

Table 12-9
Suggestions for Writing an Interoffice Memorandum.

1. Use the company form if available. If not, make your own. The headings should be:
   To: Subject:
   From: Date:

2. In the text first give a very brief introduction: What is the problem or subject, why are you writing the memo, what are you going to present? Make this adequate for a reader many years later.
3. Next give the background and the facts in your presentation.
4. Then give your proposed solution and your reasons for reaching that conclusion.
5. Always be sure that you write to the audience (your peers, your boss, his boss (?), etc.) who will read the memo, and try to answer all of the questions they will have. Do not give a "hard sell," and be aware of your (and your boss's) restrictions and limitations in the matter. Be brief and clear.
6. Proofread carefully for:
   a. a clear logic pattern
   b. appropriate data
   c. use of the third person, good grammar, spelling, punctuation, sentence structure, etc.
   d. not overstated and yet persuasive language

Table 12-10
Notes on Semi-informal Presentations.

1. Carefully adjust your talk to the time available for presentation. If you consistently overrun your time you will probably not be asked to make many reports.
2. Give an adequate introduction so that all of your audience understands why the presentation is being made, and what the problem and background are.
3. Present your facts and logic pattern clearly, and yet as briefly as possible.
4. For any talk over 5–10 minutes be sure to have visual aids.
5. Build the talk around your recommendation, and be as persuasive as possible, but do not overstate your case.
6. Be sure that your presentation is geared for, and appropriate for all of those in the meeting—especially the decision makers.

these reports until you are confident of their substance, form, grammar, spelling, and so on, is often very worthwhile. They can be that important to your career.

Informal Presentations

Next in importance to interoffice memorandums is generally semi-informal presentations to groups or departments within one's company. This can be as simple as answering questions at weekly production group meetings, or the somewhat more demanding brief presentations. Table 12-10 lists some of the considerations for making effective presentations, but each individual situation will bring out others. As with memos, engineers will be indirectly judged by the quality of these presentations, and they also can be quite important to one's advancement. The work that is done on economic analyses will be a frequent part of these presentations. When such talks are a major part of one's job some engineers have found it interesting and advisable to take speaking courses such as the Dale Carnegie courses (How to Win Friends and Influence People) or join Toastmaster Clubs. You improve your speaking skills and outside-of-industry contacts, and many managements look upon this with considerable approval.

REFERENCES

AIChE. 1988. Job referral service. 345 E. 47th St., New York, NY 10017.

Basta, Nicholas. 1987. Chemical engineering education: Are changes really called for?''. *Chemical Engineering* (June 22):23–27.

Beakley, George C. 1984. Careers in engineering and technology. Macmillan Publishing Co., New York.

Career Research Systems, Inc. 1987. Career opportunity index. P.O. Box 8969, Fountain Valley, CA 92728:1–184.

Chemical and Engineering News. 1986. (Feb. 3).

Chemical and Engineering News. 1987a. ACS committee on professional training annual report. (May 18):59–60.

Chemical and Engineering News. 1987*b*. Facts and figures for the chemical industry. (June 8):58.

Chemical and Engineering News. 1987*c*. (Dec.):5.

Chemical Week. 1984. For graduates, more jobs in 1984. (April 22):12.

Civic Data Corp. *Southern California Business Directory and Buyers Guide.* 404 S. Bixel St., Los Angeles, CA 90017.

College Placement Council. *The CPC Annual.* 62 Highland Ave., Bethlehem, PA 18017.

Curry, Michael J. 1987. Jobs for problem solvers. *Chemtech* (April):200–201.

Dale Carnegie & Associates, S. 1287 N, 6000 Dale Carnegie Drive, Houston, TX 77036.

Dennis, Ronald M. 1984. Job search resource. *Chemical Engineering* (March 19):170–184.

Donaldson, T. L. 1986. Chemical Engineering Enrollment Survey, AIChE.

Dunn & Bradstreet, Inc. *Million Dollar Directory.* 3 Century Dr., Parsippany, NJ 07054.

Easton, Thomas A. 1984. Careers in science. Dow Jones-Irwin.

Farrell, Pia. 1985. *Chemical Engineering* (April 15):22–25.

Graham, E. Earl. 1983. Surveys on the initial employment of chemical engineers. Penn. St. University.

Hagerty, D. Joseph, Earl R. Gerhard, and Charles A. Plank 1979. *Opportunities in Chemical Engineering.* VGM Career Books.

Hileman, Bette. 1987. Employment outlook. *Chemical and Engineering News* (Oct. 28):29–59; 1985 (Oct. 28):40.

Jackson, Bernard. 1987. Overcoming unemployment. *Chemical Engineering* (Aug. 17):159–160.

Jones, Susan R. 1987. The class of 1987 faces fewer jobs. *Chemical Week* (May 27):6–8.

Mathis, J. F. 1986. Chemical engineers meet a changing world. *Chemical Engineering Progress* (July):17–21.

Moody's Investor Service, Inc. *Moody's Industrial Manual.* 99 Church St., New York, NY 10007.

Peterson's Guides. 1987. *Peterson's Guide to Engineering, Science, and Computer Jobs.* P. O. Box 2123, Princeton, NJ 08540.

SRI International. 1986. *Directory of Chemical Producers.* Meno Park, CA 94025.

Schultz, Fred O., Jr. 1984. Demand for chemical engineers, 1983 survey. *Chemical Engineering Progress* (Aug.):23–25.

Shinnar, Reuel. 1987. The crisis in chemical engineering. *Chemical Engineering Progress* (June):16–21.

Standard and Poor's Corp. *Standard and Poor's Register of Corporations, Directors, and Executives.* 25 Broadway, New York, NY 10004.

Thomas Publishing Co. *Thomas Register of American Manufacturers.* One Penn Plaza, New York, NY 10001.

Times-Mirror Publishing Corp. *California Directory of Manufacturers.* Spring St., Los Angeles, CA 90012.

APPENDIX 1

EQUIPMENT COST ESTIMATES

The following section provides very rough cost estimates for a wide variety of process equipment. It must be remembered in using these charts that there is no such thing as an exact, definite, fixed price for any piece of equipment of a given size or capacity. As with buying merchandise, clothing or a car there are many styles, quality differences, optional features and designs to meet specific needs or services. Presumably charts could be made for each of these variations, but the number would be large and confusing, and for many preliminary estimates the engineer would not know exactly what he wanted at that stage of the design, so only average, representative equipment should be more useful. Again, a range of prices could be shown, but usually a single line is more practical, keeping in mind that the price could quite normally vary considerably depending upon the exact design requirements and the company policy on quality, maintenance, and so on.

With these generalities in mind, the following charts have been taken from a number of sources. Most are from cost estimating articles or books, although some are from recent vendor quotations. In case only a single source was available, that reference has been noted. However, often many sources were available and a somewhat biased consensus of opinion curve was selected. In this case the sources were not noted except for inclusion in the reference list at the end of the appendix. In case different variables were used as the sizing parameter, the most logical one in the author's opinion was selected.

All costs were factored to an early 1987 basis, or a chemical engineering index number of 320. When equations were available for the cost relationship they were listed beneath the charts, and when straight line functions existed for the costs on log–log paper a sizing exponent was given:

$$\text{cost size } 2 = \text{cost size } 1 \left(\frac{\text{size } 2}{\text{size } 1}\right)^{\text{size exponent}}$$

In a number of references various authors have estimated the fraction of the purchased equipment cost that it takes to install the equipment. This generally included freight and shipping costs, foundations, mounting, and simple electric and piping connections, such as switch gear, starters, flange connections, and so on. Unfortunately these numbers often varied widely, so the range and average are both listed when available:

installed cost = purchase price × installation factor

A similar number that also includes all of the adjacent minor equipment and connections is sometimes listed in the literature (principally by Guthrie 1975 and Ulrich 1984) covering the cost of purchase and installation of the major equipment as well as all of the supporting equipment around each major unit. This is called the module factor, and when available is also listed under the charts as the range given by different authors and the average value.

cost of the installed module = purchase price × module factor

As a final item under the equipment cost graphs, often a simple factor can be used to estimate the cost of some other material, pressure, size, or other variable for the equipment, than is shown on the graph. For instance, the cost of a stainless steel agitated tank is 1.7 times the cost of a mild steel tank (which is shown on the chart). These factors have also been listed when available, and again, sometimes as a consensus of different authors' estimates.

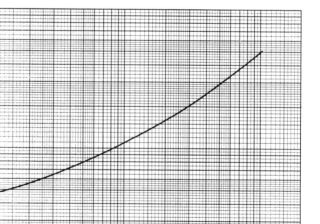

Adsorbers, Activated Carbon
Mild steel construction, including instruments and controls

Weight of carbon, 1,000 lb.

Equations:

Cost = 15,200 + 1, 100$W_C^{0.481}$ for W_C >250, < 10,000 lb.

Cost = 76,200 + 0.422$W_C^{1.2}$ for W_C >10,000, <200,000 lb.

W_C = weight of activated carbon, lb.

Agitators
Dual turbine blades; mild steel; 30–45 rpm, motor, gear reduction, shaft
Propeller; mild steel, single blade

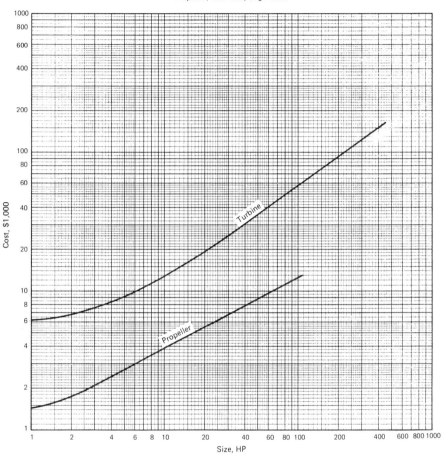

Size exponent:
Turbine: >30 HP 0.68
4–30 HP 0.56
<4 HP 0.23
Propeller: 3–100 HP 0.51
1–3 HP 0.42

Installation factor:

	Range	Average
Turbine	1.20–40	1.32
Propeller	1.12–32	1.22
Module factor	2.0	

Factors for:

	Range	Average
Turbine:		
Single blade	0.75–0.85	0.82
56–100 rpm	0.57–0.70	0.66
125–230 rpm	0.37–0.51	0.47
316 stainless	1.23–1.87	1.47
Propeller:		
Stainless steel		1.19
With seal (for closed tank)		1.32

Agitated Tanks*
Jacketed, agitated, mild steel

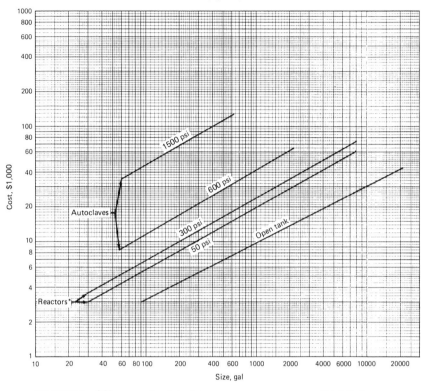

Size exponent	0.53	Installation factor:			Material factors:
Module factor	2.5		Range	Avg.	Stainless steel
		Open tank	1.41 – 66	1.58	1.2 – 2.2, avg. = 1.7
		Low pressure	1.30 – 57	1.44	Glass lined
		Autoclave	1.50 – 70	1.60	1.2 – 2.0, avg. = 1.6
					*See Reactors

Air Conditioning
Compressor, motor, controls, condenser, refrigerant

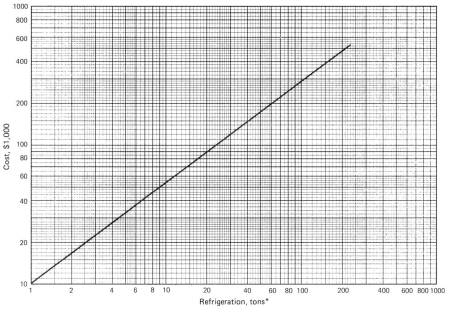

Size exponent   0.73
Installation, Module factor
   1.38–53 avg. 1.46

*One ton = 12,000 Btu

Blenders

Mild steel construction

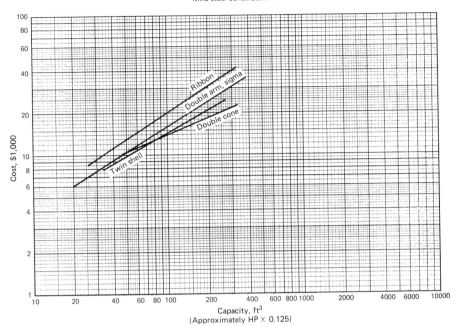

Capacity, ft^3

(Approximately HP × 0.125)

Size exponents:

Ribbon, double arm, sigma,
twin shell   0.60

Double cone 0.42

Material factor:

304 stainless steel 1.6

Installation factor 1.30

Module factor:

Ribbon   2.0

Sigma   2.8

Double arm, cone,
twin shell   2.2

Blowers
30-in. water (~1 psi) to 30 psi; cast iron, with motor

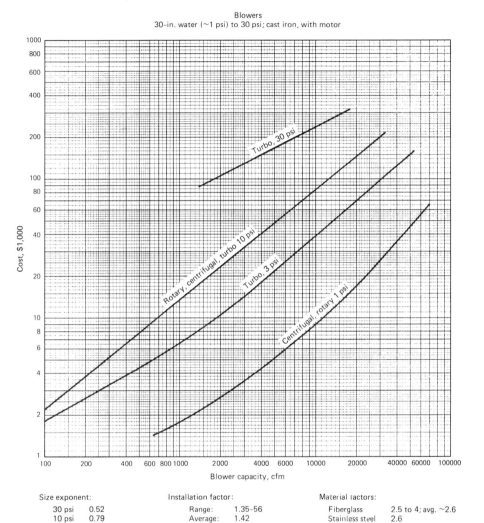

Blower capacity, cfm

Size exponent:		Installation factor:		Material factors:	
30 psi	0.52	Range:	1.35–56	Fiberglass	2.5 to 4; avg. ~2.6
10 psi	0.79	Average:	1.42	Stainless steel	2.6
<3 psi	variable				
		Module factor:			
		Rotary	2.2		
		Centrifugal	2.5		

Boilers

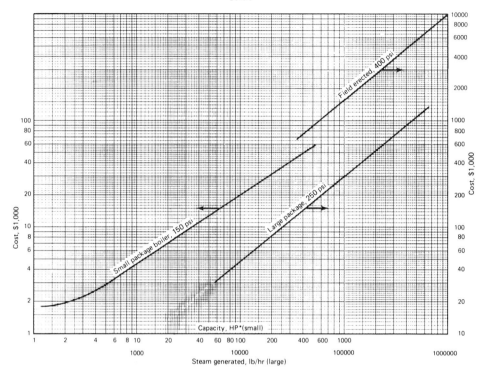

Size exponents:

Package boilers
5 - 1,000 HP   0.65
6 - 600 · 10³ lb/hr 0.77
Field erected   0.82

*One HP = 33,5000 Btu

Installation factor:
1.21 - 82 avg. 1.53

Module factor:
Package   1.8
Field erected 1.8 - 2.0 avg. 1.9

Pressure factor:
Large package
400 psi   1.31
500 psi   1.74
Field erected
1000 psi   1.35
3000 psi   1.58

Coal fired
Large package 1.61
Field erected 1.36

Boilers, Waste Heat

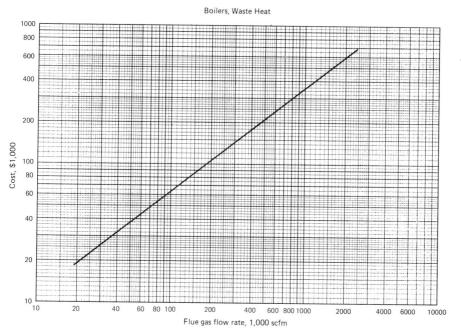

Size exponent 0.75

Installation factor:
1.40–82 avg. 1.67

Module factor 1.81

Factors:
High-temperature
operation 1.2
Finned tubes 1.5
Alloy-clad tubes 3.0
Mechanical ash
removal 1.8
Radiation section 2.0

Buildings
Office type with air conditioning, restrooms, plaster or equivalent
walls, insulation, modest architectural features

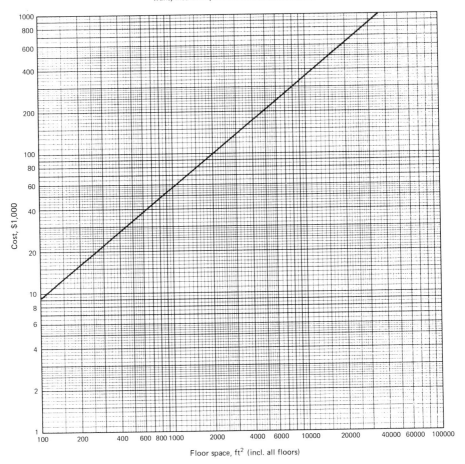

Floor space, ft² (incl. all floors)

Size exponent 0.8

Factors:

Warehouse	0.25
Laboratory	1.5
Manufacturing bldg.	0.5

Centrifuges
Solid-bowl, screen-bowl, pusher types,
316 stainless steel

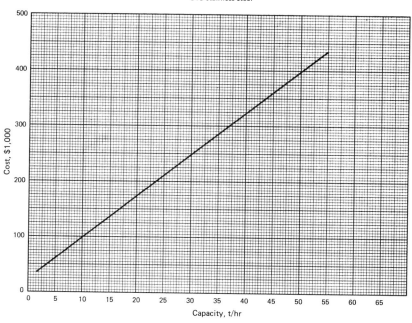

Installation factor:

    Range 1.20–2.02 avg. 1.54

Module factor: 2.0

Material factors:

    Carbon steel 0.68
    Monel   1.35
    Nickel   1.7
    Hastalloy C 2.6

Chimneys, Stacks
Carbon steel, lined, insulated, with foundations (tall); No lining (short)

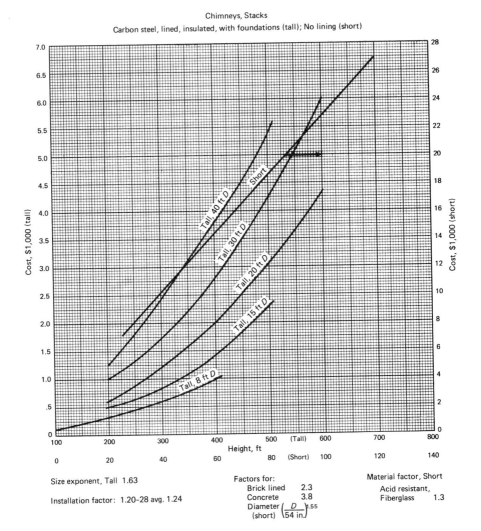

Size exponent, Tall 1.63

Installation factor: 1.20–28 avg. 1.24

Factors for:
Brick lined   2.3
Concrete   3.8
Diameter $\left(\dfrac{D}{54\ \text{in.}}\right)^{.55}$
(short)

Material factor, Short
Acid resistant,
Fiberglass   1.3

Classifier, Rake or Spiral
Mild steel construction

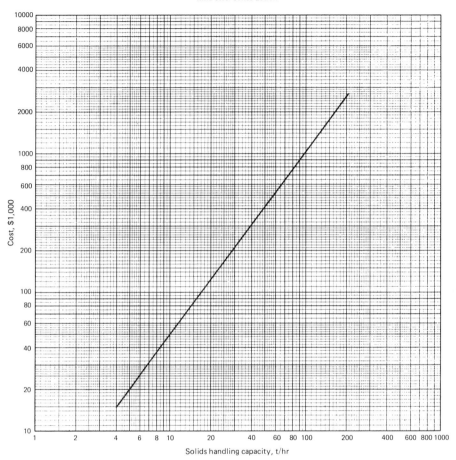

Cost, $1,000

Solids handling capacity, t/hr

Size exponent 1.32   Installation factor   Module factor 2.3
1.63-2.61 avg. 2.12

Columns, Distillation, Absorption Towers, etc.
Mild steel construction, 0–50 psi, vertical

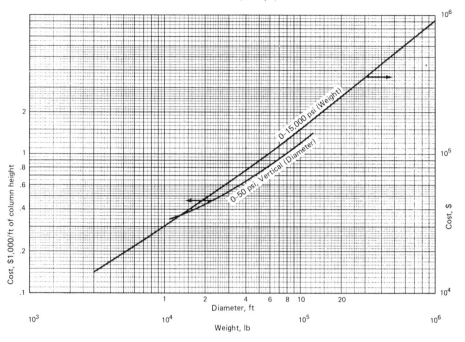

Size exponent

10^5–10^6 lb 0.78

Installation cost:

1.29–2.03 avg. 1.72

Module factor

Vertical   4.16
Horizontal 3.05

Material factors

Carbon steel	1.0
Stainless 304	1.7
Stainless 316	2.1
Monel 400	3.6
Titanium	7.7
Carpenter 20 CB-3	3.2
Nickel 200	5.4
Inconel 600	3.6
Incoloy 825	3.7

Other factors

Horizontal vessel 0.6

Pressure $\dfrac{P}{50}^{0.44}$ or see chart

Pressure factors (vertical)

psi		psi	
50	1.00	800	3.80
100	1.25	900	4.00
200	1.55	1,000	4.20
300	2.00	1,500	5.40
400	2.40	2,000	6.50
500	2.80	3,000	8.75
600	3.00	4,000	11.25
700	3.25	5,000	13.75

Column Trays
Mild steel

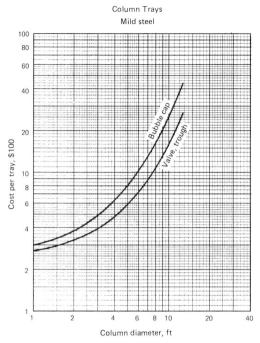

Cost per tray, $100 (y-axis)

Column diameter, ft (x-axis)

Installation factor 1.20

Number factor:		Tray type factor:		Material factor:	
25	1	Turbo grid (stamped)	0.8	Brass	1.2
20	1.05	Grid, plate, seive	1.0	304 stainless	1.5
15	1.25	Trough, valve	1.2	316 stainless	1.9
10	1.50			347 stainless	2.1
5	2.30			Inconel	3.3
1	3.0			Monel	7.7

Column Packing

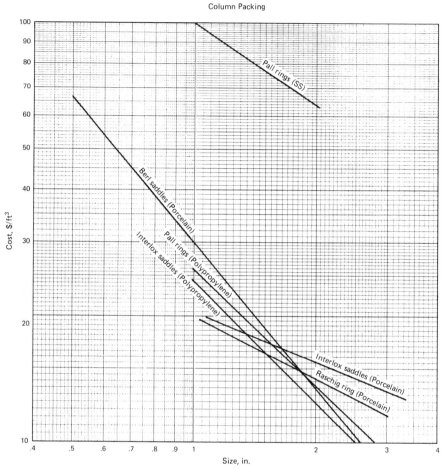

Cost, $/ft³ vs Size, in.

Material Factors; size exponents

Material	Size exponent	Ratio for other material		Material	Size exponent	Ratio for other material	
Berl saddles		1.24	Stoneware	Interlox saddles			
Porcelain	−1.16			Porcelain	−0.4	0.94	Stoneware
Pall ring SS	−0.64	0.30	Carbon steel	Polypropylene	−0.95		
Polypropylene	−0.95			Raschig rings		6.11	Stainless steel
				Porcelain	−0.5	2.35	Carbon
						1.58	Mild steel
						0.78	Stoneware

Compressors, Medium-Low Pressure

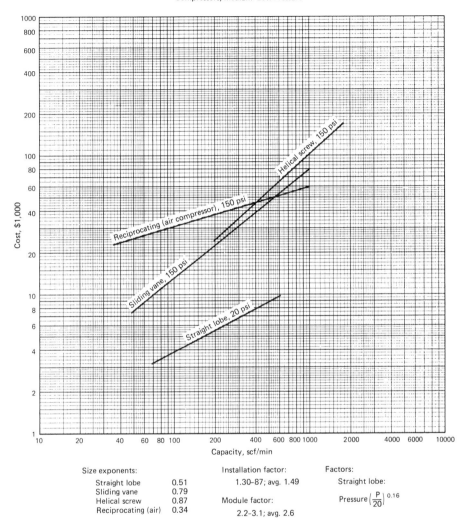

Size exponents:

Straight lobe	0.51
Sliding vane	0.79
Helical screw	0.87
Reciprocating (air)	0.34

Installation factor:

1.30–87; avg. 1.49

Module factor:

2.2–3.1; avg. 2.6

Factors:

Straight lobe:

$$\text{Pressure} \left(\frac{P}{20}\right)^{0.16}$$

Compressors, High-Capacity and/or Pressure
1,000 psi; electric motor drive, gear reducer, steel

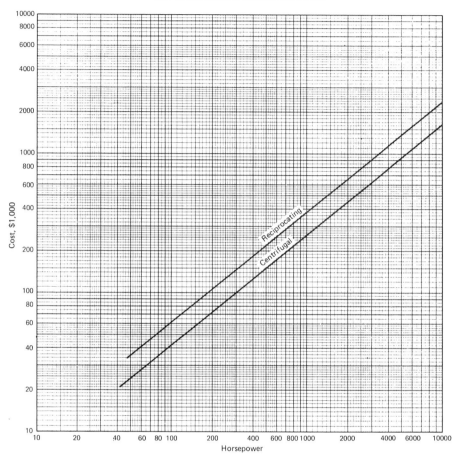

Size exponent 0.80

Equation:
(Isothermal compression)

$HP = 0.0044 P_1 Q_1 \ln P_2/P_1$

P_1 = inlet pressure, psi
P_2 = outlet pressure, psi
Q_1 = inlet flow rate, cfm

Installation factor:
1.30–87; avg. 1.49

Module factor:
2.15–3.1; avg. 2.6

Factors:

Turbine drive	1.13
Gas engine	1.41
Pressure $\left(\dfrac{P}{1000}\right)^{0.18}$	
Stainless steel	2.5
Nickle alloy	5.0

Conveyors
Mild steel construction

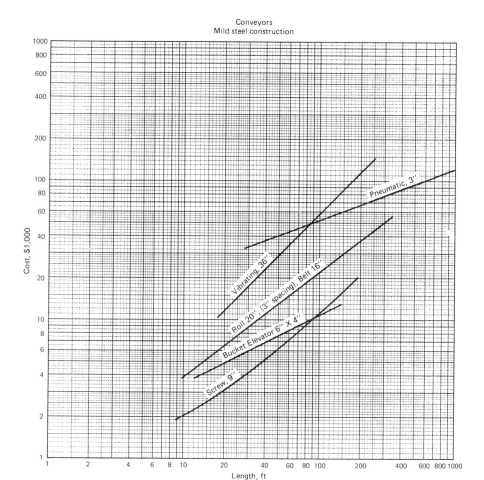

Length, ft

Cost, $1,000

Size exponent:

Screw conveyor	0.78
Belt conveyor	0.76
Bucket elevator, roll	0.5
Pneumatic conveyor	0.37
Vibrating	1.0

Installation factor:

Range 1.40–2.15 avg. 1.72

Module factors:

Screw, pneumatic, roll	2.2
Belt, bucket, vibrating	2.4

Size factors:

Screw conveyor $\left(\dfrac{diameter}{9\ in.}\right)^{1.2}$

Belt conveyor $\left(\dfrac{width}{16\ in.}\right)^{0.6}$

Bucket elevator $\left(\dfrac{bucket\ wd.\ X\ ht.}{6\ X\ 4 = 24\ in.^2}\right)^{0.37}$

Pneumatic conveyor $\left(\dfrac{diameter}{3\ in.}\right)^{0.55}$

Roll $\left(\dfrac{width}{20\ in.}\right)^{0.55}$; 4 in. spacing X 0.84

Vibrating $\left(\dfrac{width}{36\ in.}\right)^{0.57}$

Coolers, Quenchers

Mild steel construction; Cascade cooler, 2 in. diameter pipe

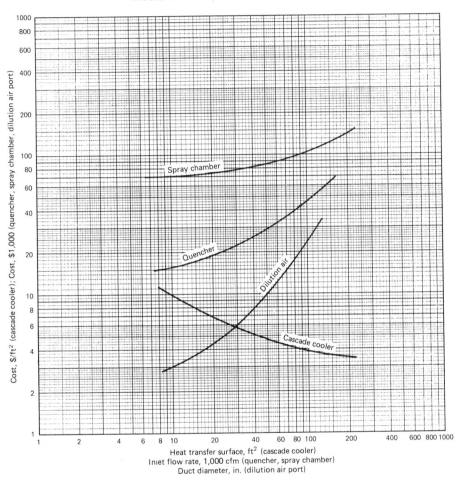

Cost, $/ft² (cascade cooler); Cost, $1,000 (quencher, spray chamber, dilution air port)

Heat transfer surface, ft² (cascade cooler)
Inlet flow rate, 1,000 cfm (quencher, spray chamber)
Duct diameter, in. (dilution air port)

Installation factor:

  1.40–1.85; avg. 1.62

Module Factor:

  Spray chamber, quencher 2.7

Equations:

  Spray chamber: $\$(358 \times M\ scf + 65,000)$
  Quencher $\$(335 \times M\ scf + 12,200)$

Factors:

  Cascade cooler
  $\left(\dfrac{pipe\ diameter}{2\ in.}\right)^{0.6}$

Cooling Tower
15° F range, 10° F approach, 82° F wet bulb

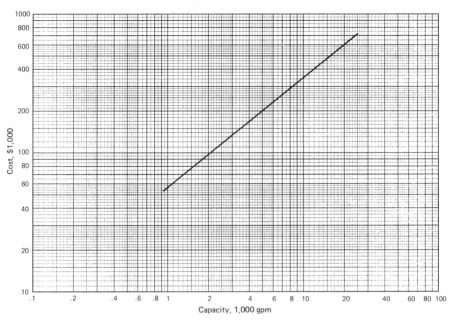

Capacity, 1,000 gpm

Size exponent   0.79

Installation factor 1.20

Module factor   1.70

Factors;

Range: $\left(\dfrac{T°F}{15}\right)^{0.57}$

Wet bulb temperature

°F	Factor	Approach, Δ°F	Factor
68	0.65	6	1.60
70	0.68	8	1.20
72	0.72	10	1.00
74	0.77	12	0.85
76	0.82	16	0.65
78	0.87	20	0.50
80	0.93	24	0.40
82	1.00		

Crystallizers
Mild steel construction

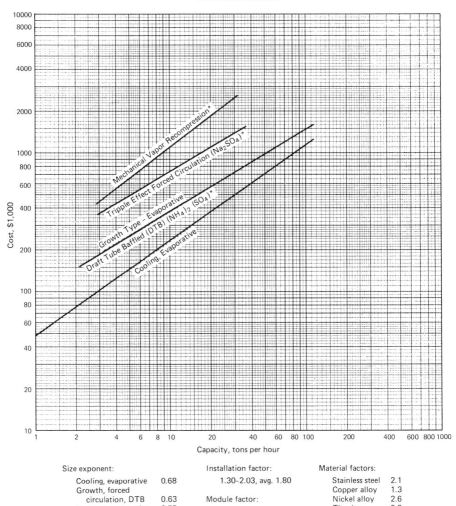

Size exponent:

Cooling, evaporative   0.68
Growth, forced
   circulation, DTB   0.63
Vapor recompression   0.75

Installation factor:

1.30–2.03, avg. 1.80

Module factor:

2.4–2.9, avg. 2.6

Material factors:

Stainless steel   2.1
Copper alloy   1.3
Nickel alloy   2.6
Titanium   6.0

*Courtesy of Swenson.

Dryers
Mild steel construction

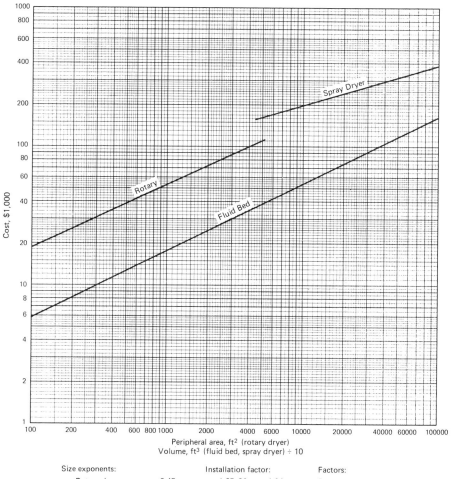

Peripheral area, ft² (rotary dryer)
Volume, ft³ (fluid bed, spray dryer) ÷ 10

Size exponents:		Installation factor:	Factors:	
Rotary dryer	0.45	1.25–96; avg. 1.64	Rotary to:	
Fluid bed	0.48		Roto-Louvre	1.25
Spray dryer	0.29	Module factor:	Vacuum shelf	0.35
		Rotary 2.3	(shelf area)	
		Fluid, spray 2.7	Materials:	
			Nickle alloy	3.7
			Brick-lined,	
			stainless steel	2.2

Ducts
Wall thickness 1/8 in.

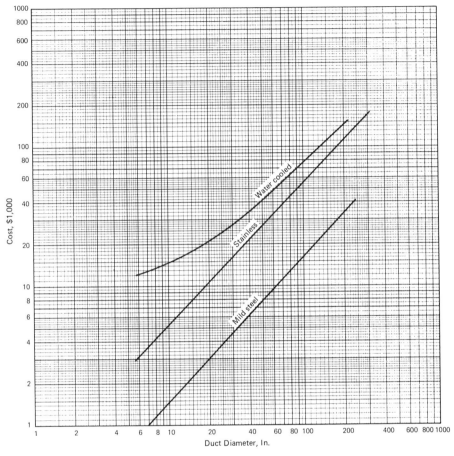

Cost, $1,000

Duct Diameter, In.

Size exponent: 1.08
Installation factor: 1.45

Equations: $/ft
Mild steel $(-2.22 + 1.66D)$
Stainless $(-6.43 + 5.84D)$
Water cooled $(79 + 6.78D)$

Factors:
Mild steel:
(wall thickness/1/8 in.)$^{0.68}$
Stainless:
(wall thickness/1/8 in.)$^{1.0}$

Dust Collectors
Mild steel construction

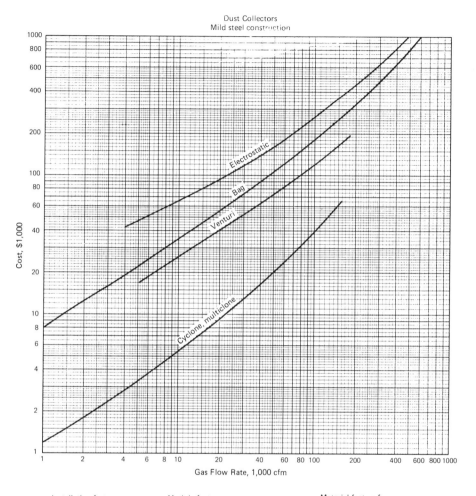

Cost, $1,000

Gas Flow Rate, 1,000 cfm

Installation factor:
1.76–2.00; avg. 1.90

Module factors:

Electrostatic precipitators	2.3
Bag filters	2.2
Venturi scrubber	2.5
Cyclone, multiclone	3.0

Material factors for
venturi, cyclone scrubbers:

High temperature with membrane, brick lining	1.6
304 Stainless	1.8
316L Stainless	2.1
316L Stainless, clad	1.9
Monel	3.0
Monel clad	2.7
Titanium	3.2

Evaporators*
Single effect; stainless steel

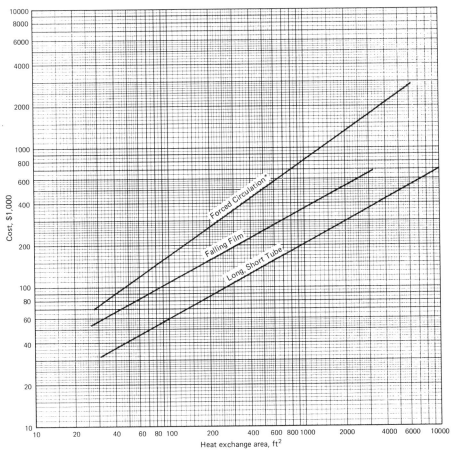

Cost, $1,000 (vertical axis)

Heat exchange area, ft² (horizontal axis)

Forced Circulation*
Falling Film
Long, Short Tube

Size exponents:		Installation factor:	Material factors:	
Forced circulation	0.7	1.5-2.50; avg. 2.09	Mild steel	0.44
Falling film, long,			Copper alloy	0.57
short tube	0.53	Module factor:	Nickle alloy	1.22
		Forced circulation 2.9	Titanium	2.93
		Falling Film 2.3		

*Also, see Cyrstallizers

Fans
Mild steel; motor, starter; 3½ in. $H_2O\Delta P$

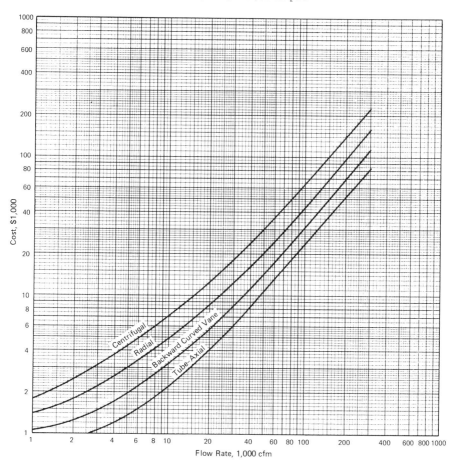

Installation factor:

Range: 1.30–2.05; avg. 1.61

Module factor: 2.2

Factors:

Pressure: $\left(\dfrac{\Delta P}{3.5}\right)^{0.3}$

Fiberglass: 1.8
Stainless steel: 2.5

Filters
Stainless steel

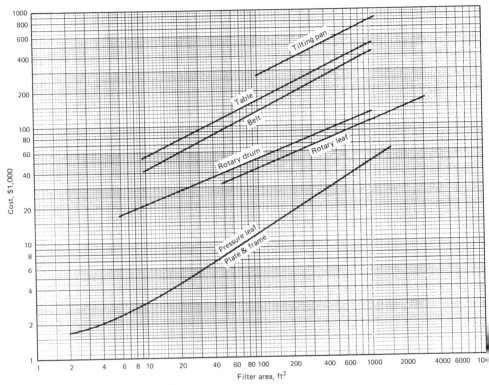

Cost, $1,000 vs Filter area, ft²

Size exponents:

Rotary vacuum drum, leaf	0.39
Vacuum table, tilting pan, belt	0.5
Pressure leaf, plate & frame	0.61

Installation factor:

1.19 - 2.21; avg. 1.69

Module factor:

Rotary table, belt, tilting pan
1.4 - 2.8; avg. 2.4
Others 2.8

Factors:

Rotary drum; belt/screw or string discharge	1.22
General/paper pulp	2.17-3.38
Mild steel/stainless steel	0.69
Vacuum table mild steel/SS	0.48
Vacuum filter auxilliaries (vac. pump, receivers, etc.), Often ~ 50% of filter cost	

Flares
Mild steel, High Btu*, with accessories

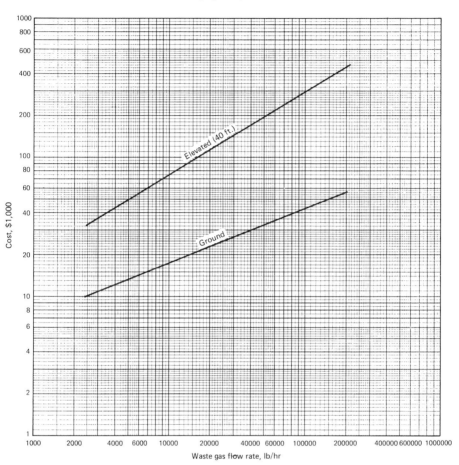

Cost, $1,000

Waste gas flow rate, lb/hr

Size exponents:

Elevated	0.59
Ground	0.39
Installation factor	1.45

Factors:

Ground: Low/High* Btu,	0.3
Elevated: Low/High* Btu,	0.8
Corrosive	2.0
Guyed (100 ft.), self-supporting (x elevated)	1.3-1.8

*High = 1,000; low = 60 Btu/ft^3

Furnaces
Mild steel tubes

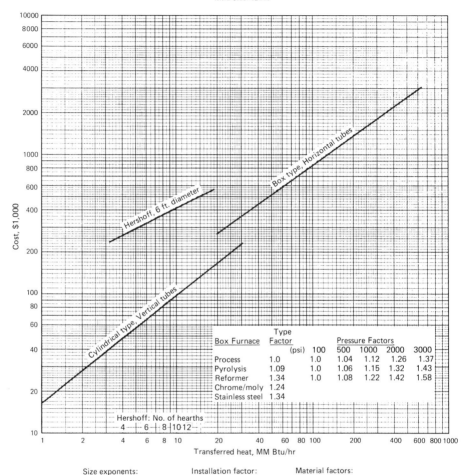

Box Furnace	Type Factor		Pressure Factors				
		(psi)	100	500	1000	2000	3000
Process	1.0		1.0	1.04	1.12	1.26	1.37
Pyrolysis	1.09		1.0	1.06	1.15	1.32	1.43
Reformer	1.34		1.0	1.08	1.22	1.42	1.58
Chrome/moly	1.24						
Stainless steel	1.34						

Size exponents:

Hershoff	0.48
Box	0.70
Cylindrical	0.78

Installation factor:

1.30-71; avg. 1.52

Module factor: 2.1

Material factors:

Cylindrical, vertical tubes:
Stainless steel	1.74
Chrome/moly	1.44
with Dowtherm	1.33

Hershoff diameter:
6-19 ft $(D/6)^{0.55}$
>19 $(D/6)^{0.65}$

Generator, Electric Power

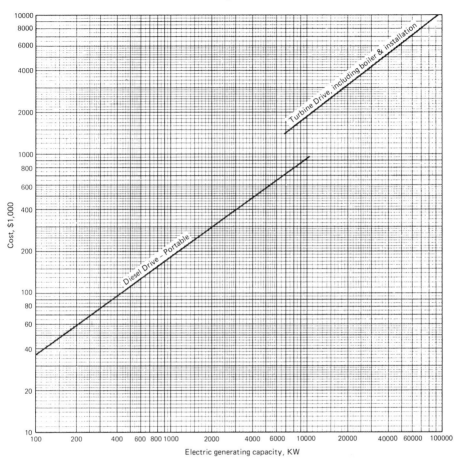

Cost, $1,000

Electric generating capacity, KW

Size exponents:		Installation factor:	Factors:	
Diesel drive	0.71	2.22–2.39	Gas/diesel engine	1.81
Turbine drive	0.76	avg. 2.31	Coal/oil, gas	
		Module factor 2.5	(turbine)	1.29

Heat Exchangers; Shell and Tube, Double Pipe, Air Cooled
Mild steel construction; Shell and tube floating head
150 psig pressure, 3/4 X 1 in. square pitch, 16 ft tubes

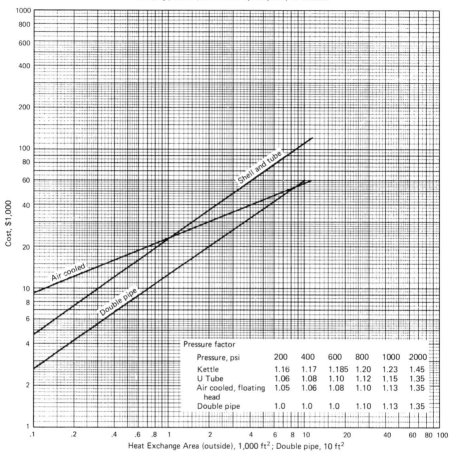

Heat Exchange Area (outside), 1,000 ft^2; Double pipe, 10 ft^2

Pressure factor						
Pressure, psi	200	400	600	800	1000	2000
Kettle	1.16	1.17	1.185	1.20	1.23	1.45
U Tube	1.06	1.08	1.10	1.12	1.15	1.35
Air cooled, floating head	1.05	1.06	1.08	1.10	1.13	1.35
Double pipe	1.0	1.0	1.0	1.10	1.13	1.35

Size exponent

Shell & tube,
Double pipe   0.68
Air cooled   0.39

Installation factor
1.23-2.10, avg. 1.61

Module factor

Shell & tube   3.2
Double pipe   1.8
Air cooled   2.2

Other factors: Exchanger type

Shell & tube to:
Kettle reboiler   2.35
U-tube   1.85
Fixed tube sheet   1.79

Shell & tube material factor
$a + (a/100)^b$

Shell	Tube	a	b
CS	SS	1.75	0.13
SS	SS	2.70	0.07
CS	monel	2.1	0.13
monel	monel	3.3	0.08
CS	Ti	5.2	0.16
Ti	Ti	9.6	0.06
CS	moly	1.40	0.05
moly	moly	1.67	0.16
CS	Admiralty	1.08	0.05

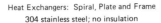

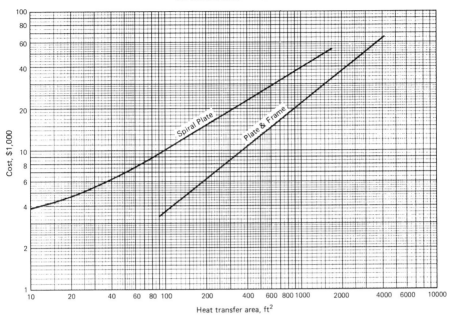

Heat Exchangers: Spiral, Plate and Frame
304 stainless steel; no insulation

Size exponent:

Plate & frame   0.78

Equations:

Spiral plate: $ = 660A^{0.59}$
Plate & frame: $ = 100A^{0.78}$

Installation Factor:

Plate & frame:
Mild steel   1.70
Stainless   1.53

Material Factor:

Mild steel   0.43
316 stainless   1.1
Nickel   1.2
Titanium   2.6

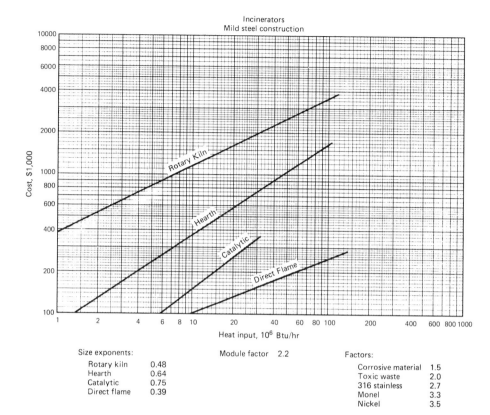

Incinerators
Mild steel construction

Size exponents:

Rotary kiln	0.48
Hearth	0.64
Catalytic	0.75
Direct flame	0.39

Module factor   2.2

Factors:

Corrosive material	1.5
Toxic waste	2.0
316 stainless	2.7
Monel	3.3
Nickel	3.5

Insulation
2 in. thickness for pipe

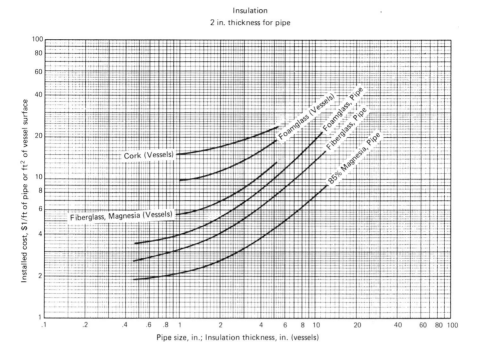

Insulation thickness factors for pipe:

3 in.	1.5
1½ in.	0.7
1 in.	0.55
½ in.	0.4

Ion Exchange*
Mild steel construction; 465 ppm removed

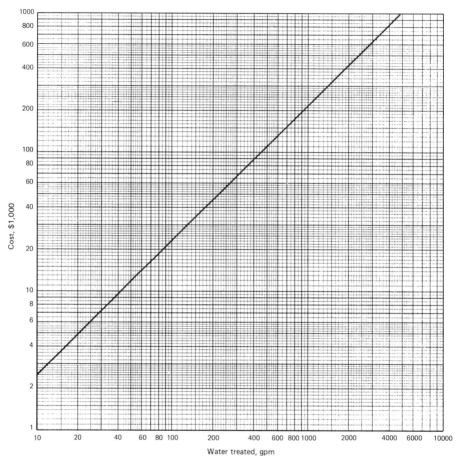

Cost, $1,000

Water treated, gpm

Size exponent: 0.97

Module factor 2.0

*See water treating.

Installation factor:
1.58–65, avg. 1.62

Factor

Ions removed $\left(\dfrac{\Delta}{465}\right)^{0.51}$

(~35 ppm typical for boiler makeup)

(~320 ppm typical for cooling tower makeup)

Mills: Hammer, Jaw, Gyratory, Roll Crushers

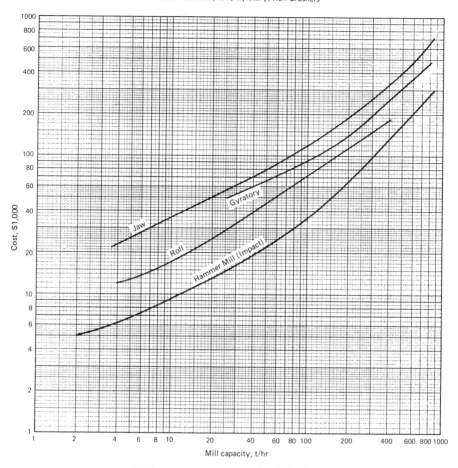

Mill capacity, t/hr

Module factor:
Hammer 2.8
Others 2.1

Installation factor:
1.30–2.15, avg. 1.83

Mills: Ball, Rod, Pebble (Wet), Jet, Rubbish

Reduction ratio 34 (i.e., ~ 1/2 in. – 65 mesh; 3/4 in. – 45 mesh)

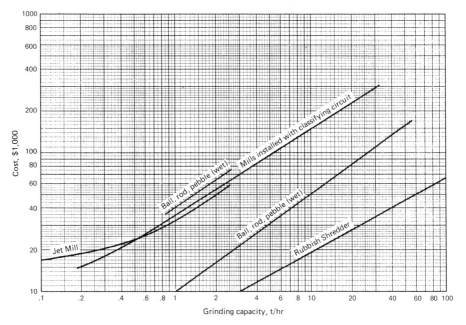

Grinding capacity, t/hr

Size exponent		Installation factor:	Factors:
Shredder	0.53	1.30 – 2.15; avg. 1.83	Ball, etc. mills
Mills; installed	0.62		Size reduction
purchased	0.70	Modular factor:	$\left(\dfrac{\text{Reduction}}{34}\right)^{1.3}$
		1.8 –2.8; avg. 2.3	
			Dry/wet = 1.25

Motors, Drives
Electric: totally enclosed, fan cooled (TEFC)

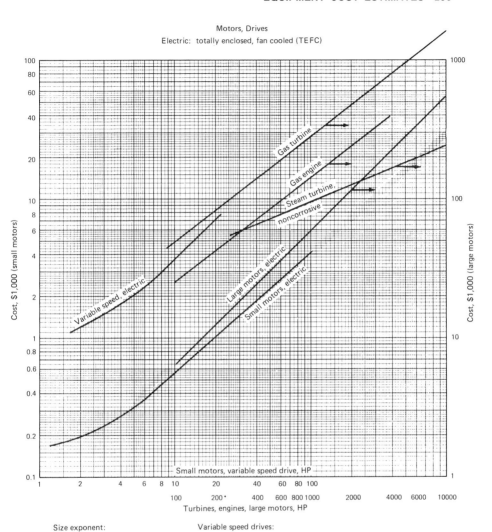

Size exponent:

Electric motors, small	0.86
Gas turbine, engine	0.76
Steam turbine	0.41

Module factor:

Electric	2.0 (1.5 on fans, pumps, compressors)
Gasoline	2.0
Gas, steam turbines	3.5

Variable speed drives:

Ratio	Factor
1.5 to 5/1	1.0
6/1	1.08

Factors: Electric motors

Speed,	1800	1.0
rpm	3600	1.04
	1200	1.6
	900	2.6
Construction:	TEFC	1.0
	Explosion proof	1.2
	Drip proof	0.74

Pipe, Pipelines
Mild steel

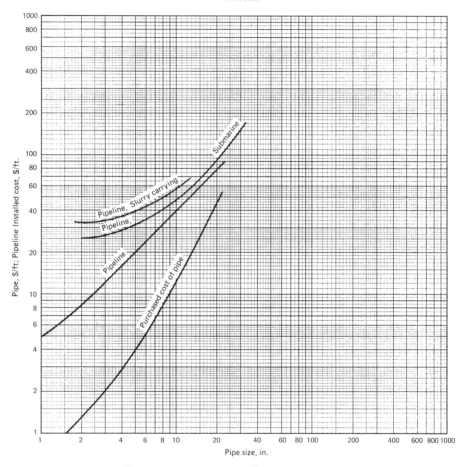

Size exponent:

Pipelines   0.99

Factors:

304 stainless, schedule 10S
Bare pipe   2.05
Traced, insulated   3.4
Fittings   18
Valves   94

Presses: Roll, Screw
Mild steel construction

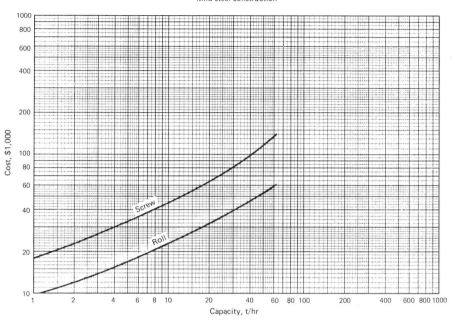

Installation factor: 2.05

Module factor: 2.4

Material factors:

Stainless steel 1.5
Nickle alloy 1.9

Pumps, Centrifugal
Cast iron, horizontal, includes motor, coupling, base

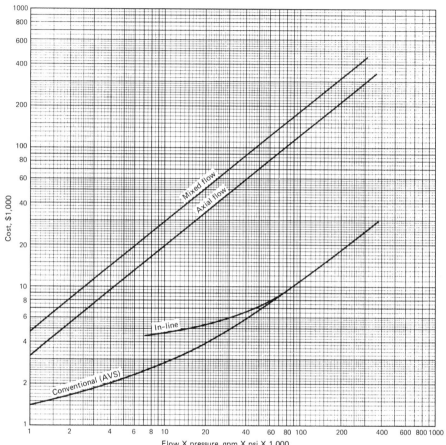

Flow X pressure, gpm X psi X 1,000
(approximately HP)

Factors:

	Conventional	In-line	Axial Flow	Mixed Flow
Size exponent			0.79	0.79
Installation	1.30	1.27	1.58	1.32
Module factor	1.5	1.75	2.05	1.70
Cast steel	1.4	1.3		
316 stainless	2.0	1.6		
Copper alloy	1.3			
Nickel alloy	3.6			
Titanium	5.7			

Pressure factor:

	to 150 psi	150– 500 psi	500– 1000 psi
In-line	1.00	1.48	1.92
Conventional	1.0	1.62	2.12

Factors:

Conventional: APS/AVS = 1.6
In-line: vertical/horizontal = 0.89
Mixed, axial flow: vertical/horizontal = 1.12

Pumps, Miscellaneous
Mild steel construction

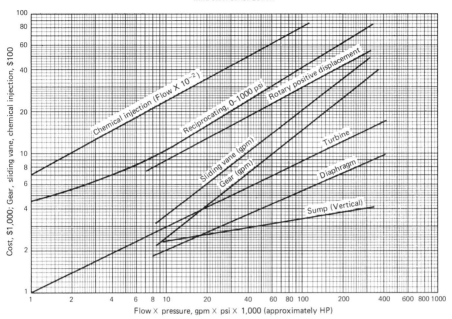

Flow X pressure, gpm X psi X 1,000 (approximately HP)

Factors

	Reciprocating	Turbine	Chemical Injection
Size exponent	0.59	0.47	0.52
Installation		1.38	1.58
Module factor	3.3	1.80	2.83
Cast iron	1.0		1.0
Cast steel	1.8		1.25
Stainless	2.4		1.95
Nickel alloy	5.0		
0–150 psi	1.0		1.0
150–500	1.32		1.37
500–1000	1.53		1.79

General installation factor:
1.25 - 2.40; avg. 1.74

Other size exponents:

Diaphragm	.43
Rotary	.52
Gear	.75
Sump	.15

Factors:
Reciprocating: ΔP (1,000 - 5,000)/(0 - 1,000) = 3.8
Sump: 3600/1800 rpm = 1.2
1200/1800 in. = 1.5
Chemical injection: Fixed/variable speed = 1.67

Pressure Vessels*
Mild steel construction

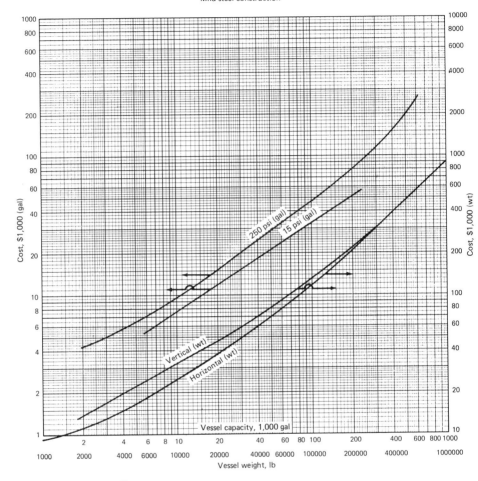

Size exponent:
15 psi (gal) 0.64

*See columns for pressure and material correction factors.

Reactors*
304 stainless steel; jacketed; no agitation

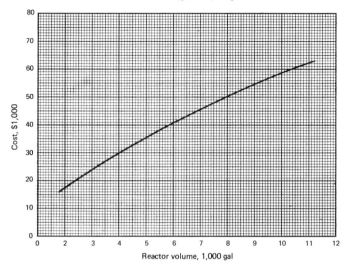

Reactor volume, 1,000 gal

Module factors:		Installation factor:	Material factors:	
Stainless	1.8	1.40 – 2.10; avg. 1.70	316 stainless	1.2
Glass lined	2.1		Glass lined	.8
Mild steel	2.3		Lead lined	.7
			Mild steel	.6

*See agitated tanks.

Refrigeration
40° F temperature

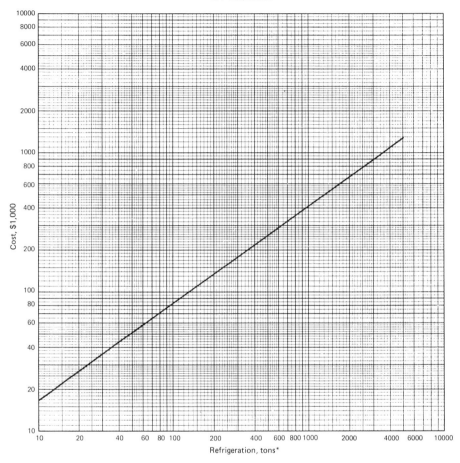

Size exponent 0.69

*One ton = 12,000 Btu

Installation, Module factor

1.38–53; avg. 1.46

Evaporative temperature factors:

+20F	1.5
0	1.9
−20	2.4
−40	3.5

Screens, Vibrating
Mild steel, single deck

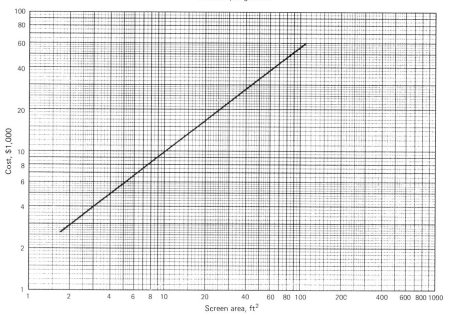

Cost, $1,000

Screen area, ft^2

Size exponent: 0.75

Module factor: 2.8

Installation factor:

1.45–2.27; avg. 1.85

Factors:

Double deck	1.6
Stainless steel	1.25
Nickel alloy	1.8

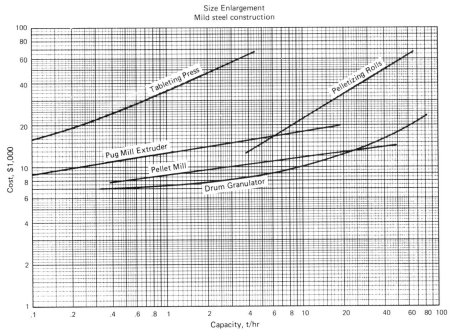

Size Enlargement
Mild steel construction

Cost, $1,000

Capacity, t/hr

Size exponent

Pug mill	0.15
Pellet mill	0.12
Pelletizing rolls	0.58

Installation factor: 2.05

Factors

	Pug mill extruder	Disk, drum granulator	Others
Module factor	2.2	2.8	2.05
Stainless steel	1.2	1.1	1.2
Nickel alloy	1.4	1.3	1.4

Screw/pug mill extruder	5.6
Disk/drum granulator	0.68

Tanks
Mild steel construction unless otherwise noted

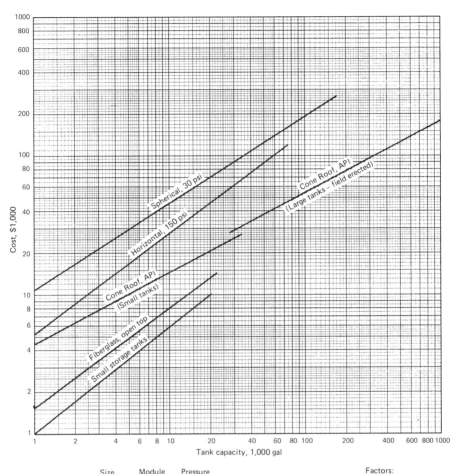

	Size exponent	Module factor	Pressure factor				
Small cone top	0.51	1.6					
Large cone top	0.51	1.9					
Horizontal, pressure	0.72	2.08	200	250 psi			
			1.18	1.38			
Sphere	0.62	1.87	50	75	100	125	200
			1.08	1.19	1.26	2.39	1.53
Fiberglass	0.71						
Small storage	0.71						

Factors:

Rubber lined	1.5
Lead lined	1.6
Stainless	2.0
Floating roof; large, field erected	1.8

Installation factor:
1.20 – 2.30. avg. 1.88

Tanks, (Small)
304 stainless steel

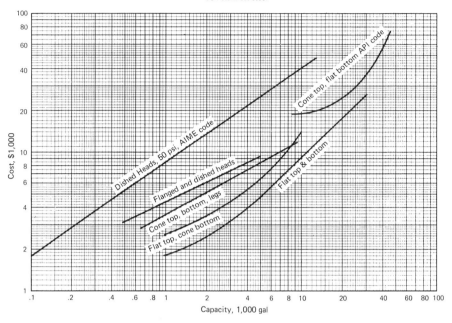

Cost, $1,000 (y-axis)

Capacity, 1,000 gal (x-axis)

Size exponent		Factor	
Dished head, 50 psi	0.68	316/304 stainless steel	1.39
Flanged, dished head	0.48		
Cone top, bottom, legs	0.57		
Flat top, bottom	0.93		

Thickeners, Clarifiers
Rake mechanism, concrete tank, drive

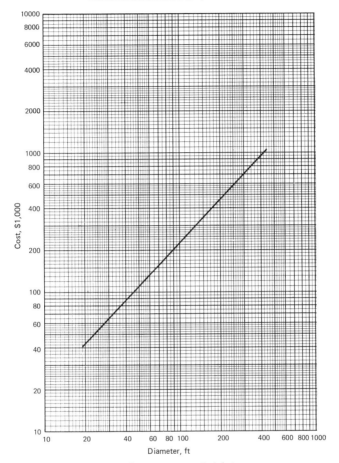

Cost, $1,000

Diameter, ft

Size exponent: 1.03
Installation factor:
 1.63–2.61; avg. 2.12

Module factor: 3.0

Tank factor:
 Concrete/steel 0.7 for
 units under 40 ft diameter

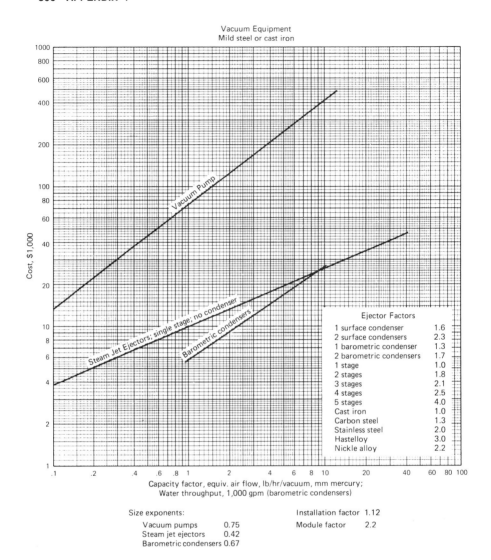

Vacuum Equipment
Mild steel or cast iron

Capacity factor, equiv. air flow, lb/hr/vacuum, mm mercury;
Water throughput, 1,000 gpm (barometric condensers)

Ejector Factors

1 surface condenser	1.6
2 surface condensers	2.3
1 barometric condenser	1.3
2 barometric condensers	1.7
1 stage	1.0
2 stages	1.8
3 stages	2.1
4 stages	2.5
5 stages	4.0
Cast iron	1.0
Carbon steel	1.3
Stainless steel	2.0
Hastelloy	3.0
Nickle alloy	2.2

Size exponents:

Vacuum pumps	0.75	Installation factor	1.12
Steam jet ejectors	0.42	Module factor	2.2
Barometric condensers	0.67		

REFERENCES

Allen, D. H., and R. C. Page. 1975. Revised techniques for predesign cost estimating. *Chemical Engineering* (March).

Alonso, J. R. F. 1971. Estimating the costs of gas cleaning plants. *Chemical Engineering*.

Axtell, Oliver, and James M. Robertson. 1986. *Economic Evaluation in the Chemical Process Industries*. John Wiley & Sons, New York.

Beckman, James, ed. 1986. *Series Design of Equipment. Vol. 1, Plant Design and Cost Estimating*. American Institute of Chemical Engineers, New York.

Bennett, Richard D. 1987. Evaporator, crystallizer costs. Swenson Process Equipment Inc., 15700 Lathrop Ave., Harvey, IL 60426; 1988, Matching Crystallizer to Material, *Chemical Engineering* (May 23): 118–127.

Blecker, H. G., H. S. Epstein, and T. M. Nichols. 1974. Wastewater Equipment. *Chemical Engineering* (Oct.).

Chase, D. J. 1970. Plant costs vs. capacity. *Chemical Engineering* (April).

Chemical Engineering, comp. & ed. 1979 and 1984. *Modern Cost Engineering: Methods and Data*. 2 vols. McGraw-Hill, New York.

Chemical Engineering, comp. & ed. 1979. *Process Technology and Flowsheets*. McGraw-Hill, New York.

Clark F. D., and S. P. Terni. 1972. Thick wall pressure vessels. *Chemical Engineering* (April).

Corripio, A. B., K. S. Chrien, and L. B. Evans. 1982. Estimate cost of heat exchangers and storage tanks via correlations. *Chemical Engineering* (Feb.): 125–127.

Desai, M. B. 1981. Preliminary cost estimating of process plants. *Chemical Engineering* (July 27).

Epstein, L. D. 1971. Costs of standard vertical storage tanks and reactors. *Chemical Engineering* (July 13):141–142.

Fang, C. S. 1980. The cost of shredding municipal solid waste. *Chemical Engineering* (April 21):151–152.

Guthrie, Kenneth M. 1974. *Process Plant Estimating, Evaluation, and Control*. Craftsman Book Co., Solana Beach, CA.

Hall, R. S., J. Mately, and K. J. McNaughton 1982. Current costs of process equipment. *Chemical Engineering* (April 5).

Happel, J., and D. G. Jordan. 1975. *Chemical Process Economics*. Marcel-Dekker, New York; 219–231.

Herkimer, Herbert. 1958. *Cost Manual for Piping and Mechanical Construction*. Chemical Publishing, New York.

Hoerner, G. M. 1976. Nomograph updates process equipment costs. *Chemical Engineering* (May).

Holland, F. A., F. A. Watson, and J. K. Wilkinson. 1974. How to estimate capital costs. *Chemical Engineering* (April).

Huff, G. A. 1976. Selecting a vacuum producer. *Chemical Engineering* (March).

Kharbanda, O. P. 1979. *Process Plant and Equipment Cost Estimation*. Craftsman Book Co., Solana Beach, CA.

Klumpar, L. V., and S. T. Stavsky. 1985. Updated cost factors: process equipment. *Chemical Engineering* (July 22):73–77.

—. 1985. Commodity materials. *Chemical Engineering* (Aug. 19):76–77.

—. 1985. Installation labor. *Chemical Engineering* (Sept. 16):85–87.

Koenig, A. R. 1980. Choosing economic insulation thickness. *Chemical Engineering* (Sept. 8).

Kumana, Jimmy D. 1984. Cost update on specialty heat exchangers. *Chemical Engineering* (June 25):169.

Lindamood, D. M. 1985. Most economical thickness, hot-pipe insulation. *Chemical Engineering* (April 1):96.

Meyer, W. S. and D. L. Kime. 1976. Cost estimation for turbine agitators. *Chemical Engineering* (Sept.).

Miller, J. S. and W. A. Kapella. 1977. Installed cost of a distillation column. *Chemical Engineering* (April).

Moselle, Gary, ed. 1979. *National Construction Estimator*. Craftsman Book Co., Solana Beach, CA.

Mulet, A., A. B. Corripio, and L. B. Evans. 1981. Estimating costs of distillation and absorption towers via correlations. *Chemical Engineering* (Dec. 28):77–82.

—. 1984. Pressure vessels. *Chemical Engineering* (Oct. 5).

Patrascu, Anghel. 1978. *Construction Cost Engineering*. Craftsman Book Co., Solana Beach, CA.

Peters, M. S., and K. D. Timmerhaus. 1980. *Plant Design and Economics for Chemical Engineers.* McGraw-Hill, New York.

Pikulik, A., and H. E. Diaz. 1977. Cost estimating for major process equipment. *Chemical Engineering* (Oct. 10):107–122.

Purohit, G. P. 1985. Cost of double-pipe and multitube heat exchangers. *Chemical Engineering* (March 4):92–96. (April 1):85–86.

Sommerville, R. F. 1970. Estimating mill costs at low production rates. *Chemical Engineering.*

—. 1972. New method gives accurate estimate of distillation cost. *Chemical Engineering* (May).

Swearingen, Judson S., and John E. Ferguson. 1983. Optimized power recovery from waste heat. *Chemical Engineering Progress* 79 (Aug):66–70.

Ulrich, G. D. 1983. *A Guide to Chemical Engineering Process Design and Economics.* John Wiley & Sons, New York.

Valle-Riestra, F. J. 1983. *Project Evaluation in the Chemical Process Industries.* McGraw-Hill, New York.

Vatavuk, William M., and Robert B. Neveril. (1980–1983). Air pollution control systems (Parts 1–16). *Chemical Engineering* (Oct.–May).

—. 1980. Pollutant capture hoods. *Chemical Engineering* (Dec. 1):111–115.

—. 1984. Practical emmissions control. *Chemical Engineering* (April 2):97–99.

—. 1984. Gaseous emmissions control. *Chemical Engineering* (April 30):95–98.

Vogel, G. A., and E. J. Martin 1983. Estimating capital costs of facility components. *Chemical Engineering* 90 (24) (Nov. 28):87–90.

—. 1984. Operating costs. *Chemical Engineering* (Jan. 9):97–100.

—. 1984. Incinerator costs. *Chemical Engineering* (Feb. 6):121–122.

APPENDIX 2

COMPLETE PLANT COST ESTIMATING CHARTS

The following charts indicate the complete cost of plants to produce various chemicals in differing tonnages. The information has been assembled primarily from four sources: (1) curves on 54 plants published by Guthrie (1974), (2) curves on 18 plants published by *Chemical Engineering* (1973/1974), (3) 33 nomographs, and about 140 single plant size-cost data notations by Kharbanda (1979), and (4) several hundred recent plant construction notices in *Chemical Engineering*'s *Construction Alert*. The first three sources are quite old, with most of the information gathered from the mid-60s through the early 70s. The last source was data from 1980 through 1987. Each source was inflation-corrected to 1987 (CE Index of 320) by means of the *Chemical Engineering* (CE) Index.

The first two references were probably quite authoratative when published, and represented contractor prices for that plant alone, plus the necessary raw material and product storage. The infrastructure for a ''grass roots'' plant, or even for minor utility and other required nonplant facilities was not included. The later two sources, on the other hand, are basically press-release information stating what the complete facility cost. This might include land, site development, and/or any of the infrastructure required to make the plant function. Costs would thus be higher, and the assembled data would be much more scattered because of each location's different requirements.

Both factors, the early data's age, and the most recent data's complete cost basis, tend to limit the accuracy of the plots. When considerable data were available, high, low and average lines were shown. Presumably the high values represent more infrastructure requirements. When only one data point (i.e., one plant cost at one size) was available, the capacity versus cost line was drawn with a slope of 0.64, the average size-cost exponent of Guthrie's 54 plants.

Normally it should be expected that the costs shown in these plots should be roughly correct, and perhaps on the high side. However, some of the data from the first three references appear to be very low, so caution should be used with all of the charts. They may be useful as a guide, but not too much confidence should be placed in their accuracy. The basis for the costs should be considered as a reasonably high value for the plant alone, plus storage, and the CE Index 320.

Plant Costs, A

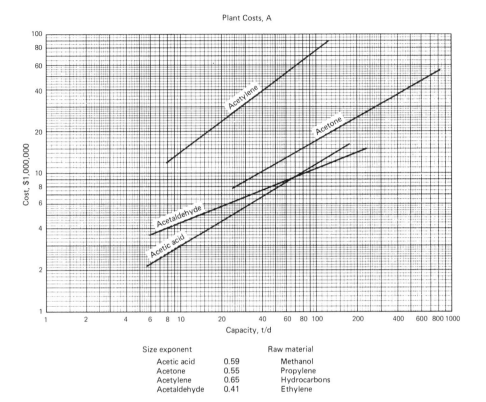

Size exponent

		Raw material
Acetic acid	0.59	Methanol
Acetone	0.55	Propylene
Acetylene	0.65	Hydrocarbons
Acetaldehyde	0.41	Ethylene

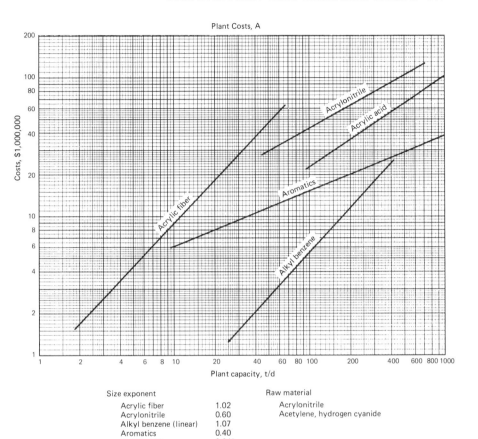

Plant Costs, A

Size exponent

Acrylic fiber	1.02
Acrylonitrile	0.60
Alkyl benzene (linear)	1.07
Aromatics	0.40
Acrylic acid	0.64*

Raw material

Acrylonitrile
Acetylene, hydrogen cyanide

*Assumed

Plant Costs, A

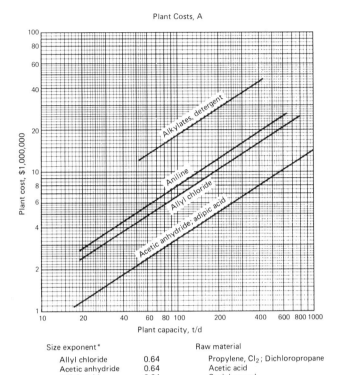

Plant cost, $1,000,000

Plant capacity, t/d

Size exponent*		Raw material
Allyl chloride	0.64	Propylene, Cl_2; Dichloropropane
Acetic anhydride	0.64	Acetic acid
Adipic acid	0.64	Cyclohexanol
Aniline	0.64	Benzene; nitric, sulfuric acids
Alylates, detergent	0.64	

*All assumed

Plant Costs, Aluminum Chemicals

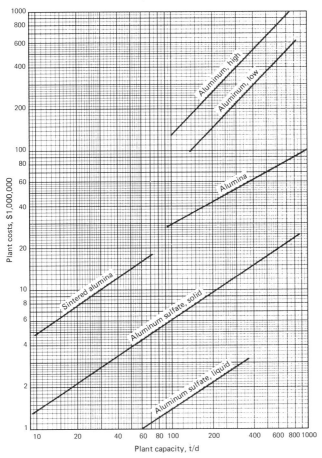

Size exponents

		Raw materials
Alumina	0.54	Bauxite
Alumina, sintered	0.64*	Alumina
Aluminum	1.0	Alumina
Aluminum sulfate	0.64*	Bauxite, H_2SO_4

*Assumed

Plant Costs, Ammonium Compounds

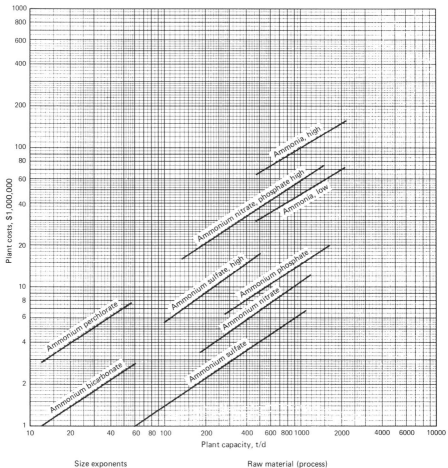

Plant capacity, t/d

Plant costs, $1,000,000

Size exponents		Raw material (process)
Ammonia	0.58	Gas, air
Ammonium nitrate	0.65	Ammonia; (prilled)
Ammonium sulfate	0.67	Ammonia, sulfuric acid, (crystalized)
Ammonium phosphate	0.64*	Ammonia, phosphoric acid, (granulated)
Ammonium perchlorate	0.64*	Ammonia, Cl_2
Ammonium bicarbonate	0.64*	Ammonia, CO_2

*Assumed

Plant Costs, B

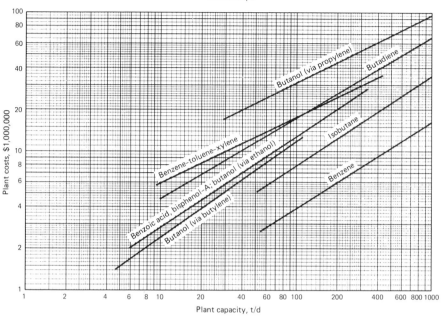

Plant capacity, t/d

Size exponent		Raw material (process)	Factors for benzene process:	
Butadiene	0.63	Butane; butylene	Detol	1.0
Butanol	0.48	Propylene	Litol	1.42
Butanol	0.69	Butylene	Pyrotol	1.48
Benzene	0.73	Toluene, H_2 (Detol)		
Benzoic acid	0.64*	Toluene		
Bisphenol A	0.64*	Acetone; phenol		
Butanol	0.64*	Ethanol		
Butane, iso	0.64*	Butane, pentane		
Benzene, toluene, xylene	0.64*	Reformate (extraction)	*Assumed	

Plant Costs, C

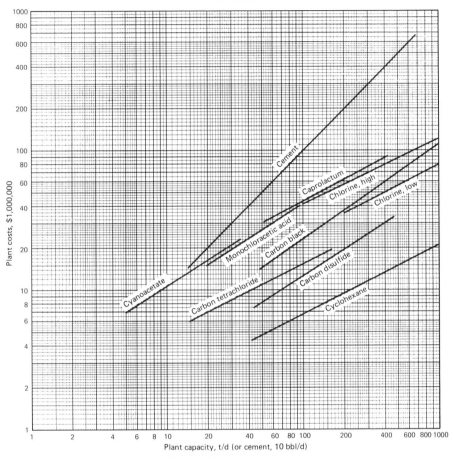

Plant capacity, t/d (or cement, 10 bbl/d)

Size exponent		Raw material (process)
Carbon black	0.67	Aromatic oils; gas
Chlorine	0.47	NaCl brine (electrolysis) (Caustic soda by-product; 1.07 lb/lb Cl_2)
Caprolactum	0.52	Cyclohexane, NH_3 (Ammonium sulfate by-product; 1.75 lb/lb caprolactum)
Cyclohexane	0.49	Benzene, H_2
Carbon tetrachloride	0.48	Propane, Cl_2 (Perchlorethylene by-product; 1.33 lb/lb CCl_4)
Carbon disulfide	0.64*	
Cement	1.0	
Cyanoacetate	0.64*	
Chloroacetic acid, mono	0.64*	

*Assumed

Plant Costs, C

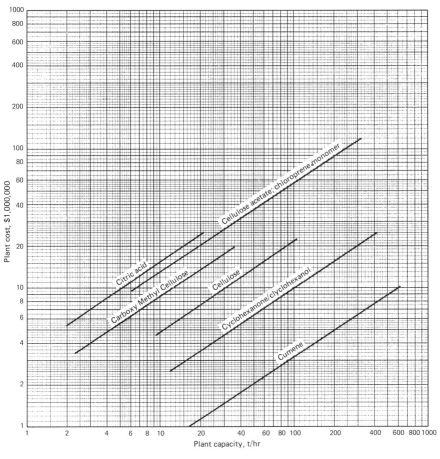

Plant capacity, t/hr

Size exponent*		Raw material (process)
Citric acid	0.64	(Submerged fermentation)
Carboxymethyl cellulose	0.64	Cellulose
Cellulose acetate	0.64	Cellulose
Cumene	0.64	Benzene, propylene
Cyclohexanone/clyclohexanol	0.64	Benzene, H_2
Chloroprene monomer	0.64	Butadiene, Cl_2

*All size exponents assumed   Factor for chloroprene raw material: Acetylene 1.57

Plant Costs, D

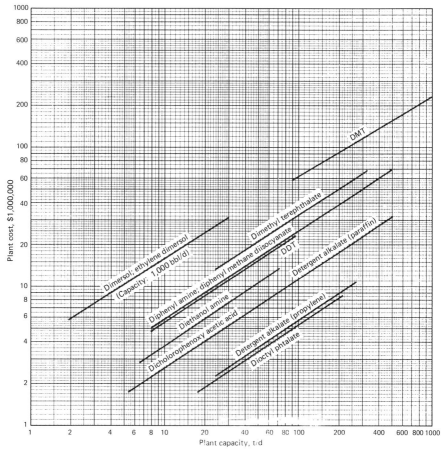

Plant cost, $1,000,000

Plant capacity, t/d

Size exponent		Raw material
DMT	0.51	Grassroots plants
Diphenyl amine	0.64*	
Dichlorophenoxy acetic acid	0.64*	Phenol
DDT	0.64*	Chloral, Chlorobenzene
Detergent alkalate	0.64*	Propylene tetramer, benzene
Detergent alkalate	0.64	n-paraffin
Diethanol amine	0.64	Ethylene oxide, ammonia
Dimethyl terephthalate	0.64*	p-xylene, methanol
Dioctyl phthalate	0.64*	Phthalic anhydride
Dimersol	0.64*	
Dimersol, ethylene	0.64	Dimerization
Diphenyl methane diisocyanate	0.64*	

*Assumed

Plant Costs, E

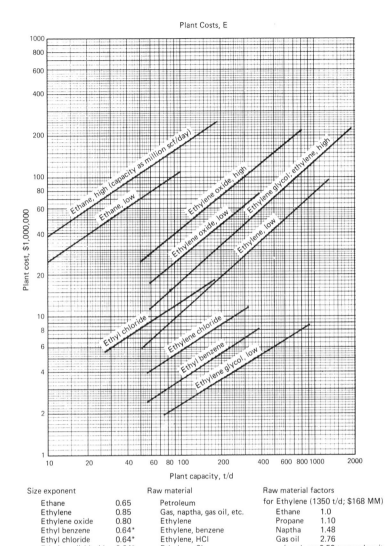

Plant capacity, t/d

Size exponent		Raw material	Raw material factors
Ethane	0.65	Petroleum	for Ethylene (1350 t/d; $168 MM)
Ethylene	0.85	Gas, naptha, gas oil, etc.	Ethane 1.0
Ethylene oxide	0.80	Ethylene	Propane 1.10
Ethyl benzene	0.64*	Ethylene, benzene	Naptha 1.48
Ethyl chloride	0.64*	Ethylene, HCl	Gas oil 2.76
Ethylene dichloride	0.64*	Ethylene, Cl$_2$	(produces 0.59 t propylene/t
Ethylene glycol	0.59	Ethylene oxide	ethylene)

*Assumed

Plant Costs, E

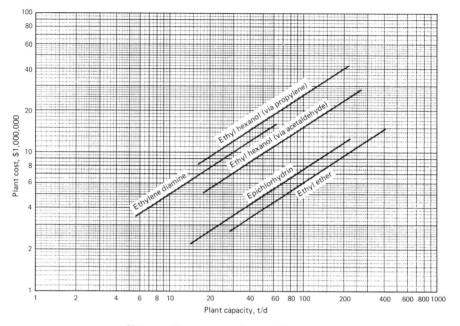

Size exponent*		Raw material
Epichlorhydrin	0.64	Allyl chloride
Ethyl ether	0.64	Propylene, synthesis gas
Ethyl hexanol	0.64	Acetaldehyde
Ethyl diamine	0.64	Ethylene dichloride

*Assumed

Plant Costs, Ethanol (Fermentation), Methanol

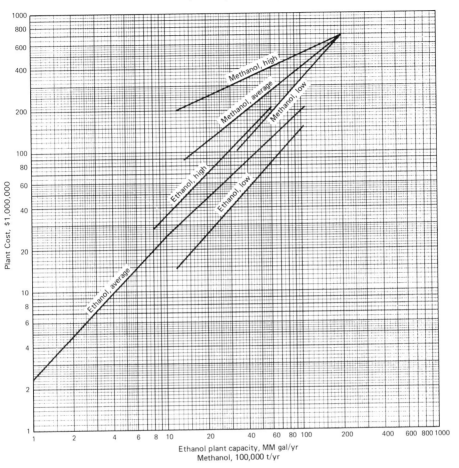

Size exponents

Methanol	0.78
Ethanol	0.90 > 10 MM gal/yr
	1.0 to 10 MM gal/yr

Raw materials (process)

Methane, CO, H_2
(Fermentation)

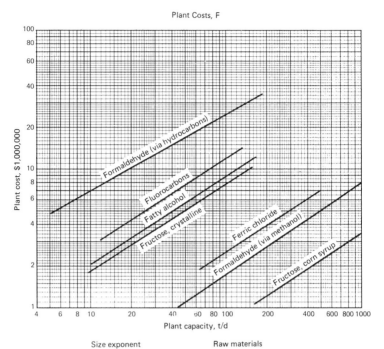

Plant Costs, F

Size exponent		Raw materials
Formaldehyde	0.55	Hydrocarbons, aqueous
	0.66	Methanol
Fatty alcohol	0.64*	Coconut oil
Fluorocarbon	0.64*	Carbon tetrachloride, HF
Ferric chloride	0.64*	Ferrous chloride, Cl_2
Fructose, crystalline	0.64	
Fructose, syrup	0.64	

*Assumed

Plant Costs, G

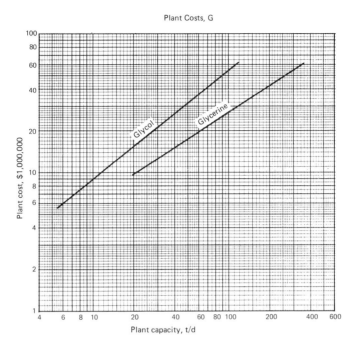

Plant cost, $1,000,000

Plant capacity, t/d

Size exponent		Raw material
Glycol	0.79	Ethylene, Cl_2
Glycerine	0.64*	Allyl alcohol, epichlorhydrin

*Assumed

Plant Costs, Gases

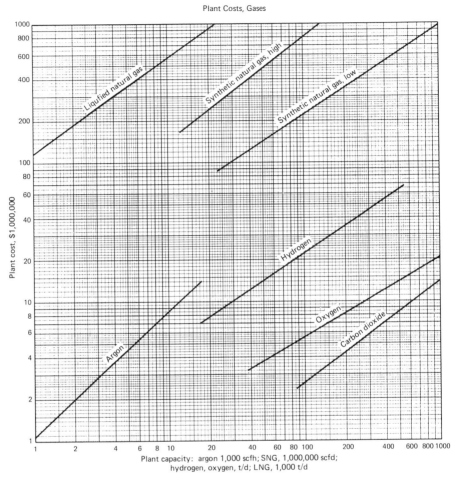

Plant capacity: argon 1,000 scfh; SNG, 1,000,000 scfd;
hydrogen, oxygen, t/d; LNG, 1,000 t/d

Size exponent		Raw material, process	Factors for SNG feedstock	
Argon	0.89	Air, liquified	Coal	1.0
Oxygen	0.59	Air, liquified	Crude oil	0.6
Hydrogen	0.65	Methane; partial oxidation; reforming	Medium, heavy gas oil,	0.5
LNG	0.68	Tealarc process	Naptha, kerosene,	0.3
SNG	0.75	Coal	light gas oil	
Carbon dioxide	0.72			

Plant Costs, Liquid Air, Hydrogen, Carbon Dioxide, Oxygen, Nitrogen

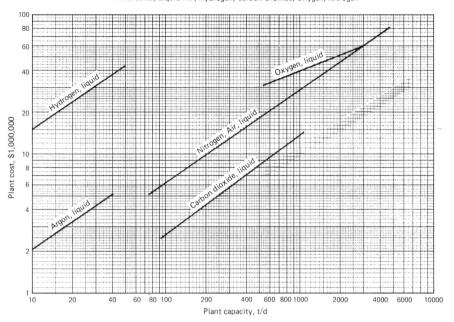

Size exponent

Carbon dioxide, liquid	0.72
Oxygen, liquid	0.37
Air, nitrogen, liquid	0.66
Argon, hydrogen, liquid	0.66 (est.)

Plant Costs, H

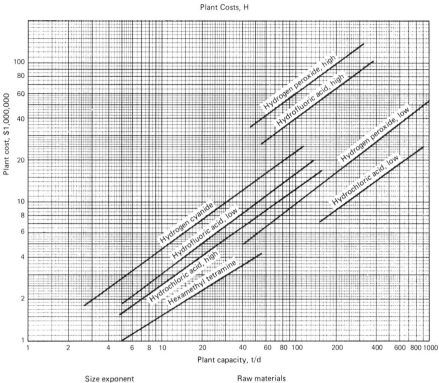

Plant capacity, t/d

Size exponent		Raw materials
Hydrochloric acid	0.69	Salt, H_2SO_4 (Na_2SO_4 by-product)
Hydrofluoric acid	0.72	CaF_2, H_2SO_4
Hexamethylene tetramine	0.64*	Methanol, ammonia
Hydrogen peroxide	0.73	Isopropylene alcohol, O_2
Hydrogen cyanide	0.70	Propane, ammonia

*Assumed

Plant Costs, I

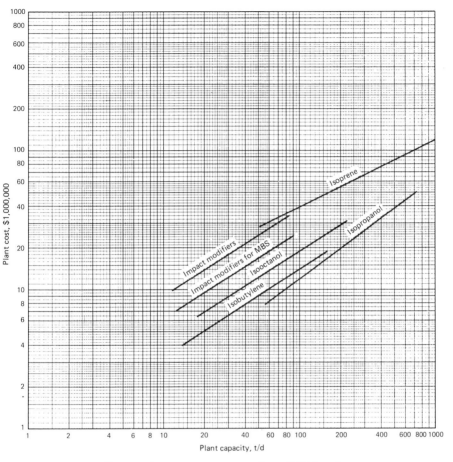

Plant capacity, t/d

Size exponent		Raw material (process)
Isoprene	0.49	Propylene, methanol, O_2
Isopropanol	0.73	Propylene
Isobutylene	0.64*	(Liquid extraction)
Isooctanol	0.64*	Heptane
Impact modifiers	0.64*	
Impact modifiers for	0.64*	
Methylmethacrylate–		
butadiene–styrene		

*Assumed

Plant Costs, L, M

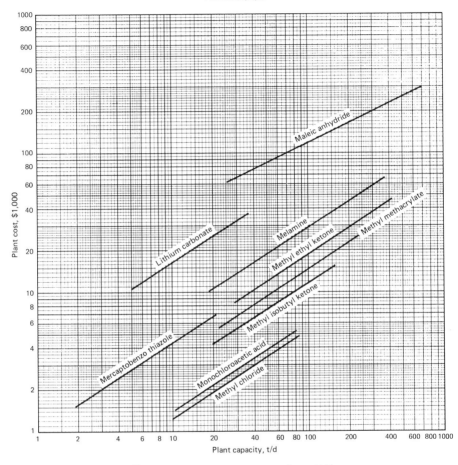

Plant capacity, t/d

Size exponent		Raw material
Lithium carbonate	0.64*	Spodumene ore
Maleic anhydride	0.48	Benzene
Melamine	0.64*	Urea, ammonia
Methyl chloride	0.64*	Methanol
Methyl ethyl ketone	0.64*	
Methyl isobutyl ketone	0.64	
Mercaptobenzo thiazole	0.64*	Aniline
Methyl methacrylate	0.64*	Acetone, HCN
Monochloroacetic acid	0.64*	Acetic acid, Cl$_2$

*Assumed

Plant Costs, M

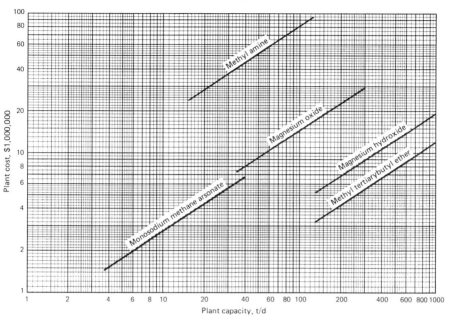

Plant cost, $1,000,000

Plant capacity, t/d

Size exponent*

		Raw material
Monosodium methyl arsonate	0.64	(coproduct sodium cocodylate − herbicides)
Magnesium oxide	0.64	Seawater; brine
Magnesium hydroxide	0.64	Seawater; brine (calcined)
Methyl tertiary butyl ether	0.64	
Methyl amine	0.64	(coproduct, 0.67 t/t dimethyl formamide)

Methanol − see page for ethanol

* Assumed

Plant Costs, N

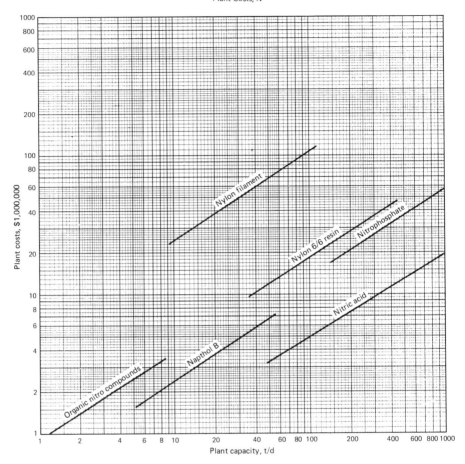

Plant capacity, t/d

Size exponent		Raw material
Nitric acid	0.59	Ammonia
Napthol B	0.64*	Napthalene
Nylon 6/6 resin	0.64*	Adipic acid
Nylon filament	0.64*	Dimethyl formamide
Nitrophosphate	0.64*	Phosphate ore, NHO_3
Nitro compounds, organic	0.64*	

*Assumed

Plant Costs, O

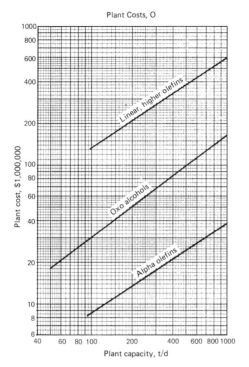

Plant cost, $1,000,000

Plant capacity, t/d

Size exponent		Raw materials
Oxo alcohols	0.74	Olefins, CO, H_2
Olefins, alpha	0.64*	Hydrocarbons; wax
Olefins, linear, higher	0.64*	

* Assumed

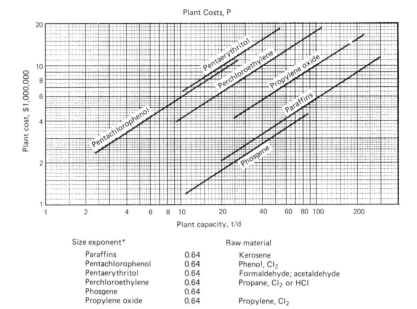

Plant Costs, P

Size exponent*		Raw material
Paraffins	0.64	Kerosene
Pentachlorophenol	0.64	Phenol, Cl_2
Pentaerythritol	0.64	Formaldehyde; acetaldehyde
Perchloroethylene	0.64	Propane, Cl_2 or HCl
Phosgene	0.64	
Propylene oxide	0.64	Propylene, Cl_2

*Assumed

Plant Costs, P

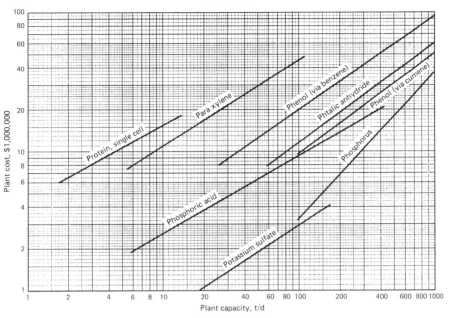

Plant capacity, t/d

Size exponent		Raw material (process)
Protein, single cell	0.64*	
Para xylene	0.61	(Crystallization)
Phenol	0.68	Benzene; toluene
	0.72	Cumene
Phosphoric acid	0.56	Phosphate rock, H_2SO_4
Phosphorus	1.06	Phosphate rock, electricity, coke
Phtalic anhydride	0.72	Napthalene; o-xylene
Potassium sulfate	0.64*	Potassium chloride, H_2SO_4

*Assumed

Plant Costs, Polymers

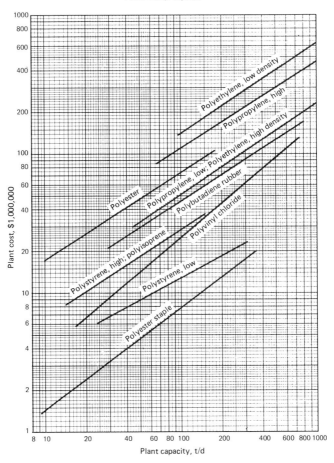

Size exponent		Raw material	Factors:
Polyethelene	0.65	Ethylene	Vinyl chloride monomer
Polypropylene	0.62	Propylene	0.82 X PVC
Polyvinyl chloride (PVC)	0.82	Ethylene, Cl	
Polypropylene	0.74	Gas, naptha, gas oil	
Polybutadiene, synthetic rubber	0.64*	Butadiene	
Polyisoprene	0.64*	Isoprene	
Polystyrene	0.53	Styrene	
Polyester; staple	0.64*	Dimethyl formamide; polyester	*Assumed

Plant Costs, Polymers

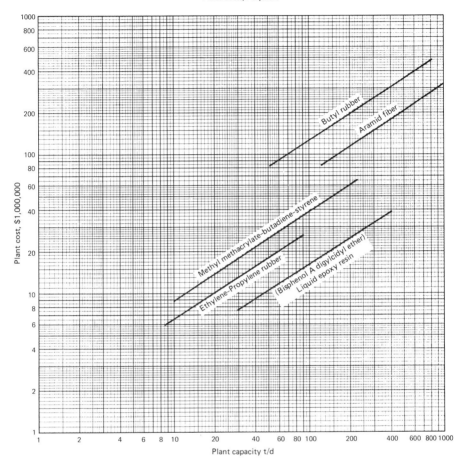

*All size exponents assumed at 0.64.

Plant Costs, Polymers

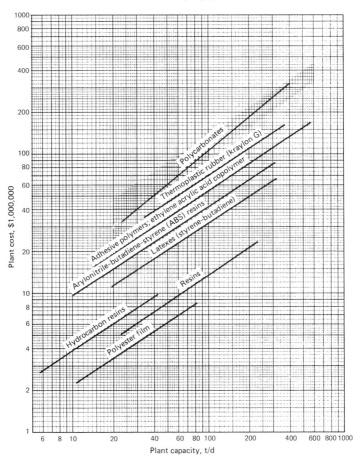

*All size exponents assumed at 0.64,
except Polycarbonate 0.79

Plant Costs, S (Organic)

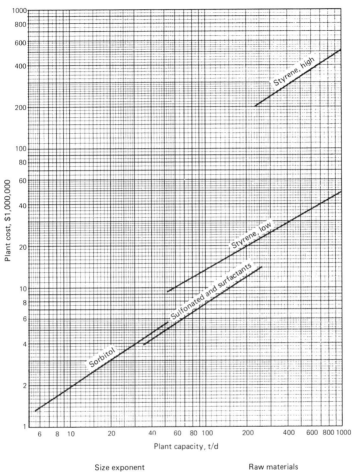

Plant capacity, t/d

Size exponent		Raw materials
Styrene	0.56	Benzene, ethylene, steam
Sorbitol	0.64*	Corn syrup
Sulfonated and sulfated	0.64*	
surfactants and detergents		

*Assumed

Plant Costs, S (Inorganic)

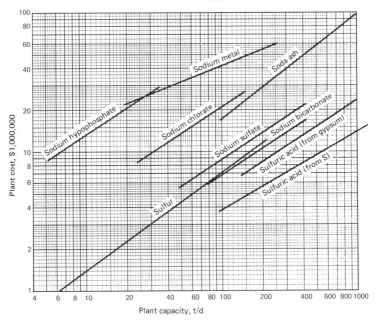

Size exponent

Sulfur	0.71
Sulfuric acid	0.56
Soda ash (Na_2CO_3)	0.74
Sodium bicarbonate	0.65
Sodium metal	0.64*
Sulfuric acid	0.64*
Sodium sulfate	0.64*
Sodium hypophosphate	0.64*
Sodium chlorate	0.64*

*Assumed

Raw materials (process)

H_2S - containing gas
Sulfur
NaCl, CO_2 (Solvey)
Soda ash or NaOH; CO_2
NaCl, electricity
Gypsum
Brine

Plant Costs, T

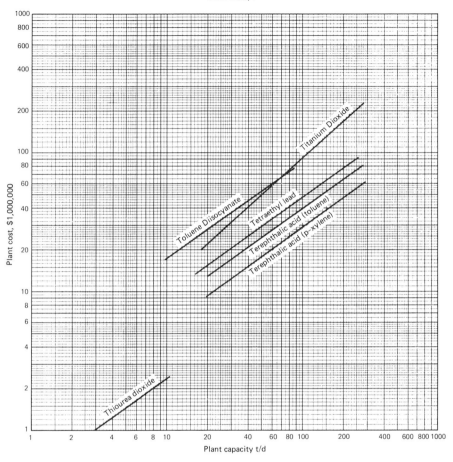

Size exponent		Raw materials
Terephthalic acid	0.64*	p–xylene
	0.64*	Toluene; benzene
Tetraethyl lead	0.64*	Ethyl chloride, Pb, Na
Titanium dioxide	0.89	Rutile ore, H_2SO_4
Toluene diisocyanate	0.64	Phosgene
Thiourea dioxide	0.64	

*Assumed

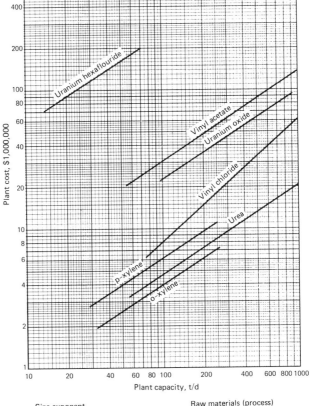

Plant Costs, U, V, X

Size exponent		Raw materials (process)
Urea	0.64	Ammonia, CO_2
Uranium oxide	0.64*	Uranium ore
Uranium hexafluoride	0.64*	Uranium ore, fluorine
Vinyl acetate	0.65	Ethylene
Vinyl chloride	0.88	Ethylene, Cl_2 or HCl
o-xylene	0.64*	Mixed xylenes (fractionation)
p-xylene	0.64*	Mixed xylenes (fractionation)

*Assumed

Plant Costs, Metals, Carbon

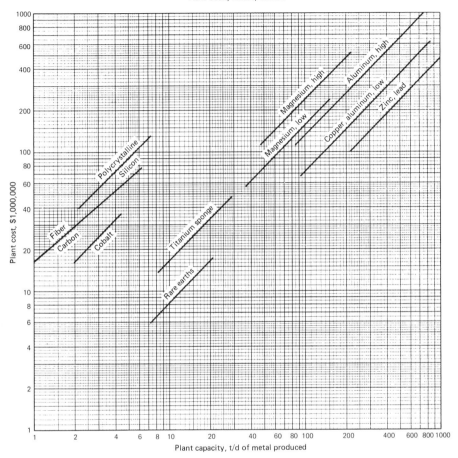

Size exponents 0.64–1.0 avg. 1.0
Carbon fibers 0.85

Plant Costs, Minerals

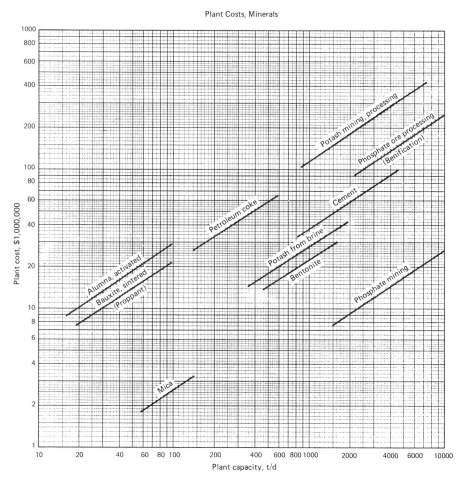

Plant cost, $1,000,000

Plant capacity, t/d

Size exponents:
Assumed to be 0.64

Plant Costs, Natural Gas Purification

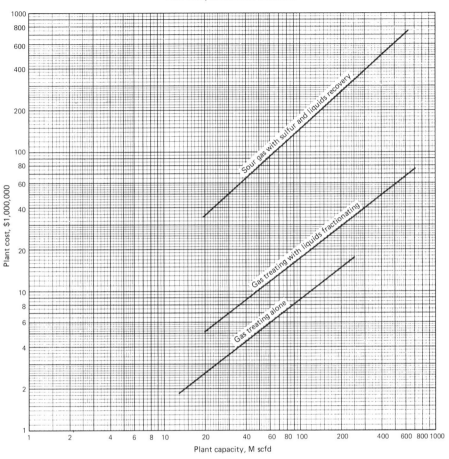

Size exponents

Gas treating alone	0.75
Gas treating with liquids fractionation	0.75
Sour gas treating with sulfur recovery and liquids fractionation	0.84

Petroleum Plant Costs, Complete Plants

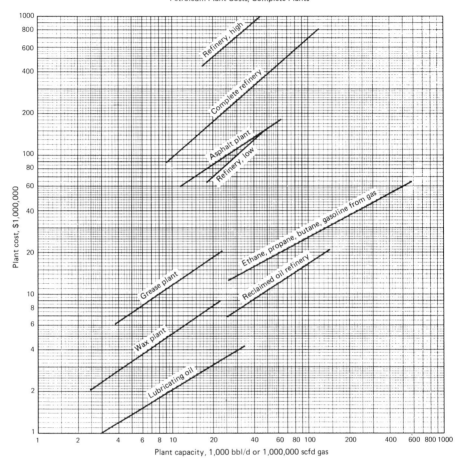

Size exponent		Raw materials
Complete refinery	0.86	
Gas processing	0.52	Recovery of "light ends"
Wax plant	0.64*	
Lube plant	0.59	
Grease plant	0.64*	
Re-refined oil	0.64*	Reclaimed motor oil

*Assumed

Petroleum Plant Costs, Cracking

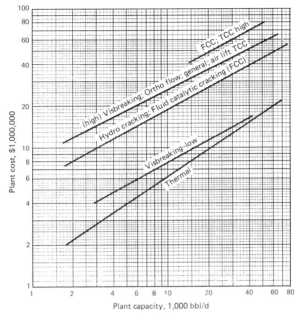

Plant cost, $1,000,000 (vertical axis)

Plant capacity, 1,000 bbl/d (horizontal axis)

Size exponent

Ortho flow; general; air lift TCC	0.49
Hydro cracking; fluid catalytic cracking (FCC)	0.53
Visbreaking	0.54
Thermal	0.65

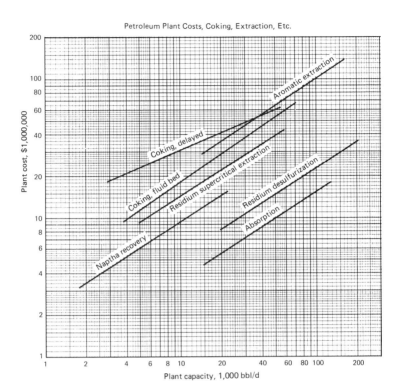

Petroleum Plant Costs, Coking, Extraction, Etc.

Plant cost, $1,000,000 (vertical axis)

Plant capacity, 1,000 bbl/d (horizontal axis)

Size exponent		Operation
Coking, delayed	0.42	Thermal cracking; coke production
Coking, fluid bed	0.64	Thermal cracking; coke production
Aromatics extraction	0.64*	Liquid extraction of aromatics
Residium supercritical extraction	0.64*	High pressure, temperation extraction
Naptha recovery	0.64*	Distillation, desulfurization, etc.
Residium desulfurization	0.64*	Hydrogenation
Absorption	0.64*	

*Assumed

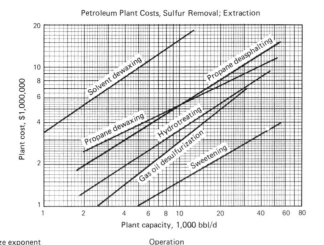

Petroleum Plant Costs, Sulfur Removal; Extraction

Size exponent		Operation
Desulfurizing		
Hydrotreating	0.64	Hydrogen treating of lube oils, naptha
Sweetening	0.57	Treatment of gasoline to remove mercaptans, sulfides
Gas oil desulfurization	0.78	Hydrogen treatment of gas oils
Extraction		
Propane deasphalting	0.61	Propane liquid extraction of vacuum distilled crudes
Propane dewaxing	0.47	Propane addition, filtration, stripping of diesel, etc. oils
Solvent dewaxing	0.66	Solvent extraction of lube oils

Petroleum Plant Costs, Gasoline Production, Distllation

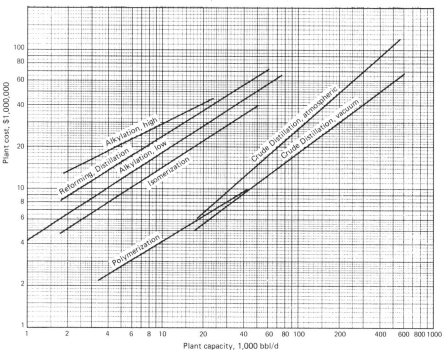

Plant capacity, 1,000 bbl/d

Size exponent		Operation
Alkylation, low	0.63	Medium weight unsat. hydrocarbons to gasoline
Alkylation, high	0.49	
Distillation, vacuum	0.73	Crude oil fractionation
Distillation, atmospheric	0.87	
Isomerization	0.64	Hydrogenation to upgrade pentane, hexane, etc.
Polymerization	0.61	Conversion of olefinic streams into higher octane
Reforming, Distillation	0.63	Dehydrogenation of paraffins, etc. into cycle compounds
		General distillation

Plant Costs, Power from Refuse, Co-generation

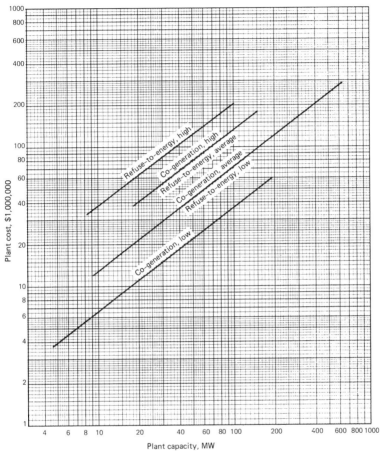

Size exponent 0.75

Water (Drinking) Preparation Plants

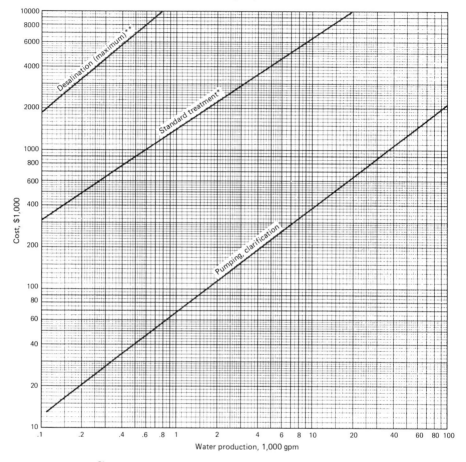

Size exponents

Desalination	0.89
Standard treatment	0.65
Pumping, clarification	0.74

*Standard treatment: floculation, clarification, filtration, chlorination.

**See Desalination graph

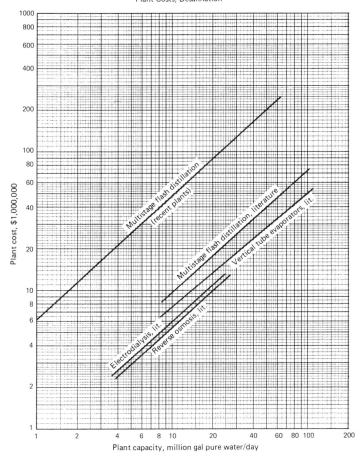

Plant Costs, Desalination

Size exponent
Multistage flash distillation,
electrodialysis, reverse osmosis   0.89
Vertical tube evaporators   0.82

Wastewater or Sewage Treatment
Secondary sewage processing: filtration, activated sludge

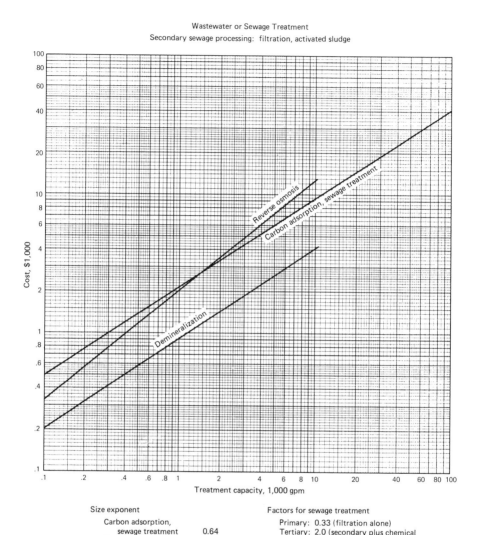

Treatment capacity, 1,000 gpm

Size exponent		Factors for sewage treatment
Carbon adsorption,		Primary: 0.33 (filtration alone)
sewage treatment	0.64	Tertiary: 2.0 (secondary plus chemical
Reverse osmosis	0.79	treatment of filtrate)
Demineralization	0.65	

REFERENCES

Chemical Engineering, ed. and comp. 1973–1974. *Sources and Production Economics of Chemical Products.* McGraw-Hill, New York: 121–180.

Chemical Engineering, ed. and comp. 1980–1988. *Construction Alert.* McGraw-Hill, New York.

Guthrie, Kenneth M. 1974. *Process Plant Estimating, Evaluation, and Control.* Craftsman Book Co., Solana Beach, CA: 125–180, 334–353, 369–371.

Guthrie, Kenneth M. 1970. Capital and operating costs for 54 chemical processes. *Chemical Engineering* (June 15):140–156.

Kharbanda, O. P. 1979. *Process Plant and Equipment Cost Estimation.* Craftsman Book Co., Solana Beach, CA.

Process Economics International. 1979–1980. Vol. 1 (2).

APPENDIX 3

MANUFACTURING COST

DATA PRESENTED

There are far less data in the literature on manufacturing cost than on the other components of cost estimating, primarily because it is a more complex and site, process, or company-specific cost. Some data do exist, however, and they are presented in the following pages. Most of the data are quite old, and difficult to easily update, although an attempt has been made to convert data to early 1987, or CE Index 320 values.

Section 1 presents manufacturing cost versus plant capacity curves of Guthrie (1974), with the percent breakdown into major cost components when available by Kharbanda (1979). The original Guthrie data were probably quite accurate as a first, general approximation, but they are old, and may have suffered badly by attempts to extrapolate them to the present time. The percent breakdown tables were undoubtedly based upon one single plant or process and location, and may be far from typical. Both sets of data at best should only be used for order-of-magnitude or "ballpark" estimates.

Section 2 gives some detailed manufacturing raw material and utility estimates from *Chemical Engineering* (1973/1974), which also probably were quite accurate when published. The processes may have changed considerably since that time, but at least these values should still be useful for conservative first approximations.

Section 3 provides more of the percent breakdowns of Kharbanda (1979), but now with the single plant size operating cost also estimated. In Section 4 Kharbanda has tabulated (or calculated) the raw materials and utilities required for many processes. As noted previously, the accuracy is probably very poor, but in many cases provides initial rough estimates that are better than nothing, and in other cases it is useful to doublecheck the figures quoted by vendors or others.

METHODS OF USE

Since each of the four sections of data overlap each other, are from different authors, and present limited lists of chemicals, each must be separately examined to make a manufacturing cost estimate. For example, ammonia is found in three of the four section's figures and tables. In cases such as this the data may not be consistent and you will have to make your best guess as to which to use and not use. This will complicate your study, but often there is some component of the information you know or feel more confident of, and this will aid in your selection. For instance, you may have heard that the average U.S. ammonia plant now uses 32 million Btu of fuel per ton of ammonia, and that the newest plants consume less than 25 million. This can allow you to somewhat adjust and evaluate the data from the three sections.

In other cases, merely knowing the current competitive selling price can allow you to adjust this data somewhat, assuming that the present manufacturers must make at least some profit on the product. This concept can lead you to further examine various alternative raw materials, processes, and producers, to see where the competitive advantages exist, which may influence and assist in your cost estimates and recommendations.

SECTION 1.   MANUFACTURING COST VS. PLANT CAPACITY (1); PERCENT COST BREAKDOWN (3)

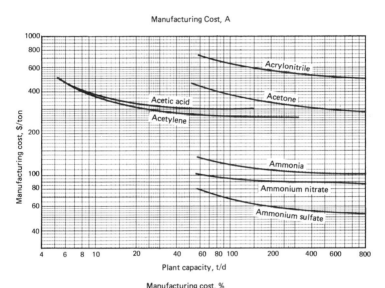

Manufacturing Cost, A

		Manufacturing cost, %			
	Plant capacity, t/d	Raw materials	Depreciation	Utilities, labor	Raw materials
Acrylonitrile	85	37	46	17	Propylene, NH₃
Acetone	70	74	6	20	Isopropanol; vapor
	70	84	6	10	; liquid
Acetic acid	70	51	24	25	Acetaldehyde, ethanol
	140	32	46	22	Methanol
Acetylene	50	15	45	40	Hydrocarbons
Ammonia	1000	36	50	14	Methane @ $1.25/MM btu
Ammonium nitrate	700	77	15	8	Ammonia

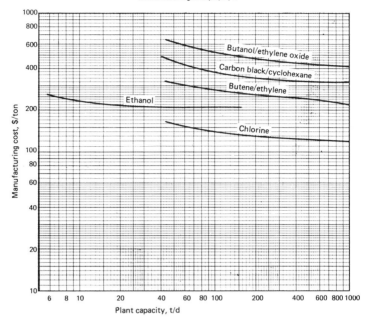

Manufacturing Cost, B, C, E

Chart axes: Manufacturing cost, $/ton (vertical, 10 to 1000) vs Plant capacity, t/d (horizontal, 6 to 1000). Curves labeled: Butanol/ethylene oxide, Carbon black/cyclohexane, Butene/ethylene, Ethanol, Chlorine.

	Plant capacity, t/d	Manufacturing cost, %			Raw materials (process)
		Raw materials	Depreciation	Utilities, labor	
Butanol	70	68	9	23	Ethanol
Ethylene oxide	85	48	33	19	Ethylene, (SD)
	85	68	18	20	Ethylene, (Shell)
Carbon black	110	27	50	23	Oil
Cyclohexane	140	92	5	3	Benzene, H_2
Butene	140	32	54	14	Butane
Ethylene	280	25	45	30	Ethane; naptha
	825	13	44	43	Propane; naptha
Ethanol	280	57	24	19	Ethylene

Manufacturing Cost, F, G, H, I, M

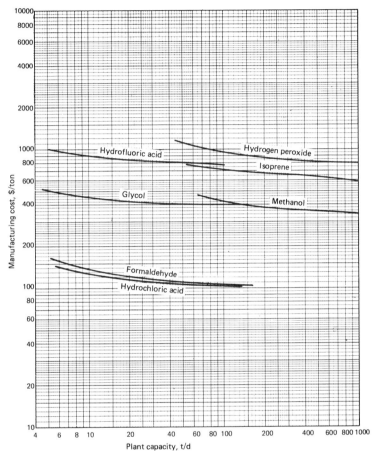

Manufacturing cost, $/ton

Plant capacity, t/d

	Plant capacity, t/d	Manufacturing cost, %			
		Raw materials	Depreciation	Utilities, labor	Raw materials
Isoprene	140	42	44	14	Propylene
	140	50	25	25	Methanol
Methanol	200	22	44	34	Methane
Glycol	110	90	4	6	Ethylene oxide
Formaldehyde	140	59	23	18	Methanol

Manufacturing Costs, N, O, P

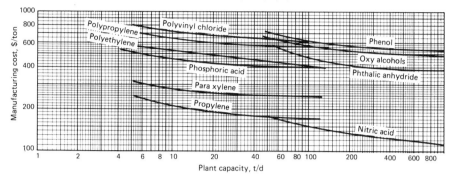

Plant capacity, t/d

	Plant capacity, t/d	Manufacturing cost, %			
		Raw materials	Depreciation	Utilities, labor	Raw materials (process)
Polyvinyl chloride	70	59	24	17	Vinylchloride; (suspension; emulsion)
Phenol	140	43	35	22	(Modified Raschig)
	140	33	35	32	Cumene
Phthalic anhydride	40	38	50	12	o-xylene
	40	53	41	6	Fluid bed; naphthalene
p-xylene	70	34	50	16	(Fractionation)
Propylene	70	77	13	10	Propane
Nitric acid	300	53	36	11	Ammonia

Manufacturing Costs, S, U, V

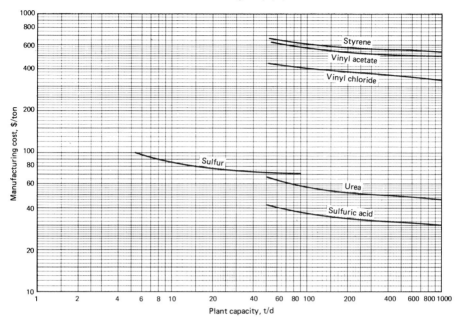

Plant capacity, t/d

	Plant capacity, t/d	Manufacturing cost, %			
		Raw materials	Depreciation	Utilities, labor	Raw materials
Styrene	140	68	14	18	Ethyl benzene
Sulfur	150	0	59	41	H_2S – rich gas
Urea	300	66	19	15	Ammonia, CO_2
Vinyl acetate	70	70	21	9	Acetic acid, acetylene
Vinyl acetate	70	49	42	9	Ethylene
Vinyl chloride	140	80	10	10	Acetylene, HCl
Vinyl chloride	140	71	15	14	Ethylene, Cl_2

Manufacturing Costs, Petroleum Plants

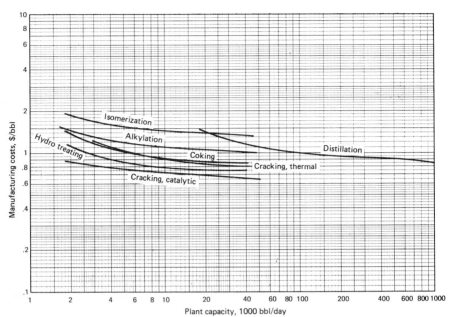

Isomerization

Hydro treating

Alkylation

Coking

Cracking, thermal

Distillation

Cracking, catalytic

Manufacturing costs, $/bbl

Plant capacity, 1000 bbl/day

Manufacturing Costs, Petroleum Plants

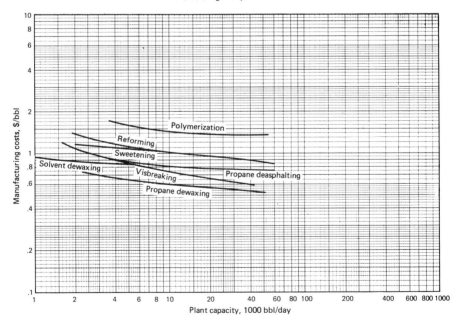

Polymerization

Reforming

Sweetening

Solvent dewaxing

Visbreaking

Propane deasphalting

Propane dewaxing

Manufacturing costs, $/bbl

Plant capacity, 1000 bbl/day

Operating Cost, Wastewater Treatment

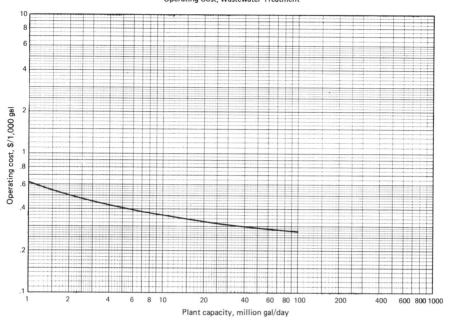

Curve: Primary, secondary treatment,
        sludge handling, chlorination

Factors
Sand filtration   0.27
Activated carbon   0.62
Electrodialysis   1.08

Manufacturing Costs

SECTION 2. DETAILED REQUIREMENTS PER TON OF PRODUCT

1. Acetaldehyde 225 t/d (75,000 t/yr) (*Hydrocarbon Process* 1967)

	One Stage (Oxygen)	Two Stages (Air)
Raw materials:		
Ethylene	1,340 lb	1,340 lb
Oxygen (99.5%), scf	9,460	—
Air, scf		54,000
HCl (as 20° Be acid), lb	30	80
Catalyst, $	2.75	2.75
Utilities:		
Electricity, KW hr	45	270
Steam (150 psig), M lb	2.4	2.4

	One Stage (Oxygen)	Two Stages (Air)
Process Water, M gal	1.7	7.2
Demineralized water, gal	120	—
Cooling tower water, M gal	48	48
Labor, operators/shift	3 to 4	3–4

2. Ammonia (2)

	Natural Gas	Naphtha
Raw materials:		
Gas, process and fuel, MMBtu	32.6	32.7
Catalyst and chemicals, $	1.13	1.23
Utilities:		
Electricity, kW hr	15	26.6
Makeup water, M gal	2.9	3.0
Labor, operators/shift	5	5

3. Benzene (Houdry Hydrodealkation Processes) (2)

Raw materials:	Detol	Pyrotol	Litol
Cyclohexane, napthenes		Approx 98.5% yields	
Hydrogen, M scf	11.4	22.6	4.9
Catalyst, $	0.47	0.63	0.32
Clay, lb	0.54	0.82	0.36
Utilities:			
Electricity, kW hr	49	47.2	41
Fuel, MM Btu;			
consumed	2.26	3.0	3.36
produced	10.0	25.0	4.9
net	+ 7.74	+ 21.8	+ 1.54
Steam, lb; consumed	88	1,940	300
produced	—	1,810	1,074
net	− 88	−130	+ 774
Boiler feed water, gal	—	0.72	3.91
Cooling water, M gal	4.05	10.43	6.86

4. Butadiene (Shell ACN Process) (2)

Raw materials:	
Butane	98.6% yield
Acetonitrile, lb	0.296
Other chemicals, $	0.44
Utilities:	
Electricity, kW hr	72
Steam (600 psig, 600°F), lb	6,460
Refrigeration (@ 40°F), Btu	71,160
Process water, gal.	11.8
Cooling water (30°F rise), M gal	31.4
Labor, operators/shift	1.5

By-products, per ton Butadiene
Butylene, ton 1.335; Light ends, lb 11
Heavy ends, lb 89

5. Caprolactum (Stamicarbon Process); 38,500 t/yr plant (2)

Raw materials:

Cyclohexane, lb	2,120
Hydrogen (>95%; 100% basis), lb	192
Ammonia, lb	632
Aqua ammonia (20% on 100% NH₃ basis), lb	962
Oleum (100% H₂SO₄ basis), lb	2,720
Sodium hydroxide, lb	200
Benzene, lb	28
Tolulene, lb	16
Phosphoric acid	10
Catalysts, $	15.10

Utilities:

Electricity, kW hr		376
Fuel (75% furnance efficiency), MM Btu		1.7
Steam, lb;	440 psig	14,380
	184	3,340
	56	7,260
Refrigeration (6°C), M Btu		126
Boiler feedwater, gal (95°C)		295
Process water, gal		760
Cooling water, gal		42,400

By-products:

Ammonium sulfate, t/t	1.75
Hydrogen (40% H₂), lb/t	12
Hydrogen (95% H₂), lb/t	28

6. Chlorine (Hooker Diaphragm Process) (2)

Raw materials:

Salt, tons	3.52
Misc. chemicals, materials, $	0.12
Diaphragm asbestos, lb	0.2
HCl, lb	10
H₂SO₄, lb	12

Utilities:

Electricity, kW hr: to cells	2,980		
other	250		
Steam, lb: evaporation	5,460		
other	700		
Labor/plant capacity, t/d	200	500	900
Supervision	One man per shift		
Operators, man hours	0.52	0.22	0.18
Cell rebuilding, man hours	0.03	0.016	0.016
Maintenance, man hours	0.32	0.17	0.12

By-products:

Caustic Soda, t/t	2.14
Hydrogen, lb/ton	56.32

7. Cyclohexane (IFP Liquid Phase Hydrogenation Process)   (2)

Raw materials:

Benzene, lb	1,870
Hydrogen, lb	130
Catalyst, $	11

Utilities:

Electricity, kW hr	8
Steam (300 psig), lb	370
Boiler feed water, gal	230
Cooling water (10°C rise), gal	4,000

By-products:

Fuel gas, lb	620
Steam (65 psig), lb	1,900

8. Cyclohexanol (Stamicarbon Process)   (2)

Raw materials:

Cyclohexane, lb	2,200
Caustic soda, lb	180
Catalysts, $	1.56

Utilities:

Electricity, kW hr	200
Steam, lbs: 454 psig	7,300
56 psig	700
Refrigeration (0°C), M Btu	66
Process water, gal	101
Cooling water, (8°C rise), M gal	101

9. Ethylene (propylene) (Lummus Naphta Pyrolysis Process)   (2)

Raw materials:	High Severity; ethane recycle	Moderate severity; no recycle
Medium range naptha, tons (117–308°F; 73°API)	5.95	8
Catalysts, chemicals, $	1.55	1.55
Utilities:		
Electricity, kW hr	34	34
Fuel, MM Btu	26	26
Boiler feed water, gal	96	96
Cooling water, M gal (25°F rise)	70	70
Labor:		
Operators, foremen/shift	8	8
Maintenance material, % of capital	2	2
By-products, lb/ton ethylene:		
Propylene	991	1,454
Butadiene	272	356
Butylenes/butanes	251	669
Hydrogen	92	77
Methane rich gas (21,630 Btu/lb)	956	1,014
Ethane	—	400
Benzene	401	358

	High Severity; ethane recycle	Moderate severity; no recycle
Toluene	199	280
C$_8$ aromatics	105	115
C$_5$ − 400°F + gasoline	404	1,104
400°F + fuel oil (17,100 Btu/lb)	280	174

10. Formaldehyde (Reichhold Formox Process) (2) (per ton at 35% solution)

Raw Materials:

Methanol, gal	130
NaOH, lb	2.2
Catalyst, lb	0.08
Ion exchange resin, ft^3	0.0002

Utilities:

Electricity, kW hr	78.5
Steam, startup (168 hr/yr), M lb (150 psig)	300
Fuel, startup (168 hr/yr), M Btu	26.88
Process feedwater, gal	238
Boiler feedwater, gal	94
Cooling water, M gal: 85°F	22.2
60°F	3.72
Instrument air, scf	300

Labor:

Operator/shift	1
Supervisor/shift	1
Laboratory, hr/wk	3
Maintenance: % of capital	3

By-product:

Steam lb/ton 150 psig	820

11. Liquefied natural gas (TEALARC Process) 1 MMM scfd (2)

Chemicals:

Monoethanol amine, lb	0.154
Antifoamant, lb	0.0006
Caustic soda, lb	0.0112
Hydrazinc, lb	0.0052
Tri sodium phosphate, lb	0.0020
Morpholine, lb	0.0024
Chlorohydric acid, lb	0.0084
Chlorine, lb (assumes seawater cooling)	0.56
Molecular sieves, lb	0.016

Utilities:

Electricity, kW hr	29
Fuel, lb (13% of feed)	260
Steam, M lb	4.1
Cooling Water, M gal	36.4

Labor:

Operators, technicians, engineers/3 shifts	40
Maintenance/3 shifts	50

12. Methanol (ICI/Kellogg Process) (2)
    Raw materials:
Natural gas, MM Btu	27
Catalyst, $	1.40

    Utilities:
Electricity, kW hr	4.8
Fuel, net, MM Btu	5.14
Boiler feedwater, gal	297
Cooling water circulation, M gal .	44
Labor: operators/shift	4
Maintenance: % of capital	3.5
By-product: steam, lb	271

13. Phenol (2)

Raw materials:	Hercules- BPCI Phenol- Acetone		Hooker
Cumene, lb	2,700	Benzene, lb	1,855
Hydrogen, lb	0.80	HCl, lb	60
NaOH, Na_2CO_3, H_2SO_4, lb	24	NaOH, lb	40
Catalyst, $	0.84	Catalyst, $	5.81
		Scrubber oil, lb	16
Utilities:			
Electricity, kW hr	228		182
Fuel, MM Btu	0.38		6.6
Steam, M lb (450 psig)	10.6		19.5
Cooling water, M gal (30°F rise)	65		74
Labor, operators/shift	4		5
others			2/day shift 1 supervisor/shift
Maintenance, % of capital	2		
By-products:			
Acetone, lb/ton	630		
Hydrocarbons, lb (18,500 Btu/lb)	220		

14. Soda ash (Na_2CO_3; Diamond Shamrock Solvey Process) 550 t/d (2)
    Raw materials:
Salt (NaCl; brine), lb	3,060–3,200
Limestone ($CaCO_3$), lb	2,080–2,800
Ammonia, lb	4–5
Sodium sulfide, lbs of S	0.6–1.2
Coke, lb (2.2 MM Btu/ton Na_2CO_3)	160–240

    Utilities:
Electricity, kW hr	54–134
Fuel, MM Btu (oil or gas)	7.2
Treated water, gal	240
Cooling water, M gal	30 (once through)
Cooling water, makeup if recycled, Mgol	4.4
Labor, operating man hr/ton	0.6
maintenance,	0.6

Supplies: operating, $	0.21
maintenance, $	1.05

15. Sodium bicarbonate   150 t/d   (2)
    Raw materials:

Caustic soda, lb	965
Natural gas, scf	5,360

    Utilities:

Electricity, kW hr	42
Steam, lb (15 psig)	69
Process water, 85°C, gal	290
Cooling water, M gal (20°F rise)	12.9
Compressed air, 100 psig, scf	590
Labor, operators/shift	2
Maintenance, % of capital	3.5

16. Synthetic natural gas (CRG/Kellogg Naphtha Reforming Process)   (2)
    Material per MM scf of SNG (993 Btu/scf)
    Raw materials:

Naphtha, M lb (20,263 Btu/lb; Dist. 365°F	47.95
Chemicals, $	8.44
Catalysts, $	63.70

    Utilities:

Electricity, kW hr	850
Fuel, M lb (20,263 Btu/lb)	4.22
Boiler feedwater make up, M gal	4.7
Cooling tower circulation, M gal (25°F rise)	9.6
Cooling tower makeup, gal	380

17. Styrene (Monsanto/Lummus Process) 900 t/d   (2)
    Raw materials:

Ethylene, lb	620
Benzene, lb	1,680
Catalysts, chemicals, $	4.4

    Utilities:

Electricity, kW hr	76
Fuel, MM Btu	4.32
Steam, M lb: 200 psig	4.8
75 psig	2.7
Cooling water, M gal	26.1
Labor, operators/shift	3
supervisors (total)	2
Maintenance, % of capital	2-3

    By-products:

Toluene, lb/ton	126
AlCl$_3$ (22% solution), lb/ton	28
Steam condensate, gal	863

18. Sulfuric acid (Monsanto Contact Process)   (2)
    Raw material:

Sulfur, lb	674

Utilities:

Process water, gal	60
Boiler feedwater, gal	324
Cooling water, circulation (25°F rise), M gal	7
Power, Kwh (steam turbine)	9
Labor, operators/shift	1
Maintenance, % of capital	5
By-product:	
Steam, M lb (225 psig)	1.7

19. Urea (Stamicarbon CO$_2$-Stripping Process; producing prills)   (2)

Raw Materials:	Biuret	0.7–0.8%	0.2–0.25%
Ammonia, lb		1140	1140
Carbon dioxide, lb		1510	1510
Utilities:			
Electricity, kW hr		109	127
Steam, M lb (368 psig)		2	2.2
Cooling water, M gal (15°C rise)		11.5	11.5
Labor: operators, supervisors/shift		3	4
Maintenance, % of capital		3	3
By-product:			
Steam, lb (60 psig)		300	700

SECTION 3. SINGLE PLANT SIZE: PERCENT COST BREAKDOWN (3)

Chemical	Plant Capacity, t/d	Mfg. Cost, c/lb	% as Raw Material	% as Depreciation	% as Utilities, Labor	Process or Raw Material
Acetaldehyde	70	25	55	29	16	Ethylene
	70	25	72	12	16	Ethanol
Acetic anhydride	70	29	78	9	13	Acetic acid
Acyrlic staple	30	175	28	26	46	Dimethyl formamide
Adipic acid	180	38	79	9	12	Cyclohexanol
Allyl chloride	40	43	47	34	19	Propylene
Aniline	55	32	65	25	10	Nitrobenzene
Benzene	350	14.5	44	38	18	Naphtha
	140	14.5	72	17	11	Toluene
Benzoic acid	7	37	22	39	39	Toluene
Bisphenol A	30	42	63	23	14	Acetone, phenol
Caprolactam	110	62	51	31	18	Cyclohexane
Carbon disulfide	140	14	42	40	18	Methane, S
Carbon tetrachloride	30	21	53	32	15	Propane, Cl$_2$
Carboxy methyl cellulose	7	83	45	33	22	Cellulose
Cellulose acetate	70	81	36	26	26	Cellulose, acidic acid
Chloroprene monomer	65	60	34	53	13	Butadiene
	65	80	28	62	10	Acetylene
Cumene	220	17	80	10	10	Benzene, propylene
Cyclohexonone-cyclohexanol	140	40	68	18	14	Cyclohexane
Dichlorophenoxy acetic acid (2, 4)	15	78	62	24	14	Phenol
DDT	30	57	53	33	14	Chloral, chloro benzene

Chemical	Plant Capacity, t/d	c/lb	% as Raw Material	% as Depreciation	% as Utilities, Labor	Principal Raw Material
Detergent alkylate	70	25	80	12	8	Propylene, benzene
Diethanolamine	175	32	83	14	3	n-Paraffin
Dimethyl terephthalate	18	23	63	28	9	Ethylene oxide, NH$_3$
	70	56	48	32	20	p-xylene, methanol
Dioctyl phthalate	55	27	84	7	9	Phthalic anhydride
Epichlorhydrin	55	57	78	11	11	Allyl chloride, Cl$_2$
Ethyl benzene	140	14	73	11	16	Ethylene, benzene
Ethyl hexanol	55	23	45	35	20	Propylene, synthesis gas
Ethylene diamine	55	23	68	20	12	Acetaldehyde
	20	78	46	27	27	Ethylene dichloride, NH$_3$
Ethylene dichloride	230	11	88	6	6	Ethylene, Cl$_2$
Ethylene glycol	110	22	90	4	6	Ethylene oxide
Fatty alcohol	35	40	82	12	6	Coconut oil, glycerine
Fluorocarbon	35	35	78	14	8	Carbon tetrachloride, HF
Glycerine	55	60	89	6	5	Epichlorhydrin
	55	60	82	8	10	Allyl alcohol
Hexamethylene tetramine	7	37	43	17	40	Methanol, NH$_3$
Hydrogen cyanide	30	33	19	38	43	Propane, NH$_3$
Iso octanol	55	29	45	24	31	Heptane
Isobutylene	40	21	30	56	14	Butane
Isopropanol	70	14	40	44	16	Propylene
Maleic anhydride	35	37	43	38	19	Benzene
Melamine	28	33	47	33	20	Urea, NH$_3$

Mercaptobenzo thiazole	7	83	57	30	13	Aniline
Methyl chloride	30	17	60	19	21	Methanol, HCl
Methyl methacrylate	55	41	75	15	10	Acetone, HCN, methanol
Monochloro acetic acid	30	37	67	21	12	Acetic acid, Cl_2
Naphthol, beta	15	73	67	17	16	Naphthalene
Nylon 6/6 resin	110	73	71	11	18	Adipic acid
Nylon filament	30	320	28	25	47	Dimethyl formamide solution
Olefins, alpha	420	17	73	15	12	Wax
Paraffins, n	70	22	82	14	4	Kerosene
Pentachlorophenol	7	37	41	22	37	Phenol, HCl
Pentaerythritol	14	47	51	35	14	Formaldehyde, acetaldehyde
Perchloroethylene	40	19	58	29	13	Propane, Cl_2
Phosgene	35	37	60	30	10	Carbon monoxide, Cl_2
Polyester staple	30	34	34	29	39	Dimethyl formamide solution.
Polystyrene	55	32	69	22	9	Styrene
Propylene oxide	70	32	66	18	16	Propylene, Cl_2
Rubber (synthetic)	140	38	21	49	30	Isoprene
Sorbitol	15	23	52	18	30	Corn syrup
Terephthalic acid (fibre)	70	53	52	29	19	p-xylene
Tetraethyl lead	55	98	64	26	10	Ethyl chloride, Pb, Na
Toluene disocyanate	28	67	32	52	16	Phosgene
o-xylene	85	11	0	50	50	Super fractionation

SECTION 4. RAW MATERIAL AND UTILITY REQUIREMENT (3)

Acetaldehyde		Ethylene 0.67 Oxygen 0.29* CW 0.3* DW 0.003* S(L) 1.3 S(H) 0.3 E 0.21
		Ethanol 1.15 [Formaldehyde 1.1 Methanol 0.65 Solvents 0.4 Acetone 0.13]
		Butane 5.53L A 2.5*
Acetic Acid		Acetaldehyde 1.1 Manganous Acetate 0.003 A 0.23*
	‡	Methanol 0.53 CO 0.47 Catalyst(s)
		Butane 0.97 A 3.8* [Other acids, alcohols and ketones]
Acetic Anhydride		Acetaldehyde 1.2 Catalyst 0.001 Diluent 1.7 A
		Acetic Acid 1.35 Catalyst(s)
Acetone		Isopropyl Alcohol 1.2
Acetylene		Calcium Carbide (85%) 3.5 W 29
		Natural Gas 8.2* S(H) 29 CW 0.1* PW 0.8L E 0.16 Solvent 0.003 [Tar 0.18 Fuel Gas 11.2*]
		Natural Gas 5.9* Oxygen (95%) 5.4 Solvent 0.003 E 1.7 S(L) 5.0 CW 0.03* (Partial Oxidation) [10.* Off Gas C Black 0.03 Acetylene Polymers 0.005]
Acrylate, Ethyl	‡	Acetylene 0.26 Ethanol 0.46 CO 0.06 Ni Carbonyl 0.09 HCl 0.18
	‡	Propylene 0.4 Oxygen 0.48 Ethanol 0.46 Catalyst(s)
	‡	Acrylonitrile 0.53 Ethanol 0.46 W 0.18 Sulphuric acid(s)
Acrylate, Methyl	‡	β-Propiolactone 0.84 Methanol 0.37 Catalyst(s)
Acrylonitrile		Propylene 1.18 Ammonia 0.48 A 6.1* Catalyst(s)
Adipic Acid		Cyclohexane feed (95%) 0.8 Nitric Acid (100%), no recycle 1.0 Air 0.6 Cu, CO Naphthenate and Am. Metavandate(s)
Amine, Amyl		Mixed Amyl Chlorides 1.25 Ammonia 0.2 Caustic Soda, solid 0.49 Ethanol(s)
Amines, Methyl (Mixed)		Methanol 1.5 Ammonia 0.43
Alkyl Aryl Sulfonate		Dodecane 0.4 Benzene 0.13 22% Oleum 0.45 NaOH (S. G. 1.21) 0.65 L Alum. Chloride 0.01
Aluminium Chloride		Alum. Scrap 0.25 Chlorine 0.88
Aluminium Sulphate		(17% Alum. Oxide) Bauxite (55% Al_2O_3) 0.34 Sulf. Acid (80%) 0.57 Black ash (70% BaS) 0.007 Flake Glue(s)
Ammonia		Natural Gas 0.8* Catalyst, Shift 0.15, Synthesis 0.25 Caustic 0.004 Monoethanolamine 0.15 Fuel Gas for driving Comp. 6.1 m. kcal. E 0.12 W 0.025*
Ammonium Chloride		Amm. Sulphate 1.3 Sod. Chloride 1.25
Ammonium Nitrate		Ammonia 0.22 Nitric Acid (100%) 0.82 E 0.055 W 0.008* S 1.8
Aniline		Nitrobenzene 1.4 Iron borings 1.6 HCl (30%) 0.13
		Chlorobenzene 1.35 Amm. Sulphate 3.7 Cuprous Oxide 0.18
		Nitrobenzene 1.35 Hydrogen 0.08* Copper Carbonate 0.007
Aspirin		Salicyclic Acid 0.77 Acetic Anhydride 0.62 [Acetic Acid 0.35]
Barium Carbonate	‡	Black Ash (65% BaS) 1.8 Carbon Dioxide 0.22
	‡	Black Ash (65% BaS) 1.3 Soda Ash (58% Na_2O) 0.54 [Sod. Sulphide, 60% 0.65]

Benzene		Naphtha 0.88 B Diethylene Glycol 0.02 Clay 0.1 Catalyst 0.003 [Toluene 0.18 B Xylene 0.18 B Raffinate 0.44 B]
		Toluene 1.23 Hydrogen 0.03 Clay 0.002 [Paraffin HC 0.24]
		Light Oil 1.98 L Sulphuric Acid (96%) 0.01 Sod. Hydroxide 0.02
Benzoic Acid		Toluene 0.93 Air 1.72* Catalyst(s)
		Benzotrichloride 1.78 PW 1.25 Catalyst 0.013
		Phthalic Anhydride 1.43 Catalyst 0.013
Benzyl Chloride		Toluene 0.78 Chlorine 0.64
Bisphenol-A		Phenol 0.88 Acetone 0.27 HCl, Lime, Methyl Mercaptan(s)
Boric Acid		Borax 1.8 Sulphuric Acid (S. G. 1.84) 0.63
Bromine		Brine (1.000 PPM Bromine) 1,000 Chlorine 0.55 S-(V)
Butadiene		Butylene 1.3
		n-Butane 1.93 (Houdry)
Butyl Acetate		n-Butanol 0.71 Acetic Acid, Glacial 0.55 Sulphuric Acid (96%) 0.003
Butyl Alcohol		Molasses 5.4L W. 0.08 Nutrients 0.004 [Acetone 0.33 Ethanol 0.05 CO$_2$H$_2$]
		Butyraldehyde 1.03 Hydrogen 0.33* Catalyst(s)
Calcium Chloride		Brine or Liquor (9% CaCl$_2$) 8.5
Calcium Phosphate		Phosphate Rock (70 BPL) 0.6 Sulphuric Acid 0.35
Caprolactam		Cyclohexanone 0.9 Ammonia 1.5 CO$_2$ 0.5 S 0.7 20% Oleum 1.4 Catalyst(s) A (V)
Carbon Black		Oil 1.4-2.8 L or Natural Gas 5.3-7.0* A 25-38*
Carbon Disulphide		Methane 0.35* Sulphur 0.9
Carbon Tetrachloride		Carbon Disulphide 0.55 Chlorine 1.15
Carboxy Methyl Cellulose		Cellulose 0.58 Monochloroacetic Acid 0.3 Caustic Soda 0.25 Water 0.4
Cellulose Acetate		Cellulose 0.7 Acetic Anhydride 2.0 Acetic Acid 3.3 Sulphuric Acid 0.1 [Acetic Acid 5.0]
Chlorine		Salt 1.8 Sod. Carb. 0.03 Sulphuric Acid 0.01 S 11.4 E 3.3 R 1.0 Graphite 0.005 [Sod. Hydrox. 1.13 Hydrogen 0.32*]
	‡	Potassium Chloride 2.1 Nitric Acid 1.8 Oxygen 0.22 [Potassium Nitrate 2.85]
	‡	Hydrogen Chloride 1.0 Oxygen 0.23
Chloracetic Acid		Acetic Acid 0.69 Chlorine 0.8
Chlorobenzene		Benzene 0.95 Chlorine 0.9 Iron Turning(s)
Chloroform	‡	Chlorine 1.8 Methane 0.19*
Chromic Acid		Sod. Dichromate Dihydrate 1.5 Sulphuric Acid (S. G. 1.84) 1.3
Citric Acid		Molasses 4.0 Nutrients 0.01 Sulphuric Acid (95%) 0.7 Lime 0.5
Copper Sulphate	‡	Copper 0.26 Sulphuric Acid (100%) 0.4
Cresol	‡	Middle Oil (V) Caustic Soda (50%) 0.8 Sulphuric Acid (100%) 0.9
Cumene		Benzene 0.8 Propylene 0.4 Phosphoric Acid (Solid)(s)
Cyclohexane		Benzene 0.9 Hydrogen 0.07 Catalyst(s)
Cyclohexanone		Cyclohexane 1.0 Air (V) Metaboric Acid 0.005 Zinc Oxide 0.001

	Phenol 1.0 Hydrogen 0.07 Catalyst(s)
Decyl Alcohols	C$_9$ Olefin 1.15 Synth Gas 0.4* Hydrogen 0.14* Co. Ni Catalyst(s)
Dibutyl Phthalate	Phthalic Anhydride 0.6 Butyl Alcohol 0.7 Sulphuric Acid (96%) 0.01
Dichlorodifluoromethane	Carbon Tetrachloride 1.6 HF 0.4 Antimony Penta-chloride(s)
Dodecylbenzene	Benzene 0.5 Dodecene 0.9 Alum-Chloride or HF Catalyst(s)
Epichlorohydrin	Propylene 0.8 Chlorine 2.4 Caustic Soda 1.2
Ethanolamines	(75% Mono, 21% Di and 4% Tri-) Ethylene Oxide 0.8 Ammonia 0.3
Ethyl Ether	Ethanol (95%) 1.4 Sulphuric Acid (96%) 0.02
Ethyl Acetate	Ethanol (95%) 0.6 Acetic Acid (100%) 0.7 Sulphuric Acid (96%) 0.04
Ethyl Alcohol (per L)	Ethylene (97%) 0.5 Phosphoric Acid(s) Sod. Hydroxide(s)
	Ethylene 0.5 Sulphuric Acid 0.07 Sod. Hydroxide 0.02 W 0.24 S 2.4 E 0.005 Fuel 186 kcal
	Molasses 2.4 L Sulphuric Acid (79%) 0.02 Amm. Sulphate 0.0002 S 6 PW 10L CW 42 L E. 0.03
Ethyl Benzene	Benzene 0.7 Ethylene 0.27 CW 17 PW 0.7 E 0.02 Fuel 9.5 kcal
Ethyl Chloride	Ethylene 0.49 HCl 0.63 Alum. Chloride(s)
	Ethanol 0.75 HCl 0.6 Catalyst(s)
Ethylene Diamine ‡	Ethylene Dichloride 1.65 Ammonia 0.57 NaOH-for neutralisation
Ethylene Dibromide	Ethylene .015 Bromine 0.86
Ethylene Dichloride	Ethylene 0.32 Chlorine 0.8 Ethylene Dibromide(s)
Ethylene Glycol	Ethylene 0.9 Air 9.5 Silver Catalyst(s)
Ethylene Diamine Mono-ethyl ether	Ethylene Oxide 0.57 Ethanol 0.6 Catalyst(s)
Ethylene Oxide	Ethylene 1.1 A 13.1 Ag 0.5 mg. E 1.9 S 0.1 W 0.2*
Ethyl (2-) Hexyl Alcohol	Propylene (92 + %) 0.74 Synth. Gas (99%) 0.96 Cocar-bonyl(s) Butyldehyde
	Butyaldehyde 1.3 Hydrogen 0.36* Ni Catalyst 0.0001
Ferrous Sulphate	Pickling Liquor 2.5 Scrap Iron 0.08
Formaldehyde (37%)	Methanol 0.47 Air 0.8*
Glycerine	Spent Lye (5% Glyc.) 22 Sod. Hydrox. 0.1 Ferric Chloride 0.06 Al. Sulph. 0.01 Act. C 0.003 S 4.0 E 0.01 [Salt 2.2]
	Propylene 0.63 Chlorine 2.0 Sod. Hydrox. 0.045 Hydrated Lime 0.045
	Propylene 0.93 Oxygen 0.23 Isopropanol 1.1 Hyd. Perox. (100%) 0.49
Hexamethylene Diamene	Adiponitrile 1.0 Ammonia 0.05 Hydrogen 0.08 Catalyst 0.0008
Hexamethylene Tetramine	Formaldehyde (37%) 3.6 Ammonia (100%) 0.55
Hydrazine	Ammonia 1.33 Sod. Hypochlorite 3.3 Glue(s)
Hydrochloric Acid (31.5%)	Salt 0.5 Sulphuric Acid (100%) 0.48 or Niter Cake 1.3 Coal 0.4 [Salt Cake 0.63 OR 1.42]
	Chlorine 0.32 Hydrogen 0.01

Hydrofluoric Acid		Fluorspar (98% CaF$_2$) 2.25 Sulphuric Acid (100%) 2.85 S 1.75 E 0.22 R (−10°C) 360 kcal Fuel 2,300 kcal PW 0.01* CW 0.12*
Hydrogen (Per*) 99.9%		Propane 0.37 L OR Natural Gas 0.25 S 5.8 Fuel 3,200 kcal CW 90 L F 0.03
		Fuel 0.36 Oxygen (95%) 0.36 E 0.07 150 kg-h
97%		Coke 0.68 S 7.2 CW 0.27 E 0.11
Hydrogen Peroxide (70%)		Ammonia 0.025 Sulphuric Acid 0.03 PW 1.15L E. 7.1 S 8.4 Platinum (VS)
25%		Oxygen 0.17* Hydrogen 0.18* W 0.75L Pd, Solvent, Ethylanthraquinone—losses only
25%		Isopropanol 0.5 Oxygen 0.2* [Acetone 0.46]
Isopropyl Alcohol 91%		Propylene 0.9 Sulphuric Acid (85%) 0.13 Mineral oil—losses only
Lead, Tetraethyl		Sodium 0.33 Lead 0.64 Ethyl Chloride 0.8 Catalyst 0.03 Ferric Chloride Sod. Thio Sulphite(s)
Lead, Tetramethyl	‡	Mg turnings 0.18 Methyl Chloride 0.75 Lead 0.78
Litharge		Pig Lead 0.95 A (V)
Maleic Anhydride		Benzene 1.34 A 19*
Melamine		Urea 3.1 Ammonia 0.46 Carbon Dioxide 0.03 Catalyst 0.008 Act. C. 0.002 CW 0.65*
Methyl Alcohol		Carbon Monoxide 1.17* Hydrogen 2.35*
Methyl Chloride	‡	Chlorine 1.4 Methane 0.45*
		Methanol 0.7 HCl, 0.8 Alumina Gel(s)
Methyl Ethyl Ketone		Butyl Alcohol (Sec-) 1.18
Methyl Isobutyl-Ketone	‡	Acetone 1.16 Hydrogen 0.23* Acid, Alkali, Catalyst(s)
Methyl Methacrylate	‡	Acetone 0.58 HCN 0.27 Methanol 0.32 Sulphuric Acid (98%) 0.98
Methyl Parathion	‡	Phosphorous Pentasulphide 0.42 Methanol 0.24 Chlorine 0.27 p.nitrophenol 0.53
Nitric Acid (100%)		Ammonia 0.29 Pt 0.00025 mg A 3.6* W 0.13* E 0.39 [Steam 1.0]
Nitrobenzene		Benzene 0.65 Mixed Acid (Sulphuric 0.72, Nitric 0.53, Water 0.11) Sod. Carb. 0.01
p-Nonylphenol		Phenol 0.62 Nonene 0.45 Catalyst(s)
Oxygen (90%)		A 3.9* S (H) 1.67 CW 4.2L E 410 kwh
Pentaerithritol		Formaldehyde (37%) 3.2 Acetaldehyde 0.38 Alkali (50%) 1.05 Acid (as formic) 0.6
Perchlorethylene	‡	Ethylene Dichloride 1.2 Chlorine 0.64 Oxygen 0.39 Catalyst(s) [Trichloroethylene 0.8]
		Propane 0.2 Chlorine 2.5 [HCl 1.35]
		Acetylene 0.19 Chlorine 1.5 Lime (hydrate) 0.45 Catalyst(s)
Phenol		Cumene 1.38 A 1.4* Sulphuric Acid, Sod. Hydrox.(s)
		Benzene 1.00 Sulphuric Acid (96%) 1.75 Caustic Soda 1.70 S 2.0 E 0.09
		Chlorobenzene 1.25 Caustic Soda 1.37 HCl (as 100%) 0.5 Toluene 1.25 A 1.06* Catalyst(s)
Phosgene		Co 0.23* Chlorine 0.72 C (Active) 0.0005
Phosphoric Acid (100%)		Phos. Rock (70 BPL) 2.5 Sand (Silica) 1.0 Coke Breeze 0.44 C Electrode 0.01 A 4.1* E 4.5 W 40* [Slag 2.3]

		Phosphorus 0.32 A 4.1* S.W-(V)
(75% H₃PO₄)		Phos. Rock (70 BPL) 1.8 Sulphuric Acid (94%) 1.7 W 0.06* S 1.35* E 3.0
Phosphorus		Rock 12.0 Sand (Silica) 2.2 Coke 1.3 C Electrode 0.03 E 14.3
Phosphorous Oxychloride		Phosp. Trichloride 0.54 Phosp. Pentoxide 0.20 Chlorine 0.28
Phosphorous Pentasulphide		Phosphorus (White) 0.30 Sulphur 0.76
Phosphorous Trichloride		Phosphorus 0.24 Chlorine 0.82
Phthalic Anhydride		Naphthalene (78°) 1.25 A 26*
		o-Xylene 0.98 A 25*
Potassium Chloride (99%)		Sylvinite ore 2.5 W 167* S 1.25 E 0.55 (crystalization)
(60% K₂O)		Sylvinite 1.15 Flotation Reagent 0.0005 (flotation)
(97%)	‡	Saturated Lake Brine 20.6
Potassium Hydroxide		Pot. Chloride 1.46 Pot. Carbonate 0.025 Sulphuric Acid (SG 1.84) 0.1 S 7.13 E 2.0 [Chlorine 0.63 Hydrogen 0.02]
Potassium Permanganate	‡	Manganese Dioxide 0.55 (0.79-70% ore) Pot. Hydroxide (100%) 0.36
Potassium Pyrophosphate	‡	Caustic Potash (100%) 0.68 Phosphoric Acid (100%) 0.60
Propylene Glycol	‡	Propylene Oxide 0.76 Water 0.24
Propylene Oxide		Propylene (100% Basis) 0.94 Chlorinie 1.6 Lime (100% CaO) 1.1
		Propylene 0.78 Isobutane 2.16 Oxygen 0.90
Sodium		Sod. Chloride 3.15 Calcium Chloride 0.006 E 16.5
Sodium Bicarbonate		Sod. Carbonate 0.69 Carbon Dioxide 0.30
Sodium Carbonate		Salt 1.5 Limestone 1.2 Coke 1.0 Coal 0.45, Solvey
(58% Na₂O)		process;
		CW 0.07* Ammonia (Make-up) 0.003 Carbon Dioxide 0.35* Sod. Sulphide 0.002
	‡	Trona Ore 1.5 Natural soda ash
	‡	Saturated Searles Lake Brine 21.5
Sodium Chlorate		Salt 0.57 HCl (100%) 0.014 Sod. Dichromate 0.0005 Barium Chloride 0.0007 Graphite 0.001 E 5.6
Sodium Chloride (99.8%)	‡	Saturated Brine (26.3% NaCl) 3.8 Soda Ash (58%) 0.004 Caustic Soda (50%) 0.4 S (Triple Effect Evap.) 1.25
Sodium Chromate		Per 1.6 kg. Sod. Chrom. Decahydrate OR 1.0 kg. Sod. Dichrom. Dihydrate
		Chromite Ore (50% Cr₂O₃) 1.1 Limestone 1.5 Soda Ash 0.8 Sulphuric Acid (S.G. 1.84) 0.5 Fuel Oil 0.54 L S 3.0 E 0.55 [Anhy. Sod. Sulphate 0.5]
Sodium Hydroxide (99%)		Salt 1.5 Sod. Carb. (58%) 0.03 Sulphuric Acid (S.G. 1.84) 0.1 S 10 E 2.75 R 0.9 [Chlorine 0.89 Hydrogen 0.28*]
Sodium Phosphate		Phosphoric Acid (45% P₂O₅) 0.44 Sod. Carb. (58% Na₂O) 0.30 Sod. Hydrox. (76% Na₂O) 0.12
Sodium Silicate (40° Be)		Sod. Carb. (Dense, 58%) 0.16 Sand 0.29 Gas (250 kcal) 0.02* W 0.67 L E 0.02
Sodium Sulphate		Natural Brine (10% Sulphate) 10 NaCl (V) Natural Gas 0.16* Salt 0.84 Sulphuric Acid (100%) 0.75 Coal 0.6 [HCl 31.5% 1.58]

Sodium Thiosulphate	‡	(Pentahydrate) Soda Ash 0.43 Sulphur Dioxide 0.26 Sulphur 0.13
Sorbitol (85%)		Dextrose 0.95 Hydrogen 0.16* Ni Catalyst 0.0001 Ac. C 0.001 Resin (V)
Styrene		Benzene 0.87 Ethylene 0.32 Al Chloride 0.01 Ethyl Chloride(s)
Sulphur		Hydrogen Sulphide (100%) 1.18 Air 1.88*
Sulphuric Acid		Sulphur 0.34 W 16.7 L S (from W. H. Boiler) 0.1 E 0.006 A 7.8* (CONTACT)
Terephthalic Acid		p-Xylene 0.68 A (V) Acetic Acid Catalyst(s)
(Dimethyl Terephthalate)		Terephthalic Acid (Tech.) 0.87 Methanol 0.34 p-Xylene 0.67 Methanol 0.40 A (V) Catalyst(s)
Titanium Dioxide (98%)		Ilmenite (50% TiO₂) 2.25 Sulphuric Acid (S.G. 1.84) 4.5 Caustic Soda(s) [Ferrous Sulphate, Sulphuric Acid] Rutile (95% TiO₂) 1.13 Chlorine 0.15 Coke 0.25 Oxygen 0.45
Toluene		Naphtha, Sulphuric Acid, Caustic Soda MEK—(V) depending on feedstock.
Toluene Diisocyanate		4-Tolydiamine 0.88 Phosgene 1.4 Solvent(s) Inert Gas (V)
Trichloroethane (1, 1, 1)	‡	Vinyl Chloride 0.47 Chlorine 0.53 Ferric Chloride(s)
	‡	Vinylidene Chloride 0.73 HCl 0.27 Ferric Chloride(s) Ethane 0.39 Chlorine 2 [HCl 1.1 Ethylene 0.07]
Trichloroethylene		Acetylene 0.22 Chlorine 1.2 Catalyst(s)
Tricresyl Phosphate		Cresol 1.0 Phosphorous Oxychloride 0.49
Urea (60% Solid, 40% Liquid)		Ammonia 2.0 Carbon Dioxide 0.9
Vinyl Acetate		Ethylene 0.35 Acetic Acid 0.7 Pd. Catalyst(s) Acetylene 0.33 Acetic Acid 0.7
Vinyl Chloride		Ethylene Dichloride 1.65 Acetylene 0.44 HCl (Anhyd.) 0.60 Mercuric Chloride 0.0001
Xylene (95%)		Mixed Xylenes (15.8% p-) 18.0 R-make up only [Mixed o- and m- 16.9]
Zinc Oxide		Zinc Metal (Spelter) 0.87 Coal (Anthracite) 0.65 Franklinite Ore (20% ZnO) 5.3 Coal 4.0 E 0.4 Zinc Sulphide (100%) 1.3 Coke 0.85 Fluxes (V)

NOTES: 1. Each Line − Different Process. Raw materials indicative of Process 2. All Numbers − Kg. per Kg. of product, except * (Cu.m.(STP), B = Bbl, L = Litre, OR = as noted. 3. ‡ = Theoretical. [ ] = By-products. m = million. s = small. V = Variable. 4. UTILITIES A = Air. CW = Cooling Water. DW = Demineralised. PW = Process. S(L) = Steam (Low Press). S(H) Steam (High Press). E = Electricity. Kwh. R = Refrigeration.

SECTION 5.   Dust Collectors, Flares; Typical Operating Costs

	Electrostatic					Flares	
	High Voltage	Low Voltage	Liquid Scrubbing	Filter, Fabric	Centri- fugal	Direct After- burner	Catalytic After- burner
Assume: Fan efficiency, 60%; Pump efficiency, 50%							
Operating power, HP/actual cfm							
Low efficiency	0.00019	0.000015	0.0013				
Med. efficiency	0.00026	0.000030	0.0035				
High efficiency	0.00034	0.000040	0.015				
Maintenance cost, $/actual cfm							
Low	0.01	0.005	0.02	0.02	0.005	0.03	0.07
Typical	0.02	0.014	0.04	0.05	0.015	0.06	0.20
High	0.03	0.02	0.06	0.08	0.025	0.10	0.35
Liquid consumption, $/(1,000 gal) (hr)							
Low			0.35				
Typical			0.50				
High			1.00				
Pressure drop, in. H_2O							
Low	0.1	0.1		2.0	0.5	0.5	0.5
Typical	0.5	0.5		5.0	3.0	1.0	1.0
High	1.0	1.0		8.0	4.0	2.0	2.0
Fuel costs, $/(actual cfm) (hr) (50% excess air)							
With heat exchanger						0.00023	0.00014
Without heat exchanger						0.00057	0.00028

Source: Alonso 1970. Excerpted by permission of *Chemical Engineering*. Copyright 1971, McGraw-Hill, NY.

REFERENCES

1. Guthrie, Kenneth M. 1974. *Process Plant Estimating, Evaluation, and Control*. Craftsman Book Co., Solano Beach, CA.
   Hydrycarbon Process. 1967. 46 (11):135.
   Jenckes, L. C. 1970. How to estimate operating costs and depreciation. *Chemical Engineering* (Dec.).
2. Chemical Engineering, ed. and comp. 1973–1974. *Sources and Production Economics of Chemical Products*. McGraw-Hill, New York; 121–180.
3. Kharbanda, O. P. 1979. *Process Plant and Equipment Cost Estimation*. Craftsman Book Co., Solano Beach, CA.
   Ohsol, E. O. 1971. Estimating marketing costs. *Chemical Engineering* (May).
   Smith and DiGregorio 1970. *Chemical Engineering*.

APPENDIX 4

SAMPLE QUESTIONS AND ANSWERS FOR EACH CHAPTER

CHAPTER 1. INTRODUCTION

1-1. What are the principal reasons for an engineer to study economics?
Answer:
   a. Many engineering decisions require an economic input to be the most practical and effective.
   b. In the course of most engineers' careers they will need to consider some, or even frequent, economic matters.
   c. Businesses are economic entities. Consequently, staff promotions are favored for those not only with ability, but also with an economic understanding.
   d. In one's personal life some skill with economics will greatly assist in budgeting, financial management, retirement security, etc.

1-2. Why not have one's supervisor or the company's engineering or financial departments handle all of the economic problems?
Answer: There are too many small or very preliminary economic problems where you need guidance. The expense, time, and authorization trouble would be too great for others to handle them. If such economic information is needed, you generally must obtain it yourself.

1-3. Do some branches of chemical engineering use economics much more frequently than others (i.e., might it not be needed in some areas of work)?
Answer: Basically, no. Perhaps in academic, governmental, or some other fields the economic *demand* will be less, but to do the best job possible, and to advance into management, all chemical engineering fields need a knowledge and practice of economics.

1-4. Does an engineer need to have an MBA (Master of Business Administration) degree to advance into management, or handle economic assignments?
Answer: Decidely no. It is generally true that management appreciates the MBA training, but usually less so than outstanding performance and a general economic knowledge (such as should be obtained from this text). With this background almost all companies will then send promising employees to economic and managerial courses for further training.

CHAPTER 2.   EQUIPMENT COST ESTIMATING

2-1. You are a production engineer in a chemical plant, and your boss asks you for a quick answer on how much a new rotary dryer of a certain size would cost. What do you tell him?
Answer:
   a. If there is time you call a vendor and ask for an immediate, over-the-phone rough estimate.
   b. If there is not time, or you cannot get a vendor quotation, look up the information in Appendix 1. Then be sure to tell your boss that it is a very rough estimate.
   c. Be sure to include the cost of transportation and/or installation (or even auxiliaries) if that is what your boss really wants.

2-2. You are a plant technical service engineer and feel that a certain high-maintenance centrifugal pump should be replaced with one of a different design or material of construction. You call the local representative of a well-known pump manufacturer for prices, knowing only the required flow rate (gpm) and pressure (feet of head). However, he won't tell you anything without being informed of the required NPSH, the pump speed, and the type of seal. What should you do?
Answer:
   a. You might immediately call several more vendors. At least one or two will probably discuss the problem with you and give you advice and price quotations without this extra information.
   b. For your own education you probably should look up or inquire as to the importance of the requested information. NPSH is net positive suction head (i.e., is the pump suction "flooded" from an adjacent tank, have to lift from a sump, under vacuum, etc.?). The impellor speed helps determine cost, output head, abrasion wear, etc. The type of seal is quite critical in determining leakage, shaft wear, etc. Each of these factors can be important in determining the pump performance, and so should be considered. When you have quantified each point you can then recontact that vendor.

2-3. a. What is the cost of a 200-gpm, 80-ft (of water) head, cast iron centrifugal pump (with motor, coupling, and baseplate)? The CE Index is 350.
   b. What is its installed cost?
   c. Module cost?
   d. What is the purchase cost if it were 316 stainless steel?
   e. What if the SS pump operated at 400-psi suction pressure?
   f. What is the size exponent in this range?
Answer:
   a. 80 ft of water is 80 ft head/2.307 ft water per psi (Appendix 5) = 34.68 psi. Therefore, 34.68 psi × 200 gpm = 6,935. Reading from the (Conventional, AVS) chart, $2,420 × 350/320 (CE Index) = $2,647 or rounded-off to $2,600.
   b. Installation factor = 1.3, so 1.3 × 2,647 = 3,440, or rounded-off to $3,400.
   c. Module factor = 3.5, so 3.5 × 2,647 = 9,264, or rounded-off, $9,300.

d. Stainless steel factor = 2.0, so 2,647 × 2.0 = 5,294, or rounded-off to $5,300.

e. The 150–500 psi pressure factor is 1.62, so 5,294 × 1.62 = 8,576, or $8,600.

f. Obtain the curve's slope from slightly smaller and larger pumps, psi × gpm 5,000 = $2,140; psi × gpm 10,000 = $2,800;

$$\frac{\log 2800 - \log 2{,}140}{\log 10 - \log 5} = \frac{3.4472 - 3.3304}{1.0000 - .6990} = \frac{0.1168}{0.301} = 0.39$$

2-4. a. What would be the cost of a mild steel rotary dryer, 6 ft diameter by 40 feet long? The CE Index is 350.

b. Using the size exponent, what would be the cost of a 4 ft diameter by 30 ft long dryer?

c. What would be the installed cost of the larger dryer?

d. The module cost?

e. What if it were a stainless roto-louver dryer?

Answer:

a. Peripheral area: $\pi 6 \times 40 = 754$ ft.^{2} From the chart, $46,200. Correcting for a CE Index of 350:

$$46{,}200 \times \frac{350}{320} = \$50{,}531, \text{ or rounded-off to } \$50{,}500$$

b. Area = $\pi 4 \times 30 = 377$. The size exponent is 0.45 so

$$\$50{,}531 \left(\frac{377}{754}\right)^{0.45} = 36{,}991, \text{ or } \$37{,}000$$

c. Installation factor (average) = 1.64: 50,531 × 1.64 = 82,869 or $82,900.

d. Module factor 2.3, so 50,531 × 2.3 = 116,221, or $116,000.

e. Roto-louvre factor 1.25; stainless factor 2.2, so 50,531 × 1.25 × 2.2 = $138,960 or $139,000.

CHAPTER 3.   PLANT COST ESTIMATES

3-1. You are an R & D engineer who has conceived of a "brilliant" new process idea. After making a very preliminary cost estimate for the concept, what accuracy do you tell your boss the estimate has?

Answer: Use your best judgment of your knowledge and confidence factor over the range ±40–100%. Often it will be in the 50–70% range.

3-2. After an initial budget estimate by a contractor for a very large project, the contractor estimates a ±25% capital cost accuracy. What comment do you give your boss?

Answer: The contractor's claimed accuracy may be true for a very well-defined and studied project, but it is probably low. Remind your boss of the recent average 31% overrun for projects up to $500 million, and 82% overrun for larger projects. Then discuss with him the environmental impact report, community relations, etc. checklist if these factors may be involved.

3-3. You work for the production department of an electronics firm and have been asked to design and estimate the capital (plant cost and total capital requirement) of a plant to treat an acidic and metal containing wastewater. Laboratory tests have shown that reaction with soda ash (Na_2CO_3) is the most effective treatment, producing a water capable of being recycled and a reasonably small amount of sludge that can be hauled to a hazardous waste dump.

You have drawn a flow sheet, made heat (not a factor) and material balances, and sized the equipment. The purchase price (from vendors) of all of the equipment, other than the 40 ft diameter, mild steel thickener and the 1 t/hr, 50 ft^2 rotary vacuum filter and its auxilliaries is $90,000. A building is available to house the bags of soda ash, along with a forklift to transport it. There is adequate utility serviced land available to locate the plant. The CE Index is 350. Show your detailed capital estimates.

Answer: First estimate the total purchase price of the plant equipment from Appendix I.

Vendor price quotations	$90,000
Thickener, 40 ft: chart $88,000 (for a concrete tank); 0.7 factor for steel; 0.7 × 350/ 320 = $67,375	
	67,000
Filter, rotary vacuum, 50 ft^2 chart = $40,000 × 350/320 = 43,750	44,000
Filter auxilliaries (vacuum pump, liquid receivers, etc.), 50%	22,000
Total purchased cost	$223,000

Next determine the appropriate multipling factors from Table 3-4.

Total equipment = $223,000	1.00
Piping	0.40
Electrical	0.15
Instrumentation	0.30
Utilities	0.30
Foundations, structure	0.07
Insulation	0
Painting, safety	0.10
Yard Improvements	0.10
Environmental	0
Buildings	0
Land	0
Subtotal	2.42
Construction, engineering	0.40 (16.5% of subtotal)
Contractors fee	0.25 (10% of subtotal)

Contingency	0.25 (10% of subtotal)
Total plant cost	$3.32^a \times 223,000 =$ $740,000
Off-site facilities	0
Plant start-up	10
Working capital	10
Total capital required	$1.20 \times 740,000 =$ $888,000, or round off to $900,000

a. This total factor is low for a mixed processing plant, but it is appropriate for a rather small addition to an existing facility.

3-4. In plant cost estimating when do you use installation factors, module factors, and direct equipment values from the charts?

Answer: Use installation factors when dealing with equipment replacement installations, or when one or a limited number of pieces of simple equipment is involved. Module factors may be needed with plant additions or modifications when a major piece (or small groups) of equipment and all of its support equipment are required. Purchased equipment cost directly from the charts is used alone in all other cases.

3-5. In plant cost estimating do you ever make subtotals of some of the cost? If so, why?

Answer: Yes. It is easier to visualize and estimate the contractors' charges and the contingency based upon the total plant cost before these charges. For rapid estimating work no such subtotal should be taken, but to more accurately analyze (and estimate) the contractor charges and contingency on many projects a subtotal plant cost is highly desirable.

3-6. When do you use a "Lang" or overall multiplier for plant cost estimating?

Answer: Only for the simplest and most rapid and preliminary estimates. A detailed factor breakdown is almost always warranted. However, in *all* cases, your total factor should be checked against the normal overall (a Lang-type) factor.

3-7. When do you use complete plant estimating charts?

Answer: Whenever the desired estimate is the same or similar to the chemical production shown on one of the plant charts. Even though not too accurate, they are probably closer than you could estimate from any but the most knowledgeable and detailed flow diagram.

3-8. Do "cost per ton of product" or "capital ratio" estimates have much value in plant cost estimating?

Answer: Generally no, but with certain types of plants the factors are in surprisingly common use. Power plants are usually quoted as $/installed kW, some paper mills as $/ton of pulp, etc. When close in size to a known plant the numbers may be useful, but since most plant costs increase as some power of plant size (such as 0.64), these numbers are only of value over a limited size range.

3-9. Are the "other components," or auxilliary costs of capital estimates always involved?

Answer: In many cases of single equipment replacement or additions, and in simple plant modifications, no. However, on new facilities or modifications of any appreciable size, start-up costs and working capital must be considered. Off-site facilities, distribution facilities, R & D, etc. are generally very specific to any project. Sometimes there is no cost involvement, but just as often it can be modest to extensive.

CHAPTER 4. MANUFACTURING COST

4-1. If a plant normally produces 100 t/d of product, and the on-stream efficiency (OSE) or operating rate is 90%, how many tons are produced per year?
Answer: 100 t/d × 365 d/yr × 90% OSE = 32,850 t/yr (or in more correct round numbers, 33,000 t/yr)

4-2. If your plant in Los Angeles wishes to purchase sulfuric acid, and the *Chemical Marketing Reporter* correctly states that 93% H_2SO_4 costs $20/ton, 100% basis, FOB Arizona, tank car lots, what is the price delivered to your plant?
Answer: You must first determine the "commodity" freight rate (a special, low freight rate that the railroad can establish for any commodity moved in large tonnages) for sulfuric acid. Either the smelter company or the railroad can tell you this rate. Let's say that it is $30/ton from the smelter to your plant. This makes the price:

$$\$20 \text{ (purchase price)} + \frac{30}{0.93} \text{ (freight)} = \$52.26/\text{ton}$$

of 100% H_2SO_4 or 20 × 0.93 + 30 = $48.60/ton of actual 93% H_2SO_4, delivered by rail (in 100-ton capacity cars) to your plant.

4-3. What would be the yearly electricity cost for a plant that had a total of 2,000 HP of installed motors? Assume as a first approximation that they are always running and drawing their full load amperage (this is usually a considerable exaggeration on both counts). Also assume that electricity costs $0.07/kW hr, and that there is a 90% OSE, with all motors off during the downtime (also an exaggeration).
Answer: 2,000 HP × 365 d/yr × 24 h/d × 0.07 $/kW hr × 0.746 kW hr/HP (Appendix 5) × 0.90 OSE = $823,405/yr (or rounded-off to $823,000).

4-4. What would be the annual operating labor cost for a continuously operating plant that needed six operators per shift, and the average labor cost was $12.00/hr?
Answer:
a. If 5 shifts were employed, and each man were paid for 40 hr/wk: 6 men × 5 shifts × $12.00/hr × 40 hr/wk × 52 wk/yr = $748,800/yr (or rounded-off to $749,000/yr).
b. If 4 shifts were employed, and as above it was assumed that they were paid (and working) during the plant downtime: 6 men × 4 shifts × 12.00/hr × 40 hr/wk × 52 wks = $599,040 for their normal hours. However, they need to work 365 × 24 = 4 shifts (52 weeks − 29 days normally off) hr/wk; average hr/wk = 45.76, or 5.76 hr/wk overtime at time and one-half pay: 6 men × 4 shifts × $12.00/hr × 1.5 overtime premium × 5.76 hr/wk OT × 52 = $129,393. The total pay is thus 599,040 + 129,393 = $728,433/yr (or rounded-off to $728,000/yr). Either answer (a). or (b). would be acceptable, just so that the basis was stated.

4-5. For the wastewater treating plant of Problem 3-3 estimate the manufacturing cost. There is a nearby facility to provide worker safety, and the supervisor and service staff can be shared. Your equipment sizing is based upon an anticipated 3 shift/day, 365 d/yr operation with a 90% on-stream efficiency (equivalent to 330 d/yr). There is adequate water storage capacity for the anticipated downtime. Your flow sheet shows a total of 80 HP motors (electricity costs $0.09/kW hr), and the skip loader will consume 1,000 gal/yr of diesel fuel, costing $0.80/gal (the diesel tank is all ready available). Tests indicate a coagulant consumption of 5 ppm of fluid in the thickener, and the coagulant will cost $8/lb. Assume that the equivalent sales value of the "product" is that of a capital or turnover ratio of 1.0. Bagged soda ash cost $131.00/ton (CMR 2-22-88), and freight is $19/ton. It will be mixed on a batch basis to form a 25% Na_2CO_3 solution weighing 10 lb/gal and used at 3 gpm. Operating labor costs average $12.00/hr. The wastewater enters at a rate of 100 gpm.

Answer: First calculate the raw materials, utilities, and operating labor required.

Raw materials: (7920 hrs/yr)

a. Soda ash: $\dfrac{3 \text{ gpm} \times 10 \times 0.25}{2000} \times 330 \times 24 \times 60 = 1,782$ t/yr; $150/ton =

$267,300/yr

b. Coagulant: $(100 \times 8.34 \times 60 \times \dfrac{7920 \times 5}{10^6} = 1,982$ lb @ $8/lb =

$15,853/yr

Sub/total $283,200

Utilities:
  a. Electricity: $80 \times 0.746 \times 7,920 \times 0.09 = 42,500$
  b. Diesel: $1,000 \times 0.80$ $= \underline{\quad 800}$

Sub/total $43,300

Labor: Assume 2 men, 5 shifts
    $2 \times 5 \times 52 \times 40 \times 12 =$ $249,600

Next calculate the factored manufacturing costs from Table 4-4

Labor related costs:
Payroll overhead	40%	
Supervisory, miscellaneous	10	
Laboratory charges	10	
Sub-total	60% =	$149,800

Capital related costs; ($740,000 plant cost + 74,000 start up = 814,000 capital cost)
Maintenance	6%	
Operating supplies	1.5	
Environmental	1.5	
Depreciation	10	
Local taxes, insurance	4	
Plant overhead	3	
Sub-total	26.0% =	$211,800

Sales related costs: (Assume capital ratio = 1)

Patents, royalties	0	
Packaging	0	
Distribution and sales	0	
Administration	10	
R & D	0.5	
Sub-total	10.5 =	$86,200
Total Manufacturing Cost		$1,023,900/yr

Total gal treated $= 100 \times 8.34 \times 7920 \times 60 = 3.963 \cdot 10^8$ gal or $2.58/1,000 gal treated

CHAPTER 5. INTEREST CALCULATIONS; PRELIMINARY PROJECT EVALUATION

5-1. Using the capitalized cost procedure compare the merits of replacing an existing pump that costs $2,000, needs replacement every 5 years, and requires $300/yr in maintenance expense with a new one that costs $4,000 but would have a 10-year life and only need $100/yr in maintenance. Assume a 10% simple, annual interest rate basis.

Answer: Let's calculate this problem in two parts: (1) the capitalized cost based upon the purchase price alone, and then add (2) the present value of the maintenance cost calculated as an annuity.

Present pump capitalized cost:

$$\$2,000 \left(\frac{1 + 1}{(1 + 0.1)^5 - 1} \right) = \$2,000 \left(\frac{1 + 1}{1.6105 - 1} \right)$$

$$= \$2,000(1 + 1.638) = \$5,276$$

Present pump maintenance (present worth of annuity)

$$\$300 \frac{(1 + 0.1)^5 - 1}{0.1 (1 + 0.1)^5} = \$300 \left(\frac{1.6105 - 1}{0.1 \times 1.6105} \right) = \$1,137$$

Old pump total capitalize cost:

$$\$5,276 + \$1,137 = \$6,413$$

New pump capitalized cost

$$\$4,000 \left(\frac{1 + 1}{(1 + 0.1)^{10} - 1} \right) = \$4,000 \left(\frac{1 + 1}{2.5937 - 1} \right)$$

$$= \$4,000(1 + 0.6275) = \$6,510$$

New pump maintenance

$$\$100 \frac{(1 + 0.1)^{10} - 1}{0.1 (1 + 0.1)^{10}} = \$100 \frac{1.5937}{0.25937} = \$614$$

New pump total capitalized cost:

$$\$6,510 + \$614 = \$7,124$$

The present pump is thus seen to be the most economical. However, at a lower interest rate they become more nearly equal, and if the value of leakage and downtime (if this is a factor) are considered, then the more expensive pump may actually be the better choice.

5-2. You work for a company which is considering building a new plant which would cost $10,000,000, including off-site installations and start-up expense. The working capital would be $1,000,000 and the depreciation $1,000,000/yr. Assume no salvage value. What would be the:
   a. ROI (return on investment)?
   b. Payback period?

   Answer:

   a. $\text{ROI} = \dfrac{1,000,000 \text{ after-tax profit}}{10,000,000 \text{ plant cost} + 1,000,000 \text{ working capital}}$

   $\text{ROI} = \dfrac{1}{11} = 9.09\%$

   b. $\text{Payback period} = \dfrac{10,000,000 \text{ plant cost}}{1,000,000 \text{ after-tax profit} + 1,000,000 \text{ depreciation}}$
   $= 5 \text{ years}$

CHAPTER 6. PROFITABILITY ANALYSIS; DISCOUNTED CASH FLOW

6-1. *DCF Calculation (with charts).* The same plant noted in Problem 5-2 is now expected to make $500,000 after-tax profit the first year, $600,000 the second, $800,000 the third, $900,000 the fourth, and $1,000,000/yr thereafter for its 10-year life. It should have a $500,000 salvage value. As a simplification still assume a $1,000,000/gr depreciation rate. Calculate its DCF by the charts.

   Answer:
   First prepare a calculation table, list the yearly cash flows, guess at an initial interest rate, and look up the discount factors. Then make trials until the present value of all cash flows is zero.

Year	Cash Flow	Table	Assume 11% Interest Discount Factor	Present Value	Assume 12% Interest Discount Factor	Present Value
0	−11.0 MM	—	1	−11 MM	1	−11 MM
0–1	1.5	6-2	0.9470	1.421	0.9423	1.414
1–2	1.6	6-2	0.8483	1.357	0.8358	1.337
2–3	1.8	6-2	0.7600	1.368	0.7413	1.334
3–4	1.9	6-2	0.6808	1.294	0.6574	1.249
4–10	2.0 × 6	6-1 and	0.664 × 0.7320 = 0.4714 × 12	5.657	0.6188 × 0.7128	5.293

			Assume 11% Interest		Assume 12% Interest	
Year	Cash Flow	Table	Discount Factor	Present Value	Discount Factor	Present Value
		6-4			= 0.4411[a] × 12	
10	1.5	6-1	0.3329	0.499	0.3012	0.452
Total present value				0.596		0.079

Extrapolate: PV Δ 11-12% 0.596 − 0.079 = 0.517
0.079/0.517 = 0.15; DCF = 12.15

a. If solved year-by-year (Table 6-2)

4-5	0.5831
5-6	0.5172
6-7	0.4588
7-8	0.4069
8-9	0.3609
9-10	0.3201
avg. =	0.4412

6-2. If the estimated total capital (including $1,000,000 working capital) for a project was $10,000,000, the after-tax profit constant at $500,000/yr, and the project life 10 years, what would be the DCF? use both Figure 6-1 and annuity-type calculations.

Answer:

a. Figure 6-1. The depreciation rate is $900,000, thus the cash flow is $1,400,000/yr. $C_0/C_n = 10/1.4 = 7.14$. Reading Figure 6-1, the DCF is about 6.5%.

b. By annuity calculations the following equation must balance:
Equation 5-13 P/C, or Figure 6-1 Co/Cn:

$$\frac{10}{1.4} = \frac{(1 + i)^n - 1}{i(1 + i)^n}$$

First assume $i = 6\%$.

$$\frac{(1 + 0.06)^{10} - 1}{0.06 (1 + 0.06)^{10}} = \frac{1.7908 - 1}{0.06 \times 1.7908} = \frac{0.7908}{0.10745} = 7.36$$

Next assume $i = 7\%$.

$$\frac{(1.07)^{10} - 1}{0.07 \times (1.07)^{10}} = \frac{1.9672 - 1}{0.13770} = 7.02:$$

Extrapolate between these two answers: Actual $C_0/C_n = 10/1.4 = 7.14$, so 7.36 − 7.02 = 0.34; 7.14 − 7.02 = 0.12; 0.12/0.34 = 0.35 or DCF = 6.65%.

6-3. If the above problem were recalculated, but now also including the return of working capital at the end of year 10 (the realistic case), what would be the DCF value (use equations only)?

Answer: The working capital's present value would be calculated as an instantaneous cash flow by means of Equation (6-2). Thus $C_{wc} = 1,000,000/(1 + i)^{10}$. Assume $i = 7\%$, $1,000,000/(1.07)^{10} = \$508,337$; If $i = 8\%$, $C_{wc} = \$463,200$. The sum of the cash flows @ 7% (calculated in 6-2) was $0.9672/0.1377 \times 1,400,000 = \$9,833,600$, making the total $508,300 + 9,833,600 = \$10,341,900$. This is more than the original \$10,000,000 investment. Recalculating the yearly cash flows at 8% gives

$$\frac{(1 + 0.08)^{10} - 1}{0.08 (1 + 0.08)^{10}} = \frac{1.08^{10} - 1}{0.08 \times 1.08^{10}} = \frac{1.1589}{0.08 \times 2.1589} = 6.71$$

$6.71 \times \$1,400,000 = \$9,394,000$; plus $\$463,200 = \$9,857,200$

Extrapolating between these two interest rates: $10,341,900 - 9,857,200 = 484,700$, or $341,900/484,700 = 0.71$, so DCF $= 7.71\%$.

6-4. *Sensitivity analysis.* You are studying a potential new process where the plant cost (including off-site equipment and start-up) is estimated to be \$50 MM. The yearly manufacturing cost calculates to be: raw materials, \$12 MM; utilities, \$5 MM; operating labor and related costs, \$5 MM; capital related costs, \$13 MM (including depreciation); and sales related costs, \$12 MM. Sales are hoped to be \$60 MM/yr. Assume a 10-year project life and depreciation period; working capital at 20% of the manufacturing cost; 40% taxes, and no salvage value.

a. Calculate the DCF.
b. Determine the DCF at ½ and 2 times the plant (production) size. Assume that all of the product can be sold, labor costs stay constant, all other costs change proportionally with plant capacity, and the capital size exponent is 0.6.
c. Determine the DCF at 10% higher and 10% lower sales price (than case a).
d. Determine the break-even production rate (for case a) and the rate at zero cash flow. Assume that only raw material and utility costs vary in proportion to the amount of sales.
Answer:

a.

Manufacturing Cost		Period	Cash Flow. $ MM	
Raw materials	$12 MM		− $50	plant cost
Utilities	5		− 9.4	working capital (20% of
Labor related costs	5	0	−59.4	$47MM)
Capital related costs	13 (26 of capital)			
Sales related costs	12 (20% of sales)	0–9	12.8 cash flow	
Total mfg. cost	$47 MM	10	22.2 (12.8 + 9.4) C.F.	
Gross profit: 60–47	$13 MM		DCF = 17.96% by a hand-held	
			calculator	
Tax (40%)	5.2		(annual compound interest)	

a. *Manufacturing Cost*   *Period*   *Cash Flow.$ MM*

Net after-tax profit	7.8
Cash flow	12.8 MM

b. $2^{0.6} = 1.516$;

	Plant size $\times 2$	$\times 1/2$		*Plant size* $\times 2$		$\times 1/2$
Plant cost	$75.79 or	32.98 MM	Gross profit			$30 − 28.07 =
			$120 − 82.71 = 37.29 MM			1.93MM
Raw materials	24	6.	After-tax profit	22.37	1.16	
Utilities	10	2.5	Depreciation	7.58	3.30	
Labor related costs	5	5	Cash flow	29.95	4.46	
Capital related costs	19.71	8.57	Working capital	16.54	5.61	
Sales related costs	24	6	10th year C.F.	46.49	10.07	
Total mfg. cost	82.71	28.07	Total Capital	92.33	38.59	
			DCF	30.57%	4.65%	

c. *Sales Change*   − 10%   + 10%

	− 10%	+ 10%
Sales	$54 MM	$66 MM
Nonsales related operating cost	35	35
Sales related	10.8	13.2
Total mfg. cost	45.8	48.2
Gross profit	8.2	17.8
After-tax profit	4.92	10.68
Cash flow	9.92	15.68
Working capital	9.16	9.64
10th year C.F.	19.08	25.32
Total capital	59.16	59.64
DCF	11.94%	23.59%

d. Let X = % of normal production
(1) Break even (zero profit)

Raw	Sales		Capital	Sales
Matl.	Util.	Related Labor	Related	Value

$(12 + 5 + 12)X + 5 + 13 = 60X$

$31X = 18; X = 0.58$, or
break even = 58% of rated capacity
(2) Zero cash flow (i.e. no depreciation or profit)

$29 X + 13 = 60X$;
$31X = 13; X = 0.42$, or

zero cash flow occurs at 42% of rated capacity.

CHAPTER 7.   ECONOMY OF THE CHEMICAL INDUSTRY

7-1. Based upon Table 7-4, in 1986 for the 18 listed chemical companies what was the:
  a. Percent that the average chemical plant was depreciated?
  b. Average debt-to-equity ratio?
  c. Average fraction of income spent on stock dividends?
  d. Average fraction of income spent on new capital additions?
  Answer:
  a. 50.0%.   b. 33.2/66.8 = 49.7%.
  c. 44.6%.   d. 7.6/5.5 = 138%.

7-2. Based upon Figures 7-3, 7-4, and 7-6, what was the 1976–1986 average chemical industry:
  a. Return on equity?
  b. Return on sales?
  c. Operating rate (1983–1987)?

Answer:
a. About 12%.
b. About 5.5%.
c. About 78%.

7-3. What were some of the major factors resulting in declining profitability for the U. S. chemical industry over the past 20 years?
Answer:
a. Increasing foreign competition.
b. Overcapacity caused by oil companies (and others) entering the market with large plants.
c. Higher energy prices.
d. Nonprogressive and high overhead management.
e. Greatly reduced acceptance of new innovations.

7-4. What are some of the factors that the U. S. chemical industry is pursuing in an attempt to increase profitability?
Answer:
a. Cost cutting with layoffs, lower overhead, and decentralization.
b. Divestitures, acquisitions, and mergers.
c. Strengthening existing production.
d. Moving to higher value-added products.
e. Increasing foreign trade and diversification.

7-5. What are some of the worrisome aspects of current CPI activities?
Answer:
a. With many companies there is much more attention to mergers, acquisitions, and divestitures than there is to strengthening their basic production.
b. Foreign competitors and financial groups are acquiring a large segment of the U. S. CPI.
c. There is an overemphasis on value-added products and not enough capital spending on more economical basic commodities, new R & D developments and plant improvements.

CHAPTER 8. ACCOUNTING AND BUDGETS

8-1. What is the difference between cash and accrual accounting?
Answer: In cash accounting entries are only considered for income tax purposes when bills are actually paid and the funds from sales are received. When using the accrual basis debt is considered to have occurred (for tax purposes) when the obligation is incurred (the purchase made, etc.), and sales are credited when the shipment is made.

8-2. What is cost accounting?
Answer: This is the name given to the accounting procedures that establish manufacturing or production costs. It usually implies their breakdown into a larger number of subaccounts, and the allocation of costs between various divisions, processes, and products.

8-3. On many operating statements costs such as sales, legal, accounting, R&D, etc. are shown as fixed (and not controllable) expenses. Why is this?

Answer: For large companies the above items, and many more, are part of the corporate budget, and are distributed to the operating units as fixed costs for each year. The production plants may also have some of these charges on their budgets for their own specific use (such as R & D for that plant alone, in addition to its fixed share of the corporate R & D), and these then become controllable charges. For more decentralized divisions or small plants these same costs become entirely under the plant's jurisdiction, and thus are more discretionary. The tendancy for efficient management is to have more locally controllable costs, and a reduced corporate fixed "G&A" (general and administrative) charge.

8-4. Why is accurate cost allocation to individual products so important?
Answer: Most plants produce a multiplicity of products, and there are numerous shared costs between them. When competition is severe (as it usually is), only with accurate cost allocation can the management know exactly where they stand on the profitability and pricing of each product.

8-5. What are some of the computer-assisted plant management tools that are currently available?
Answer: The PMS, CIM, MRPII, etc., programs to assist with scheduling, inventory control, production efficiency, etc. in plant management.

8-6. What effect might the "just-in-time" inventory management system have in the CPI?
Answer: For most chemical production it is difficult to operate with the minimum raw material inventory, and to depend upon prompt deliveries. When it can be done, however, it would assist profitability by decreasing inventory and working capital. In most cases, however, some customers use the "just-in-time" method, and this requires more CPI inventory to meet their needs.

CHAPTER 9. CORPORATE ANNUAL REPORTS

9-1. On the balance sheet of Table 9-1 what is meant by:
  a. Other current assets?
  b. Other assets?
  c. Other current liabilities?
  Answer:
  a. Other current assets are usually investments of capital reserves that can be fairly quickly converted into cash, such as stock or bonds.
  b. Other assets include investments and intangibles, such as investments in land, real estate, other ventures, etc. that cannot be quickly sold.
  c. Other current liabilities covers all of the otherwise unlisted liabilities, such as monies owed to pension funds, bonuses, profit sharing, prefered stock, dividends, etc. It is usually a large number.

9-2. On the hypothetical income statement of Table 9-2, what was the 1986:
  a. Sales and G & A expense (as % of sales)?
  b. R & D expense (as % of sales)?
  c. Income tax percentage of income before taxes?
  d. By CPI standards is this a well-run company?
  Answer:

a. $120,000/800,000 = 15\%$
b. $10,000/800,000 = 1.25\%$
c. $20,000/81,500 = 24.5\%$
d. No. The G & A is high, R & D low, and the income tax payment relatively high for such a small company.

9-3. In examining the "Cash Flow" section of Table 9-1, and the "debt ratios," what conclusions can be reached concerning acquisitions and divestitures?
Answer: The "other internal sources" figures for the source of funds, and "other applications" spending indicate an unusually high amount of selling company operations (source of funds) and purchasing others (application). At the same time the debt-to-equity ratio increased from about 0.5 to 0.6 over the period of this table. It again indicates acquisitions, since the capital expenditures were modest compared to cash flow and dividends.

CHAPTER 10. PROJECT MANAGEMENT

10-1. Why is defining the scope of work on a project so important?
Answer: The scope of work first of all provides official management approval for the project and its detailed execution. It provides the basis for coordinating the activities of all of the company and outside staffs working on the project, and allows detailed instructions and schedules to be made for all of the work each group will perform. If adequately discussed, and formulated with each group's input and approval it provides the basis for a cooperative and successful project.

10-2. You are an engineer at a gas purification plant, and have been asked to be the project manager for the installation of a foam separator–sulfur melter at its Stredford sulfur-removal operation. A contractor has done all of the engineering, design, permitting, and procurement, but your company will do the installation with your own staff.
a. Prepare a simple bar chart schedule for the installation, making your best guess of manpower, activity duration, and costs. Perform the job as rapidly as possible, allowing 2 weeks for tie-ins. The site is at present adequately prepared, and there is room for the new equipment.
b. Prepare a budgeted expenditure curve based upon the information established in (a) above (do not include the contractor's cost, i.e., equipment, permits, engineering, etc.).
c. Prepare a critical path bar chart *and* flow sheet layout for the project. What is the critical path?
d. State how (c) above can be "crashed" to reduce the elapsed time by 20%. Estimate the extra cost to do this. Did it change the critical path?
e. Chart the labor requirement for (c) above, and then perform the maximum manpower leveling that will not raise costs or delay the project.
Answer: First break the project into as many subdivision tasks as appears to be appropriate. (A limited number will be used in this book's answer to simplify the discussion.) Then estimate the sequence, time, manpower, and cost:

a. Project Sub-Tasks

Job	Required Time	Manpower	Labor	Other	Total	Job Sequence
1. Site preperation, foundations	1 mo	4 men	$12 M	$13 M	$15 M	0–1
2. Yard paving, sumps, underground facilities	2 wk	2 men	3	2	5	1–2
3. Set the equipment	2 wk	4 men	6	3	9	2–3
4. Platforms	3 wk	2 men	4.5	1.5	6	3–4
5. Install piping	1 mo	8 men	24	6	30	3–5
6. Install electrical	1 mo	4 men	12	4	16	3–6
7. Install instruments	2 wk	2 men	3	1	4	6–7
8. Install insulation	2 wk	2 men	3	2	5	5–8
9. Painting	3 wk	2 men	4.5	4.5	9	3–9
10. Cleanup, lables, etc.	2 wk	2 men	3	1	4	7, 8–10
11. Testing, tie-in	2 wk	6 men	9	3	12	7, 9, 10–11

Total $115,000
Plus overhead 100%: $230,000

a. Project Bar Chart Time Schedule

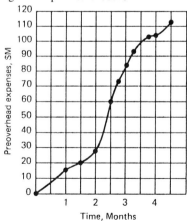

Sequence	Job	Months	0	1	2	3	4
0–1	1	Foundations					
1–2	2	Paving, etc.					
2–3	3	Set equipment					
2–4	4	Install: platforms					
3–5	5	piping					
3–6	6	electrical					
6–7	7	instruments					
3, 5–8	8	insulation					
3–9	9	Painting					
7, 8–10	10	Cleanup					
7, 9, 10–11	11	Tie-in					

b. Budgeted Expenditure Curve

c. Critical Path (Bar) Chart

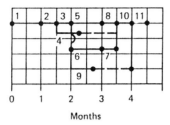

Months

Critical Path (Flow Sheet) Chart

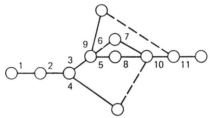

Critical path 1, 2, 3, 5, 8, 10, 11   4½ mo = 18 wks

d. 20% of 18 wk = 3.6 wk. Reduce job 1 by 2 wks, Estimated cost increase: about $1.5 M. Reduce jobs 2 and 10 by 0.3 weeks: Estimated cost $1.0M Reduce jobs 5 and 6 by 1 week; cost: 3.0 M; Total est. cost increase $5.5M (could be much more)

No change in critical path

e. Manpower Leveling

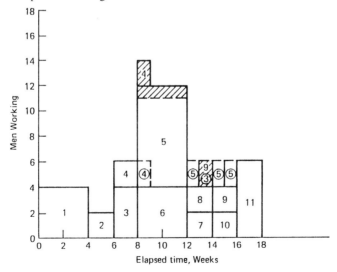

Elapsed time, Weeks

Numbers = jobs; circled numbers = jobs moved to new times;
shaded periods = men no longer working (moved to new times);
dotted lines = new job periods

10-3. What is a:
   a. PERT program?
   b. Decision tree analysis?
   Answer:
   a. A critical path program with statistical probabilities considered for the likelihood of each job duration.
   b. A form of sensitivity analysis in which all of the possible occurrences are charted to their ultimate conclusion, with present value or DCF calculations made on all of the possible possible variations.

CHAPTER 11.   PERSONAL INVESTING

11-1. You are a process engineer in a small but prosperous firm that is planning a new $10 million expansion to start one year from now, and take one year to complete. Present retained earnings plus the cash flow generated during this period will be adequate for the total financing. Your boss has asked you to outline a general investment strategy for your present funds and future cash flow to safely attempt to earn the maximum possible income until the money is needed. Please note the types of investments you recommend, and your reasons for these choices.
   Answer: Suggested investment strategy:
   a. Must have reasonable diversification, safety, and liquidity.
   b. One possible group of investments might be:
      (1) Maintain some cash (i.e., about $1 MM). This could be a money market account, but preferably short-term (30–90 day) commercial paper or CDs in a strong, sound bank.
      (2) If interest rates appear to be stable or declining, the majority of funds (i.e., about $7 MM) should be placed in a high yielding, no-load, utility stock market mutual fund with switching privileges. Interest rates must be watched carefully, and if they appear to have a possibility of rising, quickly switch to a money market fund.
      (3) Place the remaining funds in a high yielding, no-load, municipal (tax-free) bond fund, again with switching privileges so that it may be sold at any firm indication of a rise in interest rates.
      (4) Alternately to the above, with uncertain interest rates, and depending upon the economy, some investments may be placed in any or all of stocks, precious metals, overseas funds, etc. Your own company's stock might be a good investment.
11-2. a. You are single and earn $28,000/yr. How much of an IRA contribution may you deduct from your income taxes?
   b. You are married and your combined income is $49,500. How much may each person deduct from taxes for their IRA contribution?
   Answer:
   a. The single taxpayer may deduct $2,000/yr from his taxes if his income is less than $25,000, and zero if over $35,000. Thus, with $28,000 income the deduction may be $2,000 \times (7,000/10,000) = \$1,400$. Of course, the

remaining $600 may be put into the IRA account with no (actually, deferred) taxes on earnings, but not deducted from that year's income for tax purposes.

b. Married couples can each deduct $2,000 in IRA payments from their taxable income if their combined earnings do not exceed $40,000, and zero if over $50,000. The minimum deduction for married and single taxpayers is $200. Therefore, with $49,500 combined earnings, the calculation would indicate: $2,000 (500/10,000) = $100 deduction. However, the minimum deduction is $200, so this is the amount they may each utilize.

11-3. For a long-term investment in a utility stock paying originally 8% dividends, and with a dividend reinvestment program, does the stockholder have a poor investment when interest rates rise to 15%

Answer: No. The utility stock at that time will probably only be worth about $^8/_{15}$ of its original value, but the number of dollars earned each quarter is fixed (actually growing through the quarterly dividend compounding), so with dividend reinvestment one receives more shares ($^{15}/_8$) for each dividend. If the stockholder can wait until the stock value is back to its original price (or higher) before selling, he will have essentially received the always-current interest rate over the holding period, plus any increase in the dividend (many utilities increase it every year), and of course, the appreciation from compounding.

CHAPTER 12. EMPLOYMENT CONSIDERATIONS

12-1. How does the AIChE job referral service work?

Answer: It is an electronic data base that attempts to match participating members' eduction and experience with the requirements of jobs that AIChE knows are or may become open. If job and applicant match up the member will be given the name of the appropriate contact at each company. The employer will also be given the applicant's name and address if he or she wishes.

12-2. Is a multipage, elaborate resume worthwhile?

Answer: Generally no. If the resume is too pretentious it is usually a mark against the applicant. Perhaps an exception would be when applying for a managerial job, where more detail is generally needed. At that level, or with sales or other contact positions, the videotape resume is also becoming increasingly popular. It may well become commonplace for all professional positions in the coming years.

12-3. When applying for a position at an electronics (or other non-CPI) firm, what skills and areas of expertise might you list?

Answer: Many industries, such as electronics, do some (or considerable) chemical handling and processing (electroplating, etching, cleaning, semiconductors, etc.) and miscellaneous fluid flow (piping systems, clean-rooms, etc.), heat transfer (cogeneration, heating, cooling, etc.), mass transfer (scrubbers, purification, etc.), computer process control, environmental control, and other chemical engineering operations. They may not recognize each of the areas as being within the ChE training area, but they have been increasingly hiring chemical engineers to help with their problems.

12-4. Are interoffice memos really of much importance to an engineer's career?
Answer: Yes, they are one of the means by which management several layers above your boss, or in other divisions, on occasion review your work. Good memos definitely stand out, and their authors are remembered. Often you will read older memos in the file to study a problem and develop a real appreciation and respect for certain authors. They may by then be in higher positions with your firm, or with other companies, and their old memos can make a very interesting bond between you when you meet.

APPENDIX 5

CONVERSION FACTORS

Multiply	By	To Obtain
Atmospheres	76.0	Cms of mercury
Atmospheres	29.92	Inches of mercury
Atmospheres	33.90	Feet of water
Atmospheres	1.0333	Kgs sq cm
Atmospheres	14.70	Lbs sq inch
Atmospheres	1.058	Tons sq ft
Barrels—Oil	42	Gallons-Oil
British Thermal Units	0.2520	Kilogram-calories
British Thermal Units	777.5	Foot-lbs
British Thermal Units	3.927×10^{-4}	Horse-power-hrs
British Thermal Units	107.5	Kilogram-meters
British Thermal Units	2.928×10^{-4}	Kilowatt-hrs
B.T.U. min.	12.96	Foot-lbs. sec.
B.T.U. min	0.02356	Horse power
B.T.U. min	0.01757	Kilowatts
B.T.U. min	17.57	Watts
Centares (Centiares)	1	Square meters
Centigrams	0.01	Grams
Centiliters	0.01	Liters
Centimeters	0.3937	Inches
Centimeters	0.01	Meters
Centimeters	10	Millimeters
Centimtrs. of Mercury	0.01316	Atmospheres
Centimtrs of mercury	0.4461	Feet of water
Centimtrs of mercury	136.0	Kgs sq meter
Centimtrs of mercury	27.85	Lbs sq ft
Centimtrs of mercury	0.1934	Lbs sq inch
Centimeters second	1.969	Feet min
Centimeters second	0.03281	Feet sec
Centimeters second	0.036	Kilometers hr
Centimeters second	0.6	Meters min
Centimeters second	0.02237	Miles hr
Centimeters second	3.728×10^{-4}	Miles min
Cms. sec. sec.	0.03281	Feet sec. sec.
Cubic centimeters	3.531×10^{-5}	Cubic feet
Cubic centimeters	6.102×10^{-2}	Cubic inches
Cubic centimeters	10^{-6}	Cubic meters
Cubic centimeters	1.308×10^{-6}	Cubic yards
Cubic centimeters	2.642×10^{-4}	Gallons
Cubic centimeters	10^{-3}	Liters
Cubic centimeters	2.113×10^{-3}	Pints (liq.)
Cubic centimeters	1.057×10^{-3}	Quarts (liq.)

Multiply	By	To Obtain
Fathoms	6	Feet
Feet	30.48	Centimeters
Feet	12	Inches
Feet	0.3048	Meters
Feet	1/3	Yards
Feet of water	0.02950	Atmospheres
Feet of water	0.8826	Inches of mercury
Feet of water	62.43	Kgs sq m
Feet of water	0.3048	Lbs sq foot
Feet of water	0.4335	Lbs sq inch
Feet min.	0.5080	Centimeters/sec.
Feet min.	0.01667	Feet/sec.
Feet min.	0.01829	Kilometers/hr.
Feet min.	0.3048	Meters/min.
Feet min.	0.01136	Miles/hr.
Feet sec. sec.	30.48	Cms. sec. sec.
Foot-pounds	1.286×10^{-3}	British Thermal Units
Foot-pounds	5.050×10^{-7}	Horse-power-hrs
Foot-pounds	3.241×10^{-4}	Kilogram-calories
Foot-pounds	0.1383	Kilogram-meters
Foot-pounds	3.766×10^{-7}	Kilowatt-hrs
Foot-pounds min.	1.286×10^{-3}	B T Units min
Foot-pounds min	0.01667	Foot-pounds sec
Foot-pounds min	3.030×10^{-5}	Horse-power
Foot-pounds min	3.241×10^{-4}	Kg calories min
Foot-pounds min	2.260×10^{-5}	Kilowatts
Foot-pounds sec.	7.717×10^{-2}	B T Units min.
Foot-pounds sec	1.818×10^{-3}	Horse-power
Foot-pounds sec	1.945×10^{-2}	Kg calories min
Foot-pounds sec	1.356×10^{-3}	Kilowatts
Gallons	3785	Cubic centimeters
Gallons	0.1337	Cubic feet
Gallons	231	Cubic inches
Gallons	3.785×10^{-3}	Cubic meters
Gallons	4.95×10^{-3}	Cubic yards
Gallons	3.785	Liters
Gallons	8	Pints (liq.)
Gallons	4	Quarts (liq.)
Gallons, Imperial	1.20095	U S Gallons
Gallons, U.S.	0.83267	Imperial gallons

Multiply	By	To Obtain
Inches of water	0.07355	Inches of mercury
Inches of water	0.002540	Kgs sq cm
Inches of water	0.5781	Ounces sq inch
Inches of water	5.202	Lbs sq foot
Inches of water	0.03613	Lbs sq inch
Kilograms	980.665	Dynes
Kilograms	2.205	Lbs.
Kilograms	1.102×10^{-3}	Tons (short)
Kilograms	10^{3}	Grams
Kgs. meter	0.6720	Lbs foot
Kgs. sq. cm.	0.9678	Atmospheres
Kgs. sq. cm.	32.81	Feet of water
Kgs. sq. cm.	28.96	Inches of mercury
Kgs. sq. cm.	2048	Lbs sq foot
Kgs. sq. cm.	14.22	Lbs sq inch
Kgs. sq. millimeter	10^{6}	Kgs sq meter
Kiloliters	10^{3}	Liters
Kilometers	10^{5}	Centimeters
Kilometers	3281	Feet
Kilometers	10^{3}	Meters
Kilometers	0.6214	Miles
Kilometers	1094	Yards
Kilometers hr	27.78	Centimeters sec
Kilometers hr	54.68	Feet/min.
Kilometers hr	0.9113	Feet/sec.
Kilometers hr	0.5396	Knots
Kilometers hr	16.67	Meters min
Kilometers hr	0.6214	Miles hr
Kms. hr. sec.	27.78	Cms. sec. sec.
Kms. hr. sec.	0.9113	Ft sec sec
Kms. hr. sec.	0.2778	Meters sec. sec.
Kilowatts	56.92	B T Units min
Kilowatts	4.425×10^{4}	Foot-lbs. min
Kilowatts	737.6	Foot-lbs. sec.
Kilowatts	1.341	Horse-power
Kilowatts	14.34	Kg calories min
Kilowatts	10^{3}	Watts
Kilowatt-hours	3415	British Thermal Units
Kilowatt-hours	2.655×10^{6}	Foot-lbs.
Kilowatt-hours	1.341	Horse-power-hrs.
Kilowatt-hours	860.5	Kilogram-calories
Kilowatt-hours	3.671×10^{5}	Kilogram-meters

Multiply	By	To Obtain
Ounces	16	Drams
Ounces	437.5	Grains
Ounces	0.0625	Pounds
Ounces	28.349527	Grams
Ounces	0.9115	Ounces (troy)
Ounces	2.790×10^{-5}	Tons (long)
Ounces	2.835×10^{-5}	Tons (metric)
Ounces, troy	480	Grains
Ounces, troy	20	Pennyweights (troy)
Ounces, troy	0.08333	Pounds (troy)
Ounces, troy	31.103481	Grams
Ounces, troy	1.09714	Ounces (avoir.)
Ounces (fluid)	1.805	Cubic inches
Ounces (fluid)	0.02957	Liters
Ounces sq inch	0.0625	Lbs sq inch
Parts million	0.0584	Grains U S gal
Parts million	0.07016	Grains Imp gal
Parts million	8.345	Lbs. million gal
Pennyweights (troy)	24	Grains
Pennyweights (troy)	1.55517	Grams
Pennyweights (troy)	0.05	Ounces (troy)
Pennyweights (troy)	4.1667×10^{-3}	Pounds (troy)
Pounds	16	Ounces
Pounds	7000	Grains
Pounds	0.0005	Tons (short)
Pounds	453.5924	Grams
Pounds	1.21528	Pounds (troy)
Pounds	14.5833	Ounces (troy)
Pounds (troy)	5760	Grains
Pounds (troy)	240	Pennyweights (troy)
Pounds (troy)	12	Ounces (troy)
Pounds (troy)	373.24177	Grams
Pounds (troy)	0.822857	Pounds (avoir.)
Pounds (troy)	13.1657	Ounces (avoir.)
Pounds (troy)	3.6735×10^{-4}	Tons (long)
Pounds (troy)	4.1143×10^{-4}	Tons (short)
Pounds (troy)	3.7324×10^{-4}	Tons (metric)
Pounds of water	0.01602	Cubic feet
Pounds of water	27.68	Cubic inches
Pounds of water	0.1198	Gallons
Pounds of water min.	2.670×10^{-4}	Cubic ft. sec.

Unit	Multiply by	To obtain
Cubic feet	2.832×10^4	Cubic cms.
	1728	Cubic inches
	0.02832	Cubic meters
	0.03704	Cubic yards
	7.48052	Gallons
	28.32	Liters
	59.84	Pints (liq.)
	29.92	Quarts (liq.)
Cubic feet/minute	472.0	Cubic cms./sec.
	0.1247	Gallons/sec.
	0.4720	Liters/sec.
	62.43	Lbs. of water/min.
Cubic feet/second	0.646317	Million gals./day
	448.831	Gallons/min.
Cubic inches	16.39	Cubic centimeters
	5.787×10^{-4}	Cubic feet
	1.639×10^{-5}	Cubic meters
	2.143×10^{-5}	Cubic yards
	4.329×10^{-3}	Gallons
	0.03463	Liters
	0.01732	Pints (liq.)
Cubic meters	10^6	Cubic centimeters
	35.31	Cubic feet
	61,023	Cubic inches
	1.308	Cubic yards
	264.2	Gallons
	2113	Pints (liq.)
	1057	Quarts (liq.)
Cubic yards	7.646×10^5	Cubic centimeters
	27	Cubic feet
	46.656	Cubic inches
	0.7646	Cubic meters
	202.0	Gallons
	764.6	Liters
	1616	Pints (liq.)
	807.9	Quarts (liq.)
Cubic yards/min.	0.45	Cubic feet/sec.
	3.367	Gallons/sec.
	12.74	Liters/sec.
Decigrams	0.1	Grams
Deciliters	0.1	Liters
Decimeters	0.1	Meters
Degrees (angle)	60	Minutes
	0.01745	Radians
	3600	Seconds
Degrees/sec.	0.01745	Radians/sec.
	0.1667	Revolutions/min.
	0.002778	Revolutions/sec.
Dekagrams	10	Grams
Dekaliters	10	Liters
Dekameters	10	Meters
Drams	27.34375	Grains
	0.0625	Ounces
	1.771845	Grams

Unit	Multiply by	To obtain
Gallons water	8.3453	Pounds of water
Gallons/min.	2.228×10^{-3}	Cubic feet/sec.
	0.06308	Liters/sec.
	8.0208	Cu. ft./hr.
Gallons water/min.	6.0086	Tons water/24 hrs.
Grains (troy)	0.06480	Grains (avoir.)
	0.04167	Pennyweights (troy)
	2.0833×10^{-3}	Ounces (troy)
Grains U.S. gal.	17.118	Parts/million
	142.86	Lbs./million gal.
Grains/Imp. gal.	14.286	Parts/million
Grams	980.7	Dynes
	15.43	Grains
	10^{-3}	Kilograms
	10^3	Milligrams
	0.03527	Ounces
	0.03215	Ounces (troy)
	2.205×10^{-3}	Pounds
Grams/cm.	5.600×10^{-3}	Pounds/inch
Grams/cu. cm.	62.43	Pounds/cubic foot
	0.03613	Pounds/cubic inch
Grams/liter	58.417	Grains/gal.
	8.345	Pounds/1000 gals.
	0.062427	Pounds/cubic foot
	1000	Parts/million
Hectograms	100	Grams
Hectoliters	100	Liters
Hectometers	100	Meters
Hectowatts	100	Watts
Horse-power	42.44	B.T. Units/min.
	33,000	Foot-lbs./min.
	550	Foot-lbs./sec.
	1.014	Horse-power (Metric)
	10.70	Kg.-calories/min.
	0.7457	Kilowatts
	745.7	Watts
Horse-power (boiler)	33.479	B.T.U./hr.
	9.803	Kilowatts
Horse-power-hours	2547	British Thermal Units
	1.98×10^6	Foot-lbs.
	641.7	Kilogram-calories
	2.737×10^5	Kilogram-meters
	0.7457	Kilowatt-hours
Inches	2.540	Centimeters
Inches of mercury	0.03342	Atmospheres
	1.133	Feet of water
	0.03453	Kgs./sq. cm.
	70.73	Lbs./sq. ft.
	0.4912	Lbs./sq. inch
Inches of water	0.002458	Atmospheres

Unit	Multiply by	To obtain
Liters	10^3	Cubic centimeters
	0.03531	Cubic feet
	61.02	Cubic inches
	10^{-3}	Cubic meters
	1.308×10^{-3}	Cubic yards
	0.2642	Gallons
	2.113	Pints (liq.)
	1.057	Quarts (liq.)
Liters/min.	5.886×10^{-4}	Cubic ft./sec.
	4.403×10^{-3}	Gals./sec.
Lumber Width (in.) × Thickness (in.) / 12	Length (ft.)	Board Feet
Meters	100	Centimeters
	3.281	Feet
	39.37	Inches
	10^{-3}	Kilometers
	10^3	Millimeters
	1.094	Yards
Meters/min.	1.667	Centimeters/sec.
	3.281	Feet/min.
	0.05468	Feet/sec.
	0.06	Kilometers/hr.
	0.03728	Miles/hr.
Meters/sec.	196.8	Feet/min.
	3.281	Feet/sec.
	3.6	Kilometers/hr.
	0.06	Kilometers/min.
	2.237	Miles/hr.
	0.03728	Miles/min.
Microns	10^{-4}	Meters
Miles	1.609×10^5	Centimeters
	5280	Feet
	1.609	Kilometers
	1760	Yards
Miles/hr.	44.70	Centimeters/sec.
	88	Feet/min.
	1.467	Feet/sec.
	1.609	Kilometers/hr.
	0.8684	Knots
	26.82	Meters/min.
Miles/min.	2682	Centimeters/sec.
	88	Feet/sec.
	1.609	Kilometers/min.
	60	Miles/hr.
Milliers	10^3	Kilograms
Milligrams	10^{-3}	Grams
Milliliters	10^{-3}	Liters
Millimeters	0.1	Centimeters
	0.03937	Inches
Milligrams/liter	1	Parts/million
Million gals./day	1.54723	Cubic ft./sec.
Miner's inches	1.5	Cubic ft./min.
Minutes (angle)	2.909×10^{-4}	Radians

Unit	Multiply by	To obtain
Pounds cubic foot	0.01602	Grams/cubic cm.
	16.02	Kgs./cubic meter
	5.787×10^{-4}	Lbs./cubic inch
Pounds cubic inch	27.68	Grams/cubic cm.
	2.768×10^4	Kgs./cubic meter
	1728	Lbs./cubic foot
Pounds foot	1.488	Kgs./meter
Pounds inch	178.6	Grams/cm.
Pounds sq. foot	0.01602	Feet of water
	4.883×10^{-4}	Kgs./sq. cm.
	6.945×10^{-3}	Pounds sq. inch
Pounds sq. inch	0.06804	Atmospheres
	2.307	Feet of water
	2.036	Inches of mercury
	0.07031	Kgs./sq. cm.
Quarts (dry)	67.20	Cubic inches
Quarts (liq.)	57.75	Cubic inches
Quintal, Argentine	101.28	Pounds
Quintal, Brazil	129.54	Pounds
Quintal, Castile, Peru	101.43	Pounds
Quintal, Chile	101.41	Pounds
Quintal, Mexico	101.47	Pounds
Quintal, Metric	220.46	Pounds
Sq. ft. gal./min.	8.0208	Overflow rate (ft./hr.)
Temp. (°C) + 273	1	Abs. temp (°C)
Temp. (°C) + 17.78	1.8	Temp. (°F)
Temp. (°F) + 460	1	Abs. temp (°F)
Temp. (°F) − 32	5/9	Temp. (°C)
Tons (long)	1016	Kilograms
	2240	Pounds
	1.12000	Tons (short)
Tons (metric)	10^3	Kilograms
	2205	Pounds
Tons (short)	2000	Pounds
	32,000	Ounces
	907.18486	Kilograms
	2430.56	Pounds (troy)
	0.89287	Tons (long)
	29166.66	Ounces (troy)
	0.90718	Tons (metric)
Tons of water 24 hrs.	83.333	Pounds water/hour
	0.16643	Gallons/min.
	1.3349	Cu. ft./hr.
Watts	0.05692	B.T. Units/min.
	44.26	Foot-pounds/min.
	0.7376	Foot-pounds/sec.
	1.341×10^{-3}	Horse-power
	0.01434	Kg.-calories/min.
	10^{-3}	Kilowatts
Watt-hours	3.415	British Thermal Units
	2655	Foot-pounds
	1.341×10^{-3}	Horse-power-hours
	0.8605	Kilogram-calories
	3671	Kilogram-meters
	10^{-3}	Kilowatt-hours

Units of Area

Units	Square Inches	Square Feet	Square Yards	Square Miles	Square Centimeters	Square Meters
1 square inch	1	0.00694444	0.000771605	0.000000002491	6.451626	0.0006451626
1 square foot	144	1	0.1111111	0.000000358701	929.0341	0.09290341
1 square yard	1,296	9	1	0.00000032831	8361.307	0.8361307
1 square mile	4,014,489,600	27,878,400	3,097,600	1	25,899,964,703	2,589,998
1 square centimeter	0.1549969	0.001076387	0.0001195985	0.00000000000361006	1	0.0001
1 square meter	1549.9969	10.76387	1.195985	0.0000003861006	10,000	1

Units of Length

Units	Inches	Feet	Yards	Miles	Centimeters	Meters
1 inch	1	0.0833333	0.0277778	0.0000157828	2.540005	0.02540005
1 foot	12	1	0.333333	0.0001893939	30.48006	0.3048006
1 yard	36	3	1	0.000568182	91.44018	0.9144018
1 mile	63,360	5280	1760	1	160,934.72	1609.3472
1 centimeter	0.3937	0.03280833	0.010936111	0.00000621369	1	0.01
1 meter	39.37	3.280833	1.0936111	0.0006213699	100	1

Units of Liquid Measure

Units	Fluid ounces	Liquid pints	Liquid Quarts	Gallons	Milliliters	Liters	Cubic Inches
1 fluid ounce	1	0.0625	0.03125	0.0078125	29.5729	0.0295729	1.80469
1 liquid pint	16	1	0.5	0.125	473.167	0.473167	28.875
1 liquid quart	32	2	1	0.25	946.333	0.946333	57.75
1 gallon	128	8	4	1	3785.332	3.785332	231
1 milliliter	0.0338147	0.00211342	0.00105671	0.000264178	1	0.001	0.0610250
1 liter	33.8147	2.11342	1.05671	0.264178	1000	1	61.0250
1 cubic inch	0.554113	0.0346320	0.0173160	0.00432900	16.3867	0.0163867	1

INDEX